T0299313

Model-Based Monitoring and Statistical Control

Available in English for the first time, this classic and influential book by the late Kohei Ohtsu presents real examples of ships in motion under irregular ocean waves, how to understand the characteristics of fluctuations of stochastic phenomena through spectral analysis methods and statistical modeling. It also explains how to realize prediction and optimal control based on time series models.

In recent years, the need to improve safety and reduce environmental impact in ship operations has been increasing, and the statistical methods presented in this book will be increasingly needed in the future. In addition, the recent development of innovative AI technology and high-speed communications will make it possible to adapt this method not only to ship monitoring and control, but also to any field that involves irregular fluctuations, and it is expected to contribute to solving issues that have been difficult to solve in the past.

Part 1 describes classical spectral method for the analysis of stochastic phenomena. In Part 2, this book explains methods to construct time series models using the information criterion, to capture the characteristics of ship and engine motions using the model, to design a model-based monitoring system that informs navigators operating the ship and managers ashore. Furthermore, it explains statistical control method to design an autopilot system and the governor of a marine engine, while showing actual examples. Part 3 presents the basic knowledge necessary for understanding these topics of the book, namely, the basic theory of ship motion, probability and statistics, Kalman filter and statistical optimal control theory.

The late Kohei Ohtsu was a Professor Emeritus of Tokyo University of Marine Science and Technology. He has served as Vice President of Tokyo University of Mercantile Marine, Captain of the Training Ship Shioji Maru of the University, and President of Ohtsu Maritime Institute, Co. Ltd. His main research interests include statistical analysis of ship motion and optimal steering, ship maneuverability and practical application of ship-to-shore communication and ship operation data monitoring. He received awards from The Society of Naval Architects of Japan, Japan Institute of Navigation, The Society of Instrument and Control Engineers and the Japan Society of Naval Architects and Ocean Engineers, etc. He passed away in 2016 at the age of 72.

Genshiro Kitagawa is a professor emeritus of The Institute of Statistical Mathematics, and of Graduate University for Advanced Study, former President of the Research Organization of Information and Systems, and the former Director-General of the Institute of Statistical Mathematics. His primary research interests are time series modeling, nonlinear filtering and statistical modeling. He is the author of several English books on time series analysis and information criteria. He was awarded the Japan Statistical Society Prize and Ishikawa Prize, etc. and is a Fellow of the American Statistical Association.

CHAPMAN & HALL/CRC Interdisciplinary Statistics Series

Series editors: B.J.T. Morgan, C.K. Wikle, P.G.M. van der Heijden

Recently Published Titles

For more information about this series, please visit: https://www.crcpress.com/Chapman--HallCRC-Inter-disciplinary-Statistics/book-series/CHINTSTASER

Model-Based Monitoring and Statistical Control

Kohei Ohtsu
Translated by Genshiro Kitagawa and
O2 Memorial Project

CRC Press
Taylor & Francis Group
Boca Raton London New York

CRC Press is an imprint of the
Taylor & Francis Group, an **informa** business

A CHAPMAN & HALL BOOK

Designed cover image: © Kohei Ohtsu and Genshiro Kitagawa

First edition published 2024
by CRC Press
2385 NW Executive Center Drive, Suite 320, Boca Raton FL 33431

and by CRC Press
4 Park Square, Milton Park, Abingdon, Oxon, OX14 4RN

CRC Press is an imprint of Taylor & Francis Group, LLC

© 2024 © Genshiro Kitagawa, the copyright in the original text is Japanese text by Kohei Ohtsu is owned by Kinue Ohtsu.

Reasonable efforts have been made to publish reliable data and information, but the author and publisher cannot assume responsibility for the validity of all materials or the consequences of their use. The authors and publishers have attempted to trace the copyright holders of all material reproduced in this publication and apologize to copyright holders if permission to publish in this form has not been obtained. If any copyright material has not been acknowledged please write and let us know so we may rectify in any future reprint.

Except as permitted under U.S. Copyright Law, no part of this book may be reprinted, reproduced, transmitted, or utilized in any form by any electronic, mechanical, or other means, now known or hereafter invented, including photocopying, microfilming, and recording, or in any information storage or retrieval system, without written permission from the publishers.

For permission to photocopy or use material electronically from this work, access www.copyright.com or contact the Copyright Clearance Center, Inc. (CCC), 222 Rosewood Drive, Danvers, MA 01923, 978-750-8400. For works that are not available on CCC please contact mpkbookspermissions@tandf.co.uk

Trademark notice: Product or corporate names may be trademarks or registered trademarks and are used only for identification and explanation without intent to infringe.

ISBN: 978-1-032-55012-1 (hbk)
ISBN: 978-1-032-55016-9 (pbk)
ISBN: 978-1-003-42856-5 (ebk)

DOI: 10.1201/9781003428565

Typeset in CMR10 font
by KnowledgeWorks Global Ltd.

Publisher's note: This book has been prepared from camera-ready copy provided by the authors.

Contents

II   Applications of Time Series Statistical Models   87

III   For a Better Understanding   221

16 Introduction to Maneuvering, Ship Motions and Propulsion Theory   223

17 Basics of Probability   238

Preface to the English Edition

Background of the English edition

On the cold evening of January 18, 2016, during the first snowfall in Tokyo in many years, I received a phone call from Mrs. Kinue Ohtsu. She was the wife of my best friend since junior high school, Professor Kohei Ohtsu. She delivered the somber news that her husband had suddenly collapsed early in the morning while shoveling snow to free his car from the garage at home. He was rushed to the hospital by ambulance and, to our great shock, passed away later that afternoon due to a ruptured aneurysm. At the age of 72, Kohei was deeply engrossed in preparations for the establishment of a new joint company, aimed at translating his research findings into practical applications within the shipping industry.

Now, eight years after that fateful day, it brings immense satisfaction to not only Kinue but also to everyone involved in the publication of this book to finally realize her late husband's dream of releasing an English edition of his work. The aim is to make it accessible to researchers, engineers and students around the world who are interested in time-series modeling, as well as the shipping industry.

Kohei's original book in Japanese, titled "Model-Based Monitoring and Statistical Control," was published on December 15, 2012, nearly eleven years ago. Even after retiring from Tokyo University of Marine Science and Technology (TUMSAT) in 2006, he remained determined to systematically construct and leave behind the outcomes of his research in "time series modeling-based ship monitoring and control."

His research largely built upon the theories and applications of his mentor, Dr. Hirotugu Akaike, the former Director General of the Institute of Statistical Mathematics. As he expressed in the preface of his original book, Kohei's work was founded on the idea that "it is useful to employ statistical methods to address irregular phenomena such as ocean waves in rough weather." Furthermore, he believed that the necessity for such pragmatic methods would only grow in the future, contributing to improved ship safety and reduced global environmental impact. After retiring from TUMSAT, this vision prompted him to deliver special lectures as a research professor not only at TUMSAT but also at various universities and related organizations in Japan and overseas. In 2013, he established "O2 Maritime Research Institute, Inc." to accumulate empirical data to substantiate his theory and method. In this regard, this book represents the culmination of his tireless research endeavors.

Shortly after the initial book publication, I had the privilege to engage in a conversation with Kohei. Although, as a specialist in semiconductors, I couldn't fully grasp the entirety of his research at the time, I sensed a strong synergy between his theories and the capabilities inherent in semiconductor technologies. These capabilities included big data collection, storage, processing, high-speed, high-precision analysis employing AI technology and high-speed, low-latency communication technology. For this reason, I believed that his methodologies had applicability not only in ship control but also in the future growth industry of drone flight control.

It is worth noting that two years after the book's initial release, Kohei received an award from the Japan Society of Naval Architects and Ocean Engineers in 2014. According to Kinue, this recognition provided him with the confidence and made him determined to publish an English edition as soon as possible.

He was in the midst of preparations for establishing a joint venture with a marine-related engineering firm in Singapore, aiming to commercialize equipment based on the concepts within the book. However, tragedy struck when he passed away on the very eve of finalizing the investment, leading to the cancellation of the joint venture. It was an immensely regrettable turn of events.

History leading up to publication of the book

The journey toward publishing this book is a tale of determination and collaboration. In late September 2022, Kinue reached out to me, along with two of Kohei's longtime friends and university classmates, Mr. Masaru Yamamoto and Mr. Akira Fujisaki. She expressed her wish to publish an English edition of Kohei's book, a dream her late husband had held fervently. Given that her husband's dream of establishing a joint venture had been shattered due to his untimely demise, she felt that this book was the sole legacy he could bestow upon future generations. Without hesitation, the three of us readily accepted her request and committed ourselves to her cause.

Our initial steps led us to TUMSAT, Kohei's alma mater and several publishers, including the one responsible for the original book. However, in today's academic landscape, Japanese publishers face significant challenges when pursuing commercial publications. Specialized papers and academic materials are increasingly available electronically on the web, and overseas sales channels are often limited. Further, translators with expertise in marine engineering and time series analysis are hard to find. This proved to be a formidable obstacle.

Nevertheless, we refused to surrender. A significant breakthrough occurred when we discovered a CD-ROM containing the final manuscript of Kohei's original work among his belongings. In addition to Kinue and us three, we enlisted the assistance of Dr. Hiroyuki Oda Guest Professor of TUMSAT and Mrs. Michiko Oda, who later undertook the arduous tasks of proofreading and reviewing the translation. We launched the "O2 Memorial Project" for this purpose at the end of December. Shortly thereafter, our endeavor received an unexpected boon in the form of Dr. Genshiro Kitagawa, the former Director General of the Institute of Statistical Mathematics and a long-time friend and collaborator of Kohei. He eagerly agreed to support the book's publication and promptly proposed the idea to Chapman & Hall/CRC Press, which had recently published his own work. Fortunately, the company embraced the project, and Dr. Kitagawa also assumed the task of translating the book into English, which had been our most formidable challenge. Thus, the project, with the addition of Dr. Kitagawa and a total of seven members, was finally able to set sail strongly toward publication at the end of January.

First of all, Dr. Oda organized for the manuscript to be reviewed and proofread by Professor Toshihiko Nakatani of Toyama National College of Technology and Mr. Tatsuya Yashiro of YDK Technologies Co. Ltd., who had collaborated with Kohei in his research. Additionally, we secured permission from Kaibundo Publishing Co., LTD., the holder of the publishing rights for the Japanese original, to transfer the publishing rights for the English translation of the first edition to Mrs. Kinue Ohtsu, the rightful heir, free of charge. With these vital steps, the project officially commenced in January 2023, and we anticipate the English edition to be released approximately a year later.

Acknowledgments

The publication of this book is a testament to the indomitable determination of Kohei, who wished to share his work with the world before his untimely departure, and the unwavering resolve of Kinue, who made it her mission to bring this dream to fruition against all odds. It is also the result of the extraordinary cooperation of individuals who shared profound personal connections with Kohei. On behalf of the late scholar and his wife, I, as the representative of the "O2 Memorial Project," wish to convey our heartfelt gratitude to all those who contributed to the publication of this book.

In particular, I would like to extend my deepest appreciation to the remarkable individuals who made this project possible: Dr. Genshiro Kitagawa for his invaluable editing and translation expertise. Dr. Hiroyuki Oda, Mrs. Michiko Oda, Professor Toshihiko Nakatani and Mr. Tatsuya Yashiro for their meticulous peer review and proofreading. Mr. David Grubbs of Chapman & Hall/CRC Press for his support and invaluable advice on the intricate publication process. Mr. Makoto Tominaga of Kaibundo Publishing for graciously granting permission for the publication of the book in English. I would also like to express gratitude to the members of the O2 Memorial Project and to President Toshio Iseki and Professor Kiyokazu Minami of TUMSAT for their continuous guidance and support throughout the project.

I fervently hope that this book serves as a valuable resource for researchers in diverse fields and individuals engaged in the application of time series analysis and control engineering worldwide, working to address pressing issues such as safety enhancement, reduction of global environmental impact and the mitigation of societal costs.

In closing

Finally, I would like to share what I know about the fascinating personality of the author, based on my interactions with him over the past 60 years. From a young age, Kohei aspired to become a sailor, following in the footsteps of his uncle, a captain of ocean-going vessels. Simultaneously, he pursued interests in music, literature, history and even ventured into politics during his high school years. As the president of the student council, he honed his organizational and leadership skills. In line with his boyhood dream, he embarked on a path to study navigation at Tokyo University of Mercantile Marine (TUMM). However, after careful consideration, he chose to remain at the university and transitioned into academia, heeding the recommendation of his former professor.

Nonetheless, his yearning for the open sea never waned, and in addition to his research pursuits, he assumed the role of captain on a university training ship. He continued to hold a keen interest in societies and cultures worldwide. He was also a fervent enthusiast of Sherlock Holmes, visiting Baker Street whenever he found himself in London. Even up until his passing, he found solace in playing piano music by Mozart and Chopin. In the CD player in his study, we discovered Chopin's "Etude in E, Op.10 No.3" - a testament to his enduring passion for music.

   After retiring from the university, he remained confident in his physical and mental well-being and had the zest to travel across the globe when duty called. His sudden and tragic demise due to a ruptured aneurysm left everyone who knew him in shock. The fact that more than 350 people gathered to bid him farewell speaks volumes about the extent of his interactions and the richness of his personality.

November 2023

Representative of O2 Memorial Project

Yukio Tsuda

Preface

The Greek word for "helmsman" is cybernetics. Norbert Wiener, the founder of cybernetics as a discipline, wrote "The Extrapolation, Interpolation and Smoothing of Stationary Time Series" in 1942, one year before my birth. The book was so difficult to understand for us that it was secretly called "The Yellow Peril" by experts because of the color of the book cover, but it was epoch-making to enable the statistical analysis of irregular phenomena such as ocean waves in the context of correlated temporal and spatial flows.

Thirty years later in 1974, Dr. Hirotugu Akaike authored a book "Statistical Analysis and Control of Dynamic Systems," which depending on the color of the book, we researchers commonly and respectfully call "the Red Book." This book established a rational method called the Minimum AIC Method for creating statistical models of time series in the time domain, and the scope of application of this book has expanded worldwide in various fields of statistical analysis and control of irregular phenomena.

In this book, I describe a method for designing and analyzing a model-based monitoring system based on real-time data of ship motions that dynamically move on irregular ocean waves, which I have been working on for many years. In the book, I wrote how to design a model-based monitoring system and analyze the obtained results. This book also explains how to apply the identified statistical model to an automatic steering system (autopilot system) that steers the course of moving ship in irregular ocean waves, and to control the main engine's governor and the propeller revolutions. The readers can study the book referring the practical data and examples.

When dealing with irregular phenomena such as ocean waves, the statistical methods described in this book, despite their importance, I suppose, have not gained sufficient citizenship among many people studying ship and marine engineering, etc., and have been rather limited to a supplementary role to the Newtonian mechanics approach. However, I am clearly convinced that the statistical method described in this book is a realistic and proactive way to bring rich and new knowledge about how to understand and control the phenomena for engineers in the marine-related fields who are struggling under extremely strong and irregular disturbances. Recent developments at the International Maritime Organization (IMO) and elsewhere indicate that, in addition to improving ship safety, there is a growing need for realistic methods to handle ship operations in order to reduce the environmental impact of ships on the earth, and I believe that the statistical methods set out here will become more and more necessary in the future.

However, to gain deeper understanding and appreciation of the study, along with a solid observation of the phenomena, it is necessary to have a basic knowledge of linear algebra, mathematical knowledge of Fourier analysis, statistics, time series analysis, modern control theory, signal processing theory and information science, and it is necessary to spend a lot of time to gain a fair knowledge of these areas. In this book, I explain the basic knowledge covering such a wide range of fields in detail, starting from the basics, so that the reader can understand the book without referring to many other books other than the basic arithmetic knowledge of linear algebra and analysis. Therefore, the number of pages has exceeded the original planning.

I have continued and honed my research since graduating from Tokyo University of Mercantile Marine and have learned a great deal from the predecessors in this research field. In particular, Dr. Yasufumi Yamanouchi, former Director of the National Maritime Research Institute, I have learned from him how to apply time series analysis to ship motions. I am deeply grateful to Dr. Yamanouchi and he introduced me to Dr. Hirotugu Akaike, former Director of the Institute of Statistical Mathematics, who is world-renowned as the advocator of Akaike Information Criterion (AIC), which is the main criterion cited in this book. In the early stages of the research, I became acquainted with Dr. Genshiro Kitagawa of the Institute of Statistical Mathematics, and we conducted many joint research projects on the development of the autopilot and on theories related to state-space model methods in the time domain. Furthermore, I am grateful to Professor Emeritus Michio Horigome of Tokyo University of Mercantile Marine, during the development stages of the autopilot and marine governor. I learned from him a lot about modern automatic control theory and engine performance. I would like to express my deepest gratitude to these four pioneers.

Working together with hardship in full-scale experiments onboard for dedicated to the development of the research, I am particularly grateful to Dr. Hiroyuki Oda of Mitsui Shipbuilding and Engineering Akishima Laboratory Inc., Prof. Toshio Iseki of Tokyo University of Marine Science and Technology, and Dr. Masanori Ishizuka, formerly of NYK Line. I acknowledge the help of Dr. Toshihiko Nakatani, Dr. Shinji Matsuda, Dr. Tadatsugi Okazaki, Dr. Jun Kayano, Dr. Daisuke Terada, Dr. Hitoi Tamaru, Dr. Shintaro Miyoshi and others who were all my graduate students during my research years at the University, and are still actively working in this field, especially thanks them for their valuable experimental results and figures. I also deeply thank the crew of Training Ship Shioji Maru of the former Tokyo University of Mercantile Marine, who spared no effort in cooperating with us for shipboard experiments to record the data. I am also grateful to Kawasaki Kisen Co. Ltd. for providing the opportunity to conduct the shipboard experiments, Nippon Kaiji Kyokai (ClassNK) and the Members of Tokyo University of Marine Science and Technology's Ocean Broadband Research Group for their generous support, MTI Corporation for providing us with valuable ship data. I would also like to express my sincere thanks to Ms. Michiko Oda, with whom I have shared many years of hardship, for her careful proofreading assistance.

The publication of this book started on an old promise to late Mr. Kuniyoshi Tanaka, the famous editor-in-chief of Kaibundo Publishing Co. Ltd. Finally, I would like to thank Mr. Toshio Iwamoto of Kaibundo Publishing on the legacy of the late Mr. Tanaka and for his efforts in bringing this book's publication to fruition.

I hope that this book will contribute to the research in any way possible and be a useful reference for researchers and engineers in the fields of navigation and ship and ocean engineering, where this field has not yet gained citizenship, as well as for many others who are engaged day and night in the practice and research of understanding and controlling similar irregular phenomena.

November 2012

Kohei Ohtsu

List of Figures

List of Tables

1

Prologue

1.1 Introduction

When ships sail the oceans, they are subject to very strong wind and wave disturbances that are significantly different situations compared to other transportation. The intensity of the disturbance varies from mirror calm to rough seas. Between 1968 and 1970, when the author was young, serious maritime accidents happened one after another. Two large ore carriers were hit by large waves off the coast of Boso Peninsula, Japan, breaking their hulls and sinking into the ocean. In those accidents, the size of the waves expressed by the ship operators was not easily communicated to the people on land, which was frustrating for the ship operators. However, regardless of the cause of the shipwreck, the fact that the ships have broken in these accidents taught both shipbuilders and operators a great findings about the magnitude of disturbance and violent changes in the sea. Since then, the author has experienced a few times the magnitude of wind and wave disturbances and their drastic changes on several occasions during experimental voyages with large vessels in the ocean and as a captain operating small training ships. However, during that time, ship forms were improved and hull strength was strengthened, but there was no movement on the part of shipbuilders or ship owners to actively monitor ship and engine motions during navigation and make this information as a useful means of ship operation. In recent years, however, the demand for saving the earth's resources and reducing the environmental burden has extended to ship transportation, and this has led to the need to select routes for ship operations as well as to take detailed management of ships and engines. Thus, there occurred finally a movement to actively **monitor** ship and engine motions and to use these information to optimize ship operation and control.

The author has analyzed many real data of ship and engine motions and used them to design control equipment for marine systems. From this experience, the author has learned that it is important not to treat ship motion and engine motion separately in order to analyze and predict the dynamic behavior of ships and engines in real time-navigating under strong disturbances. If we can accurately predict the relationship between these two motions simultaneously, we can realize a navigation method that reduces the environmental burden while maintaining safety by optimally controlling the rudder, which controls the ship's heading, and the main engine, which propels the ship, the main control systems of a ship. The purpose of this book is to show how to develop a total statistical model of ship and engine motions using actual time series data and to clarify the effectiveness of the statistical model obtained.

The hydrodynamic approach has been the mainstream of research on ship motion. The thermodynamic approach has also been used in the study of heat engines. This was natural from the standpoint of hull form development and main engine development. However, the problem here is the actual motion and control between different systems, i.e., hull and engine, on an irregular sea surface. It is difficult to represent the dynamics using only hydrodynamic and thermodynamic approaches. In order to describe and predict these motions

DOI: 10.1201/9781003428565-1

under conditions where real data are available, it is necessary to go beyond the dynamics of these motions and challenge the flow of information, which should be called statistical dynamics, with the statistical approach[100], first advocated by Norbert Wiener. The author believes that this is currently the only way. In recent years, among these statistical approaches, the method developed in Japan, which will be described hereafter, has achieved numerous results in the field of statistical modeling of time series, and is still being developed day by day. The method of monitoring ship and engine motions used by the author here is a new model-based monitoring method in the time domain using this statistical model.

In the following, we will focus in detail on methods for building statistical models and using them to predict and control motion. Using the methods obtained here, the following two additional things can also be achieved.

1. By transmitting very few parameters obtained from the statistical model of ship and engine motions, rather than the data itself, using the recently developed ship-to-shore communication, it is possible to manage the operation of individual ships on land.

2. The accumulation of information on individual vessels by creating a database of such parameters can be applied to the problem of route selection.

1.2   Ship and Engine Motions in Ocean Waves

1.2.1   Characteristics of Ship Motion and Means of Control

The ship motions treated in this book are defined in Figure 1.1. You can find a six-degree-of-freedom motion consisting of three rotational motions and three parallel motions around the body-fixed axes. When the coordinate axes are fixed to the center of gravity of the ship, the x-axis is taken in the direction from aft to fore, the y-axis is abeam direction and the z-axis is directed from top to bottom (Figure 1.1). Table 1.1 shows the definitions of the parallel and rotational motions for each axis[63].

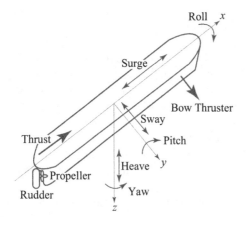

TABLE 1.1

Definition of ship motions in three axes

Axis	Parallel motion	Rotational motion
x	Surge	Roll
y	Sway	Pitch
z	Heave	Yaw

FIGURE 1.1
Coordinate of ship motion

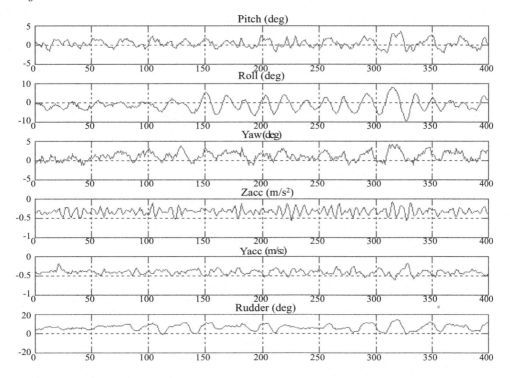

FIGURE 1.2
Time series of autopilot steering of a medium-sized container ship

Six motions listed in Table 1.1 can be broadly classified into the following two groups:

1. Group of rolling, pitching and heaving motions (motion with righting moment)

2. Group of surging, swaying and yawing motions (motion without righting moment)

In the first group with a righting moment, for example, considering a ship with rolling motion in calm water, the difference in buoyancy force between the left and right hulls causes the hull to return to its original position and repeat the motion around the vertical axis even without using any control means. In other words, there exists a **righting moment**. All motions in the first group have this righting moment.

The second group of motions without righting moment, however, requires the application of some control force to move them. First, to control the surging motion, a **main engine**[102] (propulsion system) is required. Also, considering yawing motion, this motion will not return to the original course unless the **rudder** is used, even if the deviation from the course occurs for some factors in the ship's heading.

1.2.2 Examples of Automatic and Manual Steering Records

Next, let us consider the actual ship and engine motions of a ship sailing in the ocean in the form of a time series. Figure 1.2 shows a time series recorded while a medium-sized container ship was sailing in the North Pacific Ocean with automatic steering. The sampling period is 1 second. From the top, the time series of 400 points of pitch, roll, yaw, vertical acceleration (Zacc), lateral acceleration (Yacc) and rudder angle (Rudder) are shown. Figure 1.3 shows a record of the same medium-sized container ship in rough sea when the captain decided to

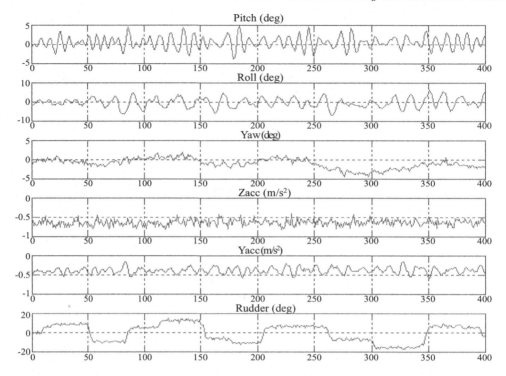

FIGURE 1.3
Time series of manual steering of a medium-sized container ship

switch from automatic steering to manual steering to keep the course constant. Comparing the time series of the two figures, we can see that the way of steering is different between the case in which a human operator takes the helm and the case in which automatic steering is used. What does skillful steering mean from the point of view of operation? How should it be evaluated? This question is a difficult and important issue for the rudder's primary task, i.e., **course-keeping control**.

1.2.3 Examples of Roll and Pitch Records Under Different Sea Conditions

Figures 1.4 and 1.5 show the roll and pitch records of a large high-speed container ship. The difference between the two figures is due to the difference in the wind conditions which the ship has; Figure 1.4 shows the case where the ship has wind from the fore direction, and Figure 1.5 shows the case where the ship has wind from the aft direction. Comparing the two figures, we can see that the ship motions are quite different even of the same ship. In particular, the period of pitch motion is quite different. In addition, the roll motion in Figure 1.5 shows a waveform peculiar to roll motion.

Figure 1.6 shows the directions in which a ship is subjected to disturbances (wind and waves). However, since the angles are not exact and standardized, the directions in the figure are only a guide for the classification in this book. How does the ship motion, and thus the engine motion, differ depending on the direction in which the ship is subjected to wind and waves? Under what conditions does the stability of the ship decrease? Ensuring this stability is a major challenge.

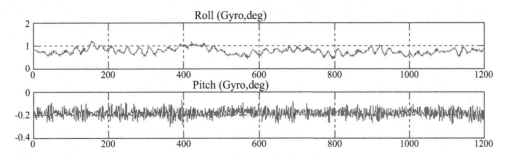

FIGURE 1.4
Roll and pitch of a large high-speed container ship having wind from the fore

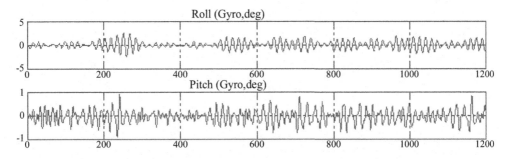

FIGURE 1.5
Roll and pitch of a large high-speed container ship having wind from the aft

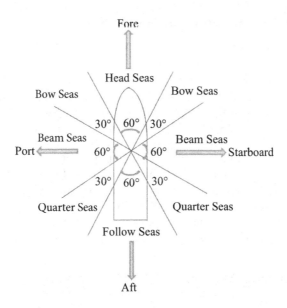

FIGURE 1.6
Direction of wind waves

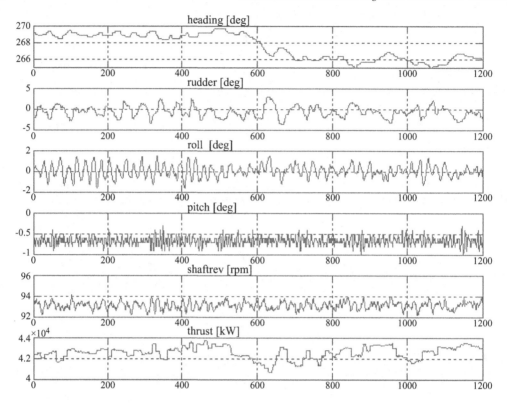

FIGURE 1.7
Ship and engine motions of a large container ship in beam seas

1.2.4   Examples of Ship and Engine Motions

Figures 1.7 and 1.8 show the time history (sampling period of 1 second) of a large container ship's motions (heading, rudder, roll and pitch, with the pitch omitted in Figure 1.8), propeller shaft revolution and thrust data, the former underway in a beam seas and the latter underway in a quarter seas. Although the records were obtained from the same ship, there appears to be a large difference depending on the weather and sea conditions. In Figure 1.7, the heading changes from 3 to 4 degrees to the left at around 600, and the roll and thrust respond to this change. These phenomena are even more pronounced in Figure 1.8. Since the ship is in a following wave, a very long-period yawing occurs, and the rudder is moving to correct it. The same movement is observed for the roll, the propeller shaft revolution and the thrust. What is the main cause of these fluctuations? How does the propeller shaft revolution change? How do pitch and roll motions affect the propeller shaft revolution and thrust? These are major challenges related to the load fluctuation of the main engine, fuel consumption, CO_2 emissions and so on.

1.2.5   Effects on Ship Motion and Engine Motion during Altering Course

Figure 1.9 shows a time series of the same large container ship altering course. In this example, the heading of the container ship changes drastically and the ship alters the heading about 20 degrees to the left at about 400 seconds because the ship took a large

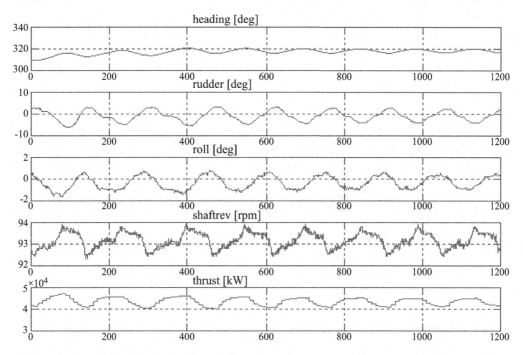

FIGURE 1.8
Ship and engine motions of a large container ship in quarter seas

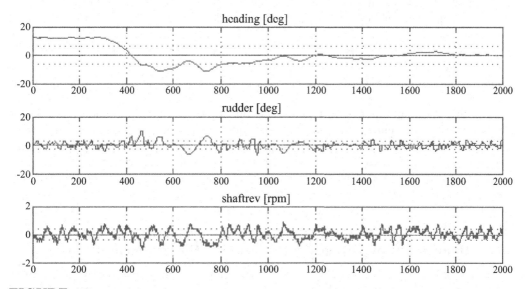

FIGURE 1.9
Time series data during course changing

port rudder angle due to a change of the desired course. Time series with such temporal changes in structure are called **nonstationary time series**. In addition, the propeller shaft revolution is also affected after 400 seconds. Thus, when steering the rudder in this way, not only the propeller shaft revolution but also other motions such as roll motion are greatly affected.

1.3   Necessity of Monitoring Ship and Engine Motions and the Equipment Requirements

The time series of ship and engine motions shown in the previous section indicate that each of the motion elements must be viewed as an intertwined and irregular phenomenon. From these figures, it will be understood that monitoring systems for ships need to be equipped to monitor ship motions and engine motions simultaneously, rather than separately monitoring navigation matters and engine matters as have been the cases in the past. In this section, we will discuss the reasons for this in detail and describe the equipment requirements of a monitoring system for ships.

1.3.1   Ensuring Safety of Voyages

Ensuring safety of navigation is the most important issue. At the time of design, although safety margins for ship performance and propulsion efficiency in calm water are taken into account, ships are exposed to dangerous situations that may cause loss of safety in heavy weather. For example, ships are exposed to various dangers such as fluctuations in ship motions due to the resonance phenomenon between the natural period of ship motions and the natural period of waves, a danger known as parametric rolling, where the period of the pitching and the rolling are in a certain relationship, and broaching during following seas. Each time the main engine is subjected to such dangers as overloading (**torque rich**), **propeller racing**, in which the propeller goes above the surface of the water and so on. However, since ships are not equipped with effective equipment for measuring motions, or even if they are equipped with such equipment, as they do not specify the current motion period, the crew cannot accurately know the motion period of the ship and engine during rough seas.

In addition, the motion of a ship depending on the intensity and direction of the ship's exposure to the surrounding sea and weather conditions is complex, and intertwined with various motion factors. Therefore, it is impossible to predict what will happen next to the main engine. Thus, to ensure the safety of the ship and engine during rough weather, a monitoring system is needed that can estimate and predict the exact period of the ship and engine motions and intuitively inform the crew of the relationship between the multiple motions[81].

1.3.2   Realization of Voyages with Reduced Environmental Load

In recent years, expectations for the realization of a low-carbon society have spread worldwide, and various emissions regulations have been discussed. Ships, which emit less CO_2 than other means of transportation, are no exception, and the 2012 revision of the MARPOL Convention by the International Maritime Organization (IMO) has introduced the EEDI (**energy efficiency design index**), an index for ship builders, as well as the EEOI (**energy efficiency operational index**), an index for shipowners. It is required that the values of these indicators do not exceed the regulation values. In the latter EEOI, the implementation of the Plan, Do, Check, Act cycle is encouraged for operations, and it is stated that it is effective to consider slow-down operation, weather routing, etc. at the planning stage. The importance of monitoring is also pointed out at the Check stage.

As mentioned above, it has been pointed out that engine condition fluctuations are caused by ship motions. Phenomena such as load fluctuations due to propeller racing, fluctuations in the engine output operating point, torque rich, etc., are factors that cause engine output to fluctuate, resulting in extra fuel consumption and higher carbon emissions. Up to now, captains who are considered to be excellent have been making speed and course selections that avoid such conditions of the main engine in close consultation with the chief engineer. In making such decisions, it is important to provide the captain and chief engineer with accurate, comprehensive and up-to-date information, and here is where the need for information provision through simultaneous monitoring of ship and engine motions arises[102].

1.3.3   Grasp of Ship Performance of Each Individual Vessel

The ship's performance changes over time after construction, resulting in changes in propulsion efficiency. This change also depends on the seasonal factors. There are various causes of long-term changes in ship performance. The main causes are the effects of hull dirt and aging. Long-term monitoring of these performances requires not only short-term monitoring on board but also long-term monitoring of each individual vessel on land. In other words, it is necessary to use ship-to-shore communications to properly classify the information received, compile it into a database, analyze it statistically and evaluate its performance. Such long-term monitoring enables the provision of information to the ship, such as weather routing according to the performance of the individual ship, which can be used for appropriate navigation in route selection, speed and course selection in actual voyages in the subsequent. Here is where the need for monitoring of ship and engine motions arises.

1.3.4   Requirements for Monitoring System

Now let us examine the actual requirements for a new monitoring system.

(1) Required signals and aggregation
   When the monitoring system described in this chapter is implemented, it will be necessary to measure environmental information such as ship's position, wind direction and wind force, and signals of the rolling and pitching angles, which have not been measured so far. In addition, time series of engine information such as torque, thrust, fuel consumption and CO_2 emissions are also important.

(2) Real-time performance and synchronization
   Real-time performance is the most important requirement for a monitoring system. In addition, the requirement for synchronization is important, as both ship and engine motions must be sampled at the same time. In this case, the recently obligated **voyage data recorder** (VDR) and the widespread of **on board LAN**, which is being standardized, will help as ship and engine motions information collection systems.

(3) Detecting the period, predicted trend and stability of ship and engine motions
   The minimum necessary statistics and parameters should be extracted from the obtained data, and the system should be equipped with a function to provide up-to-date predictive

information to operators and analysts regarding the current **trend** of the ship and engine motions, whether the system is **stable**, how long the basic **period** is, what the values are if there are missing data and what the values will be in the short term and in the long term.

(4) Relevance between the signal elements obtained

The function should provide information to the operator and analyst about the degree of influence between each variable, such as what the ship motion affecting the propeller torque and how much it is affecting the propeller torque.

(5) Gain adjustment function for rudder, governor, etc.

Autopilots, which are typical control systems for ships, and speed regulators (governors), which control the RPM of the main engine, must be equipped with a function to give gains adapted to the current state.

(6) Classification of information

It is necessary to have a function to classify the obtained information and make into a database for easy use in preparation for future utilization.

(7) Information exchange function with land

It is necessary to enhance the ship-to-shore communication function that sends information obtained at sea to land and analysis results on land back to the ship. In this case, it is impractical to send all signals obtained to land. A monitoring system that also takes into account information exchange with land should have the function of compressing ship and engine information.

1.4 Previous Research Methods on Prediction of Ship and Engine Motions

1.4.1 Research Areas on Ship Performance

The development of a monitoring system described in the previous section requires knowledge of a wide range of ship and engine motions. This section describes what are the research results and issues that have been addressed to date. Chapter 16 provides a brief summary of the theoretical treatment of ship and engine motions[63].

The performance related to the first type of motion with restoring power is called **sea keeping ability**. Control measures include anti-rolling tanks and fin stabilizers, but safety is basically ensured as long as the restoring force is sufficient. As for the second type of motion performance without restoring power, it can be divided into two categories: the yawing is controlled mainly by using the rudder (**maneuverability**) and the surging speed is controlled mainly by using propellers and other propulsion devices at the stern side of the hull (**propulsion efficiency**), which mainly controls the surging speeds. In previous studies, these three types of dynamic performance have usually been treated separately.

1.4.2 Seakeeping Qualities

Seakeeping qualities are mainly concerned with the ship's ability to withstand rolling, pitching and heaving motion during ocean navigation, especially in rough seas. When designing a ship, it is necessary to pay attention to the increase or decrease of restoring power, slamming, seawater impingement and so on. Among these ship's motions, the most dangerous motion in rough seas is rolling. In designing a ship, the rolling performance is determined based on parameters such as breadth, length overall and hull frontal shape, etc., in accordance with established rules, so that the ship's restoring force curve is sufficient. In the field of seakeeping ability research, hydrodynamic studies centering on the potential theory have progressed, and today, a method called "**strip theory**," which divides the hull into about 10 equal parts in the longitudinal direction, determines the hydrodynamic forces acting on each element (strip) and adds them to calculate the overall motion, has been developed. Strip theory considers disturbances as regular waves such as sinusoidal waves, and calculates the motion within such regular waves while changing the period. In particular, it is said to match the actual motions of longitudinal and vertical motions, and is used to calculate operational limits, etc., because these motions can be calculated given a ship type and representative regular waves (called representative waves). However, it is not suitable for monitoring systems that require realistic information because the waves cannot be measured for the real-time motion of a ship on a complex and irregular sea surface, which is the subject of this book.

1.4.3   Maneuverability

The maneuverability of a vessel is basically designed to navigate in calm sea, i.e., surface of deep sea without waves and wind, at navigational speed. The ship is equipped with a rudder as a means of maneuvering. Therefore, as the ship's maneuverability, the rudder's ability to turn the ship when the rudder is turned (**turning ability**) and the rudder's ability to keep the ship on a straight course while keeping the ship on the desired course (**course-keeping ability**) are important. The rudder area, rudder shape (especially aspect ratio) and commutation function are designed to ensure the required performance as design elements. In order to ensure good course-keeping performance, a vessel's **course stability** must be secured. Turning ability and course stability are inversely related. That is, a ship with poor course stability generally has good turning performance, and a ship with good course stability has poor turning performance. The degree of hull fatness, known as the **block coefficient**, has a significant effect on these abilities. Fat-bodied ships, such as tankers and special carriers, are known to have poor course stability.

When sailing on the ocean, maneuverability is also strongly affected by sea and weather conditions. The maneuverability is affected not only by the shape of the hull below the water's surface, but also by the size and shape of the upper structure above the water's surface. In such a disturbance, as yawing of the ship increases, the ship sails obliquely and affected by other motions. In addition, depending on the direction of the disturbance, maneuverability can change significantly, making it difficult to maintain the course. The navigator must maneuver a ship designed for performance in calm water, with limited maneuvering range and under strong and changing disturbances. Today, in order to adapt to these changes in disturbance and restriction of maneuvering space in the open sea and maintain the desired course, many ships are equipped with **autopilot** (automatic steering system) to assist the ship's operator. In addition, a **tracking system** has been developed to track the planned course. The monitoring requires prediction using many variables, such as yawing and rolling when the ship is steered under the current conditions. The current model of maneuverability based on hydrodynamic considerations and experiments is not a flexible model for the problem of predicting such a large number of variables in such changing waves, so further work is needed.

1.4.4   Propulsion Performance

The propulsion performance of a ship is the performance related to the motion of a propulsion engine, such as a propeller placed near the stern, in the bow to aft direction of the

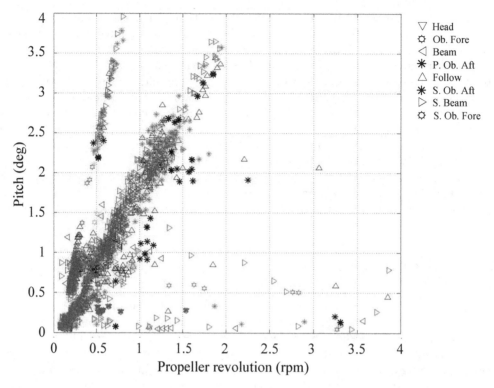

FIGURE 1.10
Scatter plot of the standard deviations of the ship's pitching and propeller revolution

hull[63]. Propulsion performance is strongly dependent on the ship type and is closely related to the development of frictional resistance-reduced vessels. The stern flow of a vessel during navigation is referred to as **wake**, and propulsion performance is strongly influenced by the shape of the hull upstream of it. Figure 1.10 shows a scatter plot of the standard deviation of the ship's pitching motion time series and the propeller revolution time series from a number of consecutive batch data set of 1200 time series, each measured every 1 second, measured continuously on a container ship. From this figure, it can be seen that the propeller shaft revolutions placed in the wake is strongly related to the ship's pitching motion. It is clear that such fluctuations in propeller shaft revolutions have an effect on the **thrust** generated by the main engine or the **torque** of the propeller.

1.5 Importance of Statistical Models in Ship and Engine Motions Analysis

1.5.1 Wiener's Theory and Statistical Estimation of the Spectrum

As mentioned in the previous section, it is impossible to accurately clarify irregular phenomena such as wave phenomena on the sea surface with the conventional framework of Newtonian mechanics. The need for a probabilistic/statistical paradigm that is different from the deterministic paradigm of Newtonian mechanics was long awaited. Therefore, the probabilistic/statistical analysis of random phenomena proposed by **Norbert Wiener** (1894-1964) emerged in the early 20th century as a product during World War II, and its applications spread to all fields of science and engineering from a relatively early in the post-war period. He himself named this theory "**cybernetics theory**,"[99][101], which means "helmsman" in Latin, because this new system of study is based on feedback. This theory was developed for the purpose of identifying the direction of fire for shooting down enemy aircraft and was born from mathematical considerations of the problem of predicting the future position of enemy aircraft based on their past and present positions. To solve this problem, it was not enough to consider aircraft dynamics alone, but rather it was necessary to introduce a probabilistic/statistical method of predicting the future based on a function called **evaluation function**, which represents the correlation between present and past positions, speeds and other factors. Although the theory was well-known for its difficulty, since it had a wide range of applications, G. Neumann[66] and W. J. Pierson Jr.[90] and others applied it to the wave growth theory in oceanography.

Wiener's theory was a mathematical theory for time series of infinite length. Therefore, statistical considerations were necessary to obtain a statistically stable spectrum for discrete time series of finite length, and the establishment of a practical statistical treatment method had to wait for the statistical method collectively referred to as the **Blackman-Tukey (BT) method**[17]. In Wiener's theory, the correlation function obtained from irregular data of infinite length is Fourier transformed, shifting the field of analysis from the field of time to the field of frequency and transforming it into a quantity called the spectrum. The spectrum is a quantity that expresses the frequency on the horizontal axis and the strength of a particular periodic component of the original time series on the vertical axis in the form of its energy. However, the spectrum calculated in this way is called the **raw spectrum** and is highly variable. In order to reduce this high variability, it was necessary to know the expected nature of the time series under consideration, recognizing that the data were sampled by chance from the underlying population and to perform effective smoothing. Blackman and Tukey found that various moving average operations, called window functions, yielded smooth and statistically reliable spectra. In this way, the seemingly irregular time series in the time domain were shifted to the frequency domain for analysis, and the regularity hidden in the irregularity could be extracted, just as light can be spectrally decomposed into various colors of different frequencies by passing it through a prism. This method was immediately applied to cross-spectrum theory, which describes the relationship between two or more time series.

In the field of marine engineering, this theory was applied to the phenomenon of irregular ship's motion in ocean waves by Dennis and Pierson [21] and Yamanouchi [103] among others. In particular, Dr. Yamanouchi pointed out the possibility of an important information leakage in the processing method of the stage called "window operation" used in the conventional Blackman-Tukey method when analyzing the rolling response waveforms to irregular wave heights generated in tank tests. Together with Dr. Hirotugu Akaike of the Institute of Statistical Mathematics, he proposed a correction method and contributed to the estimation of frequency response functions with a high degree of confidence [103] [8]. However, this method clarified the frequency structure of a time series by its spectrum in the frequency domain and was not a theory for smoothing or predicting time series in the time domain. Therefore, although it was an effective method for analyzing irregular phenomena in the frequency domain in naval architecture and nautical navigation, it was not used for active application to ship monitoring or control problems because it could not be predicted in the time domain.

1.5.2   Development of the Modeling of Time Series in Time Domain

Unlike the path followed by Wiener, attempts to predict time series in the time domain began with **G. U. Yule**[108]. A statistical model called **autoregressive model** in the time domain, which is different from the dynamical approach represented by differential equations, is introduced to predict the time series of the number of sunspot appearances. This method is called the **parametric method** because it assumes parameters in the statistical model. This idea was later developed as the **Box-Jenkins method**[18], but the method for determining the order of the autoregressive model was left to the subjective judgment of the analyst, and only low-order models could be handled, so an objective method for determining statistical models had to be established. Under these circumstances, Dr. Hirotugu Akaike of Japan developed a method for determining the order of the autoregressive model, which was used as a method for estimating parametric statistical models[20]. Dr. Akaike found that there is a bias proportional to the number of unknown parameters in the expected log-likelihood used as a measure of model certainty when the parameters of the model are estimated using the **maximum likelihood method**. By correcting for this bias, he showed that it is possible to determine which model is closer to the true model when multiple models are compared[4][5]. **Akaike Information Criterion** (**AIC**) is today not just a method for determining time series models, but has been widely applied in many fields as a method for determining statistical models[4][5][7][6].

Furthermore, Dr. Genshiro Kitagawa[45] of the Institute of Statistical Mathematics, Japan, realized interpolation, smoothing and forecasting of time series originally proposed by Wiener by making full use of the Kalman filter[38] for likelihood computation and state estimation by using the state-space representation of statistical models determined in this way. Kitagawa[45] has also developed a method for estimating states by replacing, the time transition in the Kalman filter, with the transition of particles with a statistical distribution generated by random numbers and simulating them in a computer.

1.5.3   Statistical Dynamics

The concept we use here is that of **statistical dynamics** as an alternative to physical dynamics. Statistical dynamics is a statistical concept that attempts to capture changing time series events in the context of their current and past statistical relationships. For example, even in a seemingly irregular time series such as the one shown in Section 1.2, regular relationships with a certain period are found on average. Statistical dynamics divides the temporal structure of such a fluctuating time series into two parts: the part **statistical model** that represents the regular fluctuations of the time series and the part called **noise** or **random fluctuations** that cannot be fully represented by this statistical model. In other words, the

$$\text{Actual Time Series Phenomenon} = \text{Statistical Model} + \text{Noise}.$$

Then, as a time function that represents this dynamic relationship over time, we introduce a quantity called the **covariance function** (or correlation function). Here, the covariance function is a statistic about temporal information that includes the statistical correlation between one's own past values and current values, temporal correlations with other variables, feedback relationships and so on. The statistical model is constructed from the regular relationship of this covariance function, and the part that cannot be fully expressed from the regular relationship and is completely uncorrelated with the past is considered to constitute the noise. This way of thinking is a broader concept that can be applied not only to dynamic phenomena such as ship motions, which will be discussed later, but also to the explanation of various social phenomena.

1.6 Purpose and Structure of This Book

1.6.1 Purpose of This Book

The purpose of this book is to develop statistical models of partial or total parametric time series of ship and engine motions, using the concept of statistical dynamics introduced in the previous section, for time series of ever-changing ship and engine motions from the point of view of ship operators, and to explain the control methods using the results of the models and the design methods of model-based monitoring systems that are effective for safe operations and voyages with reduced environmental impact by utilizing the information provided by this statistical model, and how to control vessels using the results of this system.

The concept of statistical dynamics described in this book is an attempt to discover new knowledge of ship and engine motions by introducing probabilistic and statistical methods. The following sections of the book are organized as follows.

1.6.2 Structure of This Book

This book as a whole is divided into three parts. Chapters 2 through 6 describe methods for analyzing time series in the frequency domain, i.e., methods for estimating the spectrum. Chapter 2 introduces the concepts of covariance function and spectrum of time series data, which play a central role in the analysis of time series. Chapter 3 explains the concept and properties of the Fourier transform, which plays an important role in spectrum analysis, and Chapter 4 explains how to estimate the spectrum with statistical reliability using the phenomena derived from its properties. Next, in Chapter 5, as a preparation for processing multi-dimensional data obtained from ships, methods for analyzing linear systems are presented and concepts such as impulse response functions and frequency response functions are introduced. Chapter 6 shows how to calculate cross-correlation functions from time series signal to obtain cross-spectra and explains how to obtain reliable frequency response functions in the frequency domain. Examples of applications to tank tests and full-scale actual experiments are then shown.

While the previous chapters have focused on analysis methods for spectra in the frequency domain, Part 2 describes methods for analyzing and controlling time series by making statistical models in the time domain. In other words, it is an explanation of the method of constructing statistical models of time series in the time domain, which has been rapidly developed after the discovery of Akaike information criterion (AIC). As a preparation for this, Chapter 7 describes various phenomena that occur when phenomena in the continuous domain are sampled and converted to discrete time series. In Chapter 8, we introduce a statistical model called the autoregressive model for stationary time series and show how to identify this statistical model using a method called the **minimum AIC estimation method** for constructing this model. We then show how to estimate the period that gives the major peak of the spectrum from the statistical model obtained in this way and explain its application to ship operations. In Chapter 9, we show how to fit a statistical model using AIC to a batch time series of data cut into intervals in order to satisfy the need for the designed monitoring system to always provide an up-to-date statistical model. This chapter is important because it shows in detail how to determine the optimal model for an actual monitoring system. The following two chapters explain the method of time series analysis using the state-space method of **Kalman filter**, which is a new topic developed by Dr. Kitagawa. In Chapter 10, we explain the method and show its application to ships, and show how to predict time series, estimate missing values, etc. Chapter 11 describes applications to ships, such as trend analysis, seasonal adjustment and estimation of the instantaneous spectrum, taking advantage of the flexibility of the state-space method, as well as a method to simulate the motion occurring at sea using a statistical model. The conversion of the statistical model obtained from the method in Chapter 9 into a state-space representation opens the horizon for the development of a monitoring system based on a new prediction method. In Chapter 12, we explain the clustering (classification) method for classifying time series models into databases. In Chapter 13, we explain how to perform simulations using statistical models. We show how to reproduce on land the phenomena occurring at sea.

In Chapter 14, the statistical model for one-dimensional time series is extended to a multi-dimensional time series model, and the unbiased impulse response function, frequency response function and relative noise contribution obtained from the model are used to analyze the relationship between ship and engine motions in a combined analysis. In particular, the noise contribution ratio, which expresses the relationship between variables, is shown to be a new and effective concept.

In Chapter 15, as an extension of the application to the analysis of ship and engine motions, we explain how to apply these statistical models to the optimal control system of a ship. The control problems addressed here include automatic steering (autopilot) systems, roll reduction system by using rudder, governor control systems and tracking (course following) systems.

Four chapters in Part 3 are appendices for readers who are not necessarily familiar with the concepts used in this book. That is, Chapter 16 provides an overview of ship and engine motion theory, Chapter 17 explains the basics of probability and statistics, Chapter 18 describes the method of the Kalman filter and Chapter 19 explains how to derive optimal control laws by dynamic programming.

1.6.3   On Reading This Book

The concept of statistical dynamics, which is the objective of this book, is not necessarily well understood in nautical science and naval architecture[106][107][80][84]. Therefore, in order to explain statistical dynamics, the structure of this book is such that the theoretical explanation is as detailed as possible at the beginning of each chapter, so that the reader can understand the theoretical background of the subject without referring to other books. As a result, the descriptions in this book do not necessarily describe actual monitoring system design methods in detail, so the reader is left to his/her own imagination when actually designing a model-based monitoring system using this book, but it is thought that this gives the reader more freedom in designing the system. At the end of each chapter, examples of applications obtained by the authors using actual ship and engine motion data are included to the extent possible.

Part I

Concept of Spectrum and its Estimation Methods

2

Covariance Function and Spectrum

A time series that varies continuously with time, such as ship motion, is called a continuous time series. A time series sampled by dividing this continuous time series into fixed intervals is called a discrete time series. The discrete time series thus obtained can be regarded statistically as a sample from the population. In this chapter, given samples of such a discrete time series, we will learn how to represent them in various figures to grasp their characteristics, and how to express the time series as a weighted sum of known functions[35].

2.1 How to Look at the Time Series

2.1.1 Frequency Distribution and Scatter Plot of Time Series

Let the continuous time series $x(t)$ be sampled for each small interval Δt as in Figure 2.1 and numbered in order from the sampling start, and we obtained N time series $\{x(s), s = 1, 2, \cdots, N\}$ as shown in Figure 2.2.

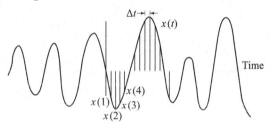

FIGURE 2.1
Discretization of time series

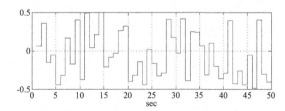

FIGURE 2.2
Example of discretized time series

Such a time series is called a **discrete time series**. In contrast, the original time series is called a **continuous time series**. The lower plot in Figure 2.3 shows the frequency of the values in the time series shown above, which is called the **frequency distribution** or **histogram**.

DOI: 10.1201/9781003428565-2

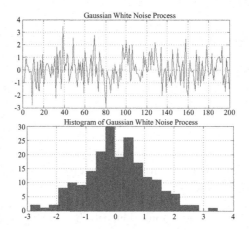

FIGURE 2.3
Histogram of a time series

2.1.2   Scatter Plot of Time Series

In the analysis of time series, the order of events that occurred is important. To understand this, it is useful to draw **scatter plot** of a time series with the interval between observations of the time series as a parameter. A scatter plot of a time series is a diagram in which the quantity occurring at a certain time is plotted on the horizontal axis, the value occurring l time ahead is plotted on the vertical axis. Here, l is the parameter and is referred to as **lag**.

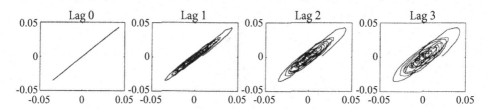

FIGURE 2.4
Scatter plot of rolling

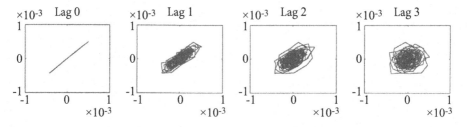

FIGURE 2.5
Scatter plot of encountering wave height

Figures 2.4 and 2.5 show scatter plots of the time series of rolling and encountered wave heights for a certain large container ship, with lags changing from left to right as $l = 0, 1, 2, 3$. For example, for $l = 1$, $x(s)$ is drawn on the horizontal axis and $x(s + 1)$ on the vertical axis.

To see what we can learn from these diagrams, let us consider the extreme cases. The waveform (a) in the upper panel of Figure 2.6 is a time series drawn by using pseudo-random

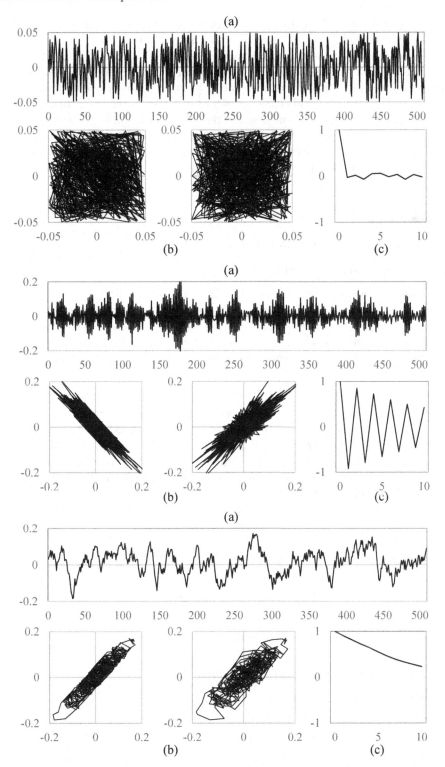

FIGURE 2.6
Various visualizations of time series, (a) waveform, (b) scatter plot with $l = 1$ and 2, (c) correlogram with $l = 0, \ldots, 10$. Top: random series, Middle: time series in which the sign changes every time, Bottom: time series in which the sign persists for a while

numbers to generate a random time series with mean 0 that is completely uncorrelated to the values of the past time series, and (b) shows a scatter plot for $l = 1, 2$. It can be seen that the scatter plots are widely scattered at $l = 1$ and 2. (c) is explained in the next section. On the other hand, waveform (a) in the middle panel of Figure 2.6 is an example of a time series showing quantities with almost alternating positive and negative signs around a mean value of 0. The scatter plot of this time series for $l = 1, 2$ is shown in (b) of the same figure. The scatter plot has a negative slope because the probability of a negative value following a positive value is large. Finally, waveform (a) in the lower panel of Figure 2.6 is an example of a time series in which the sign of the positive and negative values is not changed for a fairly long interval. Its scatter plot (b) has a positive slope.

2.2   Concept of Covariance Function

2.2.1   Variance and Covariance of Time Series

The previous section shows that referring to a scatter plot, it is possible to visually know whether or not there is a relationship between what occurred at a given time and what occurred before or after an interval of lag l. Taking this as a cue, we can use the product of the time series value $x(s)$ at time $s\Delta t$ and the value $x(s + l)$ at $l\Delta t$ time ahead of it to find out the information about them and consider the mean of the product,

$$\big(x(s + l) - m_x\big) \times \big(x(s) - m_x\big), \tag{2.1}$$

where m_x is the mean value of the time series over $N\Delta t$ time interval. N is the number of data sampled in the whole interval. In equation (2.1), if we fixed $l = 1$, i.e., $(N - l)$ times of neighbor multiplications are computed and the average is computed, it will approach zero, since in the case of random time series, positive and negative values are taken with the same probability. Figure 2.7(a) shows this situation, and it can be seen that it reaches zero at an early stage. Figure 2.7(b) shows the simulation results for the case of alternating positive and negative values, and Figure 2.7(c) shows the simulation results for the case of time series with the same sign for a certain period, showing that the former converges to negative values and the latter to positive values as the number of N increases.

In probability and statistics, the mean value of equation (2.1);

$$\frac{1}{N} \sum_{t=1}^{N-l} \big(x(s + l) - m_x\big)\big(x(s) - m_x\big) \tag{2.2}$$

is called **covariance** for $l \neq 0$ and **variance** for $l = 0$.

2.2.2   Autocovariance Function of Time Series

In time series analysis, considering lag l as a parameter

$$\begin{cases} R_{xx}(0) = \dfrac{1}{N} \sum_{s=1}^{N} \big(x(s) - m_x\big)^2, \quad l = 0 \\[4mm] R_{xx}(l) = \dfrac{1}{N} \sum_{s=1}^{N-l} \big(x(s + l) - m_x\big)\big(x(s) - m_x\big), \quad l = 1, 2, \cdots, L \end{cases} \tag{2.3}$$

is called the **autocovariance function**.

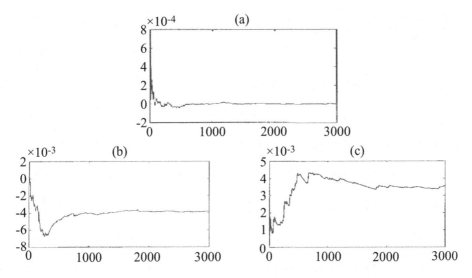

FIGURE 2.7
Convergence of sum of products as N increases. (Horizontal axis: $N\Delta t$)

Figure 2.6(c), depicted earlier, shows the autocorrelation function for each time series for lags $l = 0, 1, 2, \cdots, 10$. First, the time series in the upper panel are for random time series and are zero except for lag 0. This means that the past values are uncorrelated with the current values. The time series in the middle panel often alternates between positive and negative values. In this case, the correlation function takes its highest value at lag 0, and thereafter its absolute value decreases while alternating between positive and negative values. In the case of the lower panel, the sign of the time series is constant to some extent, and the correlation function reaches its highest value at lag 0 and gradually decreases to 0. This decrease with larger lags indicates that the values at a certain point in time do not correlate with the values in the past or distant future far from that point in time.

2.2.3　Definition and Properties of the Correlogram

Now, if $l \neq 0$ in equation (2.3), then $|C_{xx}(l)| < C_{xx}(0)$ holds (see Section 17.6, "Properties of correlation coefficient 1"). That is, the autocovariance function is maximal at lag 0. Therefore, the **autocorrelation function**, the quantity normalized by the value at lag=0;

$$\rho_{xx}(l) = \frac{R_{xx}(l)}{R_{xx}(0)} \tag{2.4}$$

is maximum at lag 0 and its value is 1. The diagram of $\rho_{xx}(l)$ is called the **correlogram**. The symmetry of the correlogram with respect to lag 0;

$$\rho_{xx}(l) = \rho_{xx}(-l) \tag{2.5}$$

will be immediately found in Figure 2.8.

Figure 2.9 shows a correlogram of time series of 1200 points of encounter wave height, Yacc at bow, rolling and pitching of a container ship observed every 1 second. From these correlograms, the followings are seen;

1.　Especially in the case of rolling, pitching and heaving, the correlograms decay while repeating between positive and negative at a certain period.

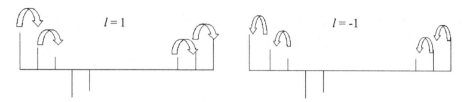

FIGURE 2.8

Sum of products in forward direction (left) and sum of products in backward direction (right)

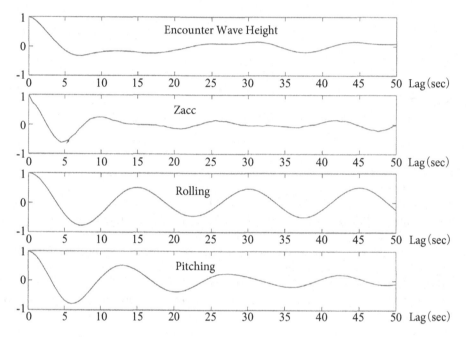

FIGURE 2.9

Correlation function of time series of ship motion

2. The rate of decay for one cycle of oscillation is greater in the case of pitching.

3. For encounter wave heights, there is no regular relationship approaching zero after $l = 5\,\text{sec}$.

2.3   Fourier Series and Time Series Analysis

2.3.1   Fourier Series

As shown in Figure 1.7, the time series of a ship's motion is apparently irregular, unlike a regular sine wave. Let us consider the problem of expressing such irregular waveforms by the weighted sum of sine and cosine waves of various periods. In this section, we deal with discrete time series as shown in Figure 2.2.

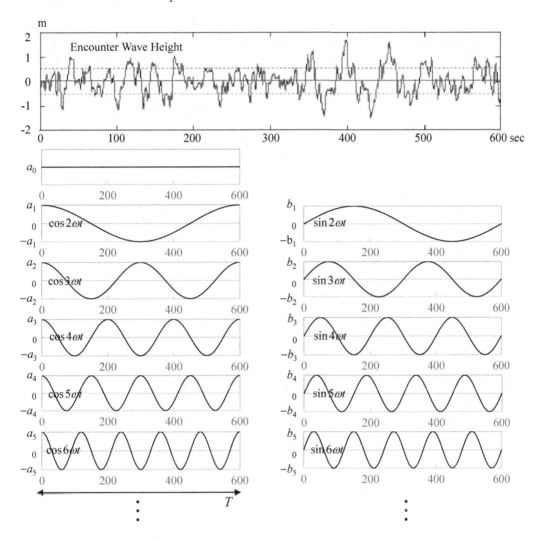

FIGURE 2.10
Fourier decomposition of time series

J. B. J. Fourier (1768–1830) constructed a function that passes through all the values at $t = 0, \pm T/2n, \pm 2T/2n, \cdots, \pm(n-1)T/2n, T/2$ of $N = 2n$ time series taken at equal intervals in the interval $[-T/2, T/2)$. It is possible to express the function as shown in Figure 2.10 by combining sine and cosine waves of various periods. That is, $x_T(t)$ can be represented by n sine and cosine waves such as $f_1 = 1/T, 2f_1, \cdots$ by adjusting the amplitude a_m and b_m, respectively. Here, f_1 is the longest period as $\cos \omega t$ and $\sin \omega t$ in Figure 2.10 and is called **fundamental frequency**. Also, $\omega = 2\pi f_1$ is called **fundamental angular frequency**. In this case, if the sampling period is set to the original time series by $\Delta t = T/2n$, the rth time point from the beginning of the time series can be expressed as $t = r\Delta t$, so $x(r) = x(r\Delta t)$ can be written as follows;

$$x_T(r) = a_0 + 2\sum_{m=1}^{n-1}\left(a_m\cos\frac{2\pi mr}{N} + b_m\sin\frac{2\pi mr}{N}\right) + a_n\cos\frac{2\pi nr}{N}. \tag{2.6}$$

By the way, the sine and cosine functions have an important property called the **orthonormality**, namely, the total sum of the products with different functions is zero;

$$\sum_{r=-n}^{n-1} \sin \frac{2\pi kr}{N} \cos \frac{2\pi mr}{N} \;=\; 0, \quad k, m = \text{integers} \tag{2.7}$$

$$\sum_{r=-n}^{n-1} \sin \frac{2\pi kr}{N} \sin \frac{2\pi mr}{N} \;=\; \begin{cases} 0 & k \neq m \\ N/2 & k = m \neq 0, n \\ 0 & k = m = 0, n \end{cases} \tag{2.8}$$

$$\sum_{r=-n}^{n-1} \cos \frac{2\pi kr}{N} \cos \frac{2\pi mr}{N} \;=\; \begin{cases} 0 & k \neq m \\ N/2 & k = m \neq 0, n \\ N & k = m = 0, n. \end{cases} \tag{2.9}$$

This formula can be easily proved using the additive theorem for trigonometric functions. In order to extract a_m and b_m from equation (2.6) using this relationship, we add up equation (2.6) by multiplying it by $\cos(\frac{2\pi mr}{N})$ or $\sin(\frac{2\pi mr}{N})$,

$$\begin{cases} a_m = \dfrac{1}{N} \displaystyle\sum_{r=-n}^{n-1} x_T(r) \cos \dfrac{2\pi mr}{N} \\ b_m = \dfrac{1}{N} \displaystyle\sum_{r=-n}^{n-1} x_T(r) \sin \dfrac{2\pi mr}{N}, \end{cases} \tag{2.10}$$

where a_m and b_m are referred to as the **Fourier cosine coefficient** and the **Fourier sine coefficient**, respectively[85].

Equation (2.6) can also be expressed in the following polar form:

$$x_T(r) = R_0 + 2 \sum_{m=1}^{n-1} R_m \cos \left(\frac{2\pi mr}{N} + \phi_m \right) + R_n \cos \frac{2\pi nr}{N}, \tag{2.11}$$

where R_m, ϕ_m, a_m and b_m are given by

$$R_m = \sqrt{a_m^2 + b_m^2}, \qquad \phi_m = \tan^{-1} \left(-\frac{b_m}{a_m} \right)$$
$$a_m = R_m \cos \phi_m, \qquad b_m = -R_m \sin \phi_m.$$

Next, consider the mean of the squares of the original time series;

$$\frac{1}{N} \sum_{r=-n}^{n-1} x_T^2(r).$$

In this case, again, using equations (2.7)–(2.9), we immediately obtain

$$\frac{1}{N} \sum_{r=-n}^{n-1} x_T(r)^2 = R_0^2 + 2 \sum_{m=1}^{n-1} R_m^2 + R_n^2. \tag{2.12}$$

This relationship is called the **Parseval's identity**. When a time series of length N is obtained, assuming that the mean value is R_0, the variance of this time series is given by

$$\sigma^2 = \frac{1}{N} \sum_{r=-n}^{n-1} (x_T(r) - R_0)^2 = 2 \sum_{m=0}^{n-1} R_m^2 + R_n^2.$$

In other words, considering the Perceval's identity physically, we can say that *the average power or the variance of a given time series is equal to the sum of the squares of the Fourier coefficients.* Furthermore, the element R_m^2 of each right-hand term represents the square of the amplitude at each frequency of the Fourier expansion of the original signal $x_T(r)$. The graphical representation of this R_m^2 on the basis of the corresponding frequency is called the **Fourier line spectrum**. In other words, it can be defined that *the Fourier line spectrum is the square of the Fourier coefficient.*

2.3.2 Fourier Integral

So far, we have treated the time series defined in N equally spaced points ($t = r\Delta t$) in the interval $-\frac{T}{2} \leq t < \frac{T}{2}$. To generalize the results of the previous section, first consider reducing the interval Δt[30]. Since T is fixed, making Δt smaller is achieved by making N larger. Suppose now that $N \to \infty$ while holding $N\Delta t = T$. In this case, since the equation (2.10) can be written as

$$a_m = \frac{1}{N\Delta t} \sum_{r=-n}^{n-1} x_T(r) \cos \frac{2\pi m r \Delta t}{N\Delta t} \Delta t,$$

and $x_T(r\Delta t)\Delta t \to x_T(t)dt$, it can be converted to the following integral representation:

$$a_m = \frac{1}{T} \int_{-T/2}^{T/2} x_T(t) \cos \frac{2\pi m t}{T} dt, \tag{2.13}$$

where dt is the infinitecimal time in the limit of the time interval Δt. The same can be said for b_m and we obtain

$$b_m = \frac{1}{T} \int_{-T/2}^{T/2} x_T(t) \sin \frac{2\pi m t}{T} dt. \tag{2.14}$$

Further, equation (2.6) can be expressed as

$$x_T(t) = a_0 + 2 \sum_{m=1}^{\infty} \left(a_m \cos \frac{2\pi m t}{T} + b_m \sin \frac{2\pi m t}{T} \right). \tag{2.15}$$

In addition, the Perseval's identity (2.12) can be rewritten as

$$\frac{1}{T} \int_{-T/2}^{T/2} x_T(t)^2 dt = R_0^2 + 2 \sum_{m=1}^{\infty} R_m^2 = a_0^2 + 2 \sum_{m=1}^{\infty} (a_m^2 + b_m^2). \tag{2.16}$$

So far, we have observed the time series $x(t)$ spread over the infinite interval $[-\infty, \infty]$ restricted in the finite interval $[-\frac{T}{2}, \frac{T}{2})$. If we express this precisely in terms of a function, we can say that we opened a window $x_T(t)$ of height 1 as shown in Figure 2.11 and observed the original time series spread over an infinite time axis,

$$x_T(t) = \begin{cases} x(t), & -\frac{T}{2} \leq t < \frac{T}{2} \\ 0, & \text{other areas.} \end{cases} \tag{2.17}$$

That is, $x_T(t)$ is equal to $x(t)$ for the finite interval $[-\frac{T}{2}, \frac{T}{2})$, and zero for the other areas.

Next, consider extending T to infinite. This does not change equations (2.13) and (2.14), only that $x_T(t)$ on the left-hand side in equation (2.15) gradually approaches $x(t)$. To

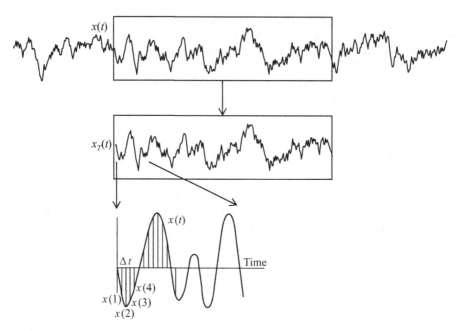

FIGURE 2.11
Sampling of time series

make the trigonometric part of the preparation easier to see, consider the Fourier series representation of the trigonometric functions in complex representation. For that, to express equation (2.15), we apply the well-known formula:

$$\sin\eta = \frac{1}{2i}(e^{i\eta} - e^{-i\eta}), \qquad \cos\eta = \frac{1}{2}(e^{i\eta} + e^{-i\eta}), \tag{2.18}$$

where $i = \sqrt{-1}$. Since

$$2a_m \cos\frac{2\pi mt}{T} = a_m\left(e^{i\frac{2\pi mt}{T}} + e^{-i\frac{2\pi mt}{T}}\right)$$

$$2b_m \sin\frac{2\pi mt}{T} = -b_m i\left(e^{i\frac{2\pi mt}{T}} - e^{-i\frac{2\pi mt}{T}}\right),$$

if we make the sum of two expressions, we obtain

$$(a_m - b_m i)e^{i\frac{2\pi mt}{T}} + (a_m + b_m i)e^{-i\frac{2\pi mt}{T}}. \tag{2.19}$$

On the other hand, since from equations (2.13), (2.14) and (2.18),

$$a_m = \frac{1}{2T}\int_{-T/2}^{T/2} x_T(t)e^{i\frac{2\pi mt}{T}} dt + \frac{1}{2T}\int_{-T/2}^{T/2} x_T(t)e^{-i\frac{2\pi mt}{T}} dt$$

$$b_m i = \frac{1}{2T}\int_{-T/2}^{T/2} x_T(t)e^{i\frac{2\pi mt}{T}} dt - \frac{1}{2T}\int_{-T/2}^{T/2} x_T(t)e^{-i\frac{2\pi mt}{T}} dt,$$

we obtain

$$a_m - b_m i = \frac{1}{T}\int_{-T/2}^{T/2} x_T(t)e^{-i\frac{2\pi mt}{T}} dt \equiv X_m$$

$$a_m + b_m i = \frac{1}{T}\int_{-T/2}^{T/2} x_T(t)e^{i\frac{2\pi mt}{T}} dt \equiv X'_m.$$

Thus, equation (2.15) can be written as

$$x_T(t) = a_0 + \sum_{m=1}^{\infty} \left(X_m e^{i\frac{2\pi mt}{T}} + X'_m e^{-i\frac{2\pi mt}{T}} \right). \tag{2.20}$$

However, since we have

$$\sum_{m=1}^{\infty} X'_m e^{-i\frac{2\pi mt}{T}} = \sum_{m=-\infty}^{-1} X'_{-m} e^{i\frac{2\pi mt}{T}}$$

and

$$X'_{-m} = \frac{1}{T} \int_{-T/2}^{T/2} x(t) e^{-i\frac{2\pi mt}{T}} dt = X_m,$$

the equation (2.20) is reexpressed as

$$x_T(t) = \sum_{m=-\infty}^{\infty} X_m e^{i\frac{2\pi mt}{T}} \tag{2.21}$$

$$X_m = \frac{1}{T} \int_{-T/2}^{T/2} x_T(t) e^{-i\frac{2\pi mt}{T}} dt, \quad -\infty < m < \infty. \tag{2.22}$$

Substituting equation (2.22) into (2.21) yields

$$x_T(t) = \sum_{m=-\infty}^{\infty} \left(\frac{1}{T} \int_{-T/2}^{T/2} x_T(t) e^{-i\frac{2\pi mt}{T}} dt \right) e^{i\frac{2\pi mt}{T}}. \tag{2.23}$$

Here, let δf_m be the difference in frequencies between neighbors, then

$$\delta f_m = f_{m+1} - f_m = \frac{m+1}{T} - \frac{m}{T} = \frac{1}{T}.$$

Therefore, equation (2.23) becomes

$$x_T(t) = \sum_{m=-\infty}^{\infty} \left(\int_{-T/2}^{T/2} x_T(t) e^{-i\frac{2\pi mt}{T}} dt \right) \delta f_m e^{i\frac{2\pi mt}{T}}.$$

It follows that by expanding the domain to $-\infty < t < \infty$, $x_T(t)$ becomes $x(t)$ and that $\frac{m}{T} \to f$ as shown in Figure 2.12, and $\delta f_m \to df$. Also the term in parentheses is the T times X_m from equation (2.22), i.e., it approaches to the limit value of TX_m,

$$\lim_{T \to \infty} TX_m = X(f). \tag{2.24}$$

As a result, equations (2.23) and (2.22) are extended as

$$x(t) = \int_{-\infty}^{\infty} X(f) e^{i2\pi ft} df \tag{2.25}$$

$$X(f) = \int_{-\infty}^{\infty} x(t) e^{-i2\pi ft} dt. \tag{2.26}$$

Equation (2.26) is called the **Fourier transformation** of the function $x(t)$, and equation (2.25) is referred to as the **Fourier inverse transformation**[30][86].

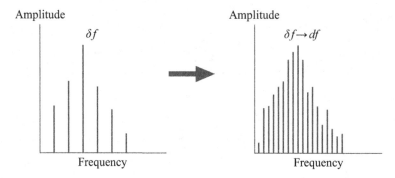

FIGURE 2.12
Change of frequency interval

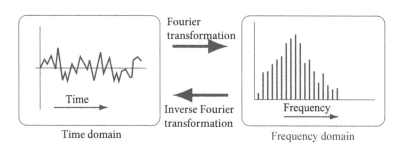

FIGURE 2.13
Transition between time domain and frequency domain

From equation (2.16), Parceval's identity can be written as

$$\int_{-\infty}^{\infty} x^2(t)dt = \lim_{T\to\infty} \sum_{m=-\infty}^{\infty} |TX_m|^2 \frac{1}{T} = \int_{-\infty}^{\infty} |X(f)|^2 df. \qquad (2.27)$$

As shown in Figure 2.13, the above operations mean that the viewpoint of the time series is shifted from the function $x(t)$ in the **time domain** to the function $X(f)$ in the **frequency domain** by Fourier transformation, and conversely from the frequency domain to the time domain by Fourier inverse transformation.

By the way, such a relation would not be possible without the existence of the Fourier integral. Now consider the absolute value of $X(f)$. In this case, since $\left|e^{-i2\pi ft}\right| = 1$, the following inequality holds;

$$|X(f)| = \left|\int_{-\infty}^{\infty} x(t)e^{-i2\pi ft}dt\right| \leq \int_{-\infty}^{\infty} \left|x(t)e^{-i2\pi ft}\right| dt \leq \int_{-\infty}^{\infty} |x(t)|dt. \qquad (2.28)$$

That is, for the Fourier integral to exist, the integral of the absolute value of the original time series must approach $x(t) \to 0$ as $|t| \to \infty$. However, this condition is not satisfied in real time series. This makes spectral analysis difficult, and suggest that some other quantity needs to be derived from the time series.

The clue of this can be found in Figure 2.9. That is, the correlation function shown in Figure 2.9 gradually decreases as the lag increases. If we restrict ourselves to the time series with correlation function that decreases as the lag increases, then their Fourier transforms exist.

TABLE 2.1

Calculation method as per formula (2.32)

$x_{-3}x_{-3}$	$x_{-3}x_{-2}$	$x_{-3}x_{-1}$	$x_{-3}x_0$	$x_{-3}x_1$	$x_{-3}x_2$
$x_{-2}x_{-3}$	$x_{-2}x_{-2}$	$x_{-2}x_{-1}$	$x_{-2}x_0$	$x_{-2}x_1$	$x_{-2}x_2$
$x_{-1}x_{-3}$	$x_{-1}x_{-2}$	$x_{-1}x_{-1}$	$x_{-1}x_0$	$x_{-1}x_1$	$x_{-1}x_2$
x_0x_{-3}	x_0x_{-2}	x_0x_{-1}	x_0x_0	x_0x_1	x_0x_2
x_1x_{-3}	x_1x_{-2}	x_1x_{-1}	x_1x_0	x_1x_1	x_1x_2
x_2x_{-3}	x_2x_{-2}	x_2x_{-1}	x_2x_0	x_2x_1	x_2x_2

2.3.3 Periodogram and Covariance Function

Arthur Schuster defined the quantity

$$I(f_m) = N\Delta t |R_m|^2 = N\Delta t \left|a_m^2 + b_m^2\right|, \tag{2.29}$$

where a_m and b_m at frequency f_m are defined as in equation (2.10), and called the plot of $I(f_m)$ versus frequency as the **periodogram**. Next, let us examine the relationship between this periodogram and the covariance function. For this purpose, using the Fourier cosine coefficient a_m and Fourier sine coefficient b_m at frequency f_m, consider the following complex number[18],

$$d_{f_m} = a_m - ib_m, \quad i = \sqrt{-1}. \tag{2.30}$$

Using this quantity, the periodogram can be written as

$$I(f_m) = N\Delta t \left|a_m^2 + b_m^2\right| = N\Delta t d_{f_m}\bar{d}_{f_m} \tag{2.31}$$

where $\bar{d}_{f_m}$ is the complex conjugate of d_{f_m}. Substituting the defining expressions for a_m and b_m into d_{f_m}, we obtain, using equation (2.18)

$$
\begin{aligned}
d_{f_m} &= \frac{1}{N}\sum_{r=-n}^{n-1} x_T(r)\cos(2\pi f_m r\Delta t) - \frac{1}{N}\sum_{r=-n}^{n-1} ix_T(r)\sin(2\pi f_m r\Delta t) \\
&= \frac{1}{N}\sum_{r=-n}^{n-1} \left(x_T(r)e^{-i2\pi f_m r\Delta t}\right).
\end{aligned}
$$

Using this result to calculate $I(f_m)$, we obtain

$$
\begin{aligned}
I(f_m) &= \frac{\Delta t}{N}\sum_{r=-n}^{n-1} x_T(r)e^{-2\pi f_m r\Delta t}\sum_{q=-n}^{n-1} x_T(q)e^{2\pi f_m q\Delta t} \\
&= \frac{\Delta t}{N}\sum_{r=-n}^{n-1}\sum_{q=-n}^{n-1} x_T(r)x_T(q)e^{-2\pi f_m(r-q)\Delta t}.
\end{aligned}\tag{2.32}
$$

By the way, if we compute this double sum according to the definition, for example, for $n = 3$ ($N = 6$), equation (2.32) for $\{x(-n), x(-n+1), \cdots, x(0), x(1), \cdots, x(n)\}$ becomes as shown in Table 2.1. In this table, $x(r)$ is written as x_r and only the product terms with respect to x_r are shown. The actual calculation is performed by multiplying each row from left to right by the corresponding exponential term while summing.

2.3.4 From Periodogram to Spectrum

In the case of Table 2.1, the same results can be obtained by fixing x_i in the multiplication of two variables $x_i x_j$, varying the lag l in the vertical direction to determine x_j and calculating

TABLE 2.2
Calculation method with different order of sums

1	2	3	4	5	6	$|l|$	$\exp^{-i2\pi\frac{l}{N\Delta t}(\cdot)\Delta t}$
$x_{-3}\,x_2$						5	(5)
$x_{-3}\,x_1$	$x_{-2}\,x_2$					4	(4)
$x_{-3}\,x_0$	$x_{-2}\,x_1$	$x_{-1}\,x_2$				3	(3)
$x_{-3}\,x_{-1}$	$x_{-2}\,x_0$	$x_{-1}\,x_1$	$x_0\,x_2$			2	(2)
$x_{-3}\,x_{-2}$	$x_{-2}\,x_{-1}$	$x_{-1}\,x_0$	$x_0\,x_1$	$x_1\,x_2$		1	(1)
$x_{-3}\,x_{-3}$	$x_{-2}\,x_{-2}$	$x_{-1}\,x_{-1}$	$x_0\,x_0$	$x_1\,x_1$	$x_2\,x_2$	0	(0)
$x_{-3}\,x_{-2}$	$x_{-2}\,x_{-1}$	$x_{-1}\,x_0$	$x_0\,x_1$	$x_1\,x_2$		$|-1|$	(−1)
$x_{-3}\,x_{-1}$	$x_{-2}\,x_0$	$x_{-1}\,x_1$	$x_0\,x_2$			$|-2|$	(−2)
$x_{-3}\,x_0$	$x_{-2}\,x_1$	$x_{-1}\,x_2$				$|-3|$	(−3)
$x_{-3}\,x_1$	$x_{-2}\,x_2$					$|-4|$	(−4)
$x_{-3}\,x_2$						$|-5|$	(−5)

as in Table 2.2. Column 7 in the table shows the absolute value of the lag. This l is multiplied directly into $(\cdot)$ in the exponential part of each row of column 8, and finally summed.

This means that in equation (2.32), we define $l = r - q$, and further we set $x(r + |l|) = 0$ if $r + |l|$ is n or more, i.e.,

$$X_T(s) = \{\cdots, 0, x(-n), x(-n+1), \cdots, x(0), x(1), \cdots, x(n), 0, \cdots\}, \quad T = N\Delta t. \quad (2.33)$$

Then $I(f_m)$ can be written as

$$I(f_m) = \frac{\Delta t}{N} \sum_{l=-N+1}^{N-1} \sum_{r=-n}^{n-1} x_T(r)x_T(r + |l|)e^{-2\pi f_m l \Delta t}.$$

Hereafter, when we use subscripts such as $X_T(s)$, it will mean in the sense of equation (2.33), and therefore $x_T(s) \in X_T(s)$ means $x_T(s)$ is an element of the set $X_T(s)$.

Thus, for $f_m = \frac{m}{N\Delta t}$, the periodogram is given by

$$
\begin{aligned}
I(f_m) &= \frac{\Delta t}{N} \sum_{l=-(N-1)}^{N-1} \sum_{r=-n}^{n-1} x_T(r)x_T(r + |l|)e^{-i2\pi\frac{m}{N\Delta t}l\Delta t} \\
&= \Delta t \sum_{l=-(N-1)}^{N-1} \sum_{r=-n}^{n-1} \left(\frac{x_T(r)x_T(r + |l|)}{N}\right) e^{-i2\pi\frac{r}{N\Delta t}\Delta t l}, \quad x_T(s) \in X_T(s).
\end{aligned}
$$

Here, the sum of the operations in the inner parentheses

$$\frac{1}{N} \sum_{r=-n}^{n-1} x_T(r)x_T(r + |l|), \quad x_T(s) \in X_T(s)$$

is given as follows, from the definition of the covariance function,

$$R_{xx}(l) = \frac{1}{N} \sum_{r=-n}^{n-1} x_T(r)x_T(r + |l|), \quad x_T(s) \in X_T(s).$$

(The $|l|$ in the above equation corresponds to l considering backward sums of products in Table 2.2). Thus, there is a relation between the periodogram $I(f_m)$ and the covariance function as

$$I(f_m) = \Delta t \sum_{l=-(N-1)}^{N-1} R_{xx}(l) e^{-i2\pi f_m l \Delta t}. \tag{2.34}$$

This relation is called the discrete type **Wiener-Khinchin theorem**. This is also the definition of **sample spectrum**, which inherits the idea of periodograms. In other words, we can say that *"The sample spectrum is the Fourier transform of the covariance function."* The sample spectrum of $x_T(s)$ is usually denoted by $S_{xx}(f)$. Using the symmetry of the autocovariance function and the relation $e^{i\theta} = \cos\theta + i\sin\theta$, from equation (2.34) we obtain

$$S_{xx}(f) = \Delta t \left\{ R_{xx}(0) + 2 \sum_{l=1}^{N-1} R_{xx}(l) \cos(2\pi f l \Delta t) \right\}. \tag{2.35}$$

Since the frequency f of this definition is given by discrete time series data, a finite number of sample spectra are given only at discrete points. The lag of the covariance function has a total length of $2N-1$ from $1-N$ to $N-1$. Thus, the fundamental frequency is

$$f_1 = \frac{1}{2N\Delta t}.$$

Since the sample spectrum is given in terms of integer multiples of this fundamental frequency, equation (2.35) becomes

$$S_{xx}(f_m) = \Delta t \left\{ R_{xx}(0) + 2 \sum_{l=1}^{N-1} R_{xx}(l) \cos 2\pi \frac{m}{2N} l \right\}. \tag{2.36}$$

It should be noted here that in the above equation, the upper bound of the summing term l on the right-hand side is $N-1$. This implies that for discrete time series, we can find out such as;

1. The highest frequency that can be computed for a discrete time series is $\frac{1}{2\Delta t}$ with $k = N$ in the frequency term $\frac{k}{2N\Delta t}$. We call $\frac{1}{2\Delta t}$ the **Nyquist frequency**.

2. When calculating $R_{xx}(l)$, the number of multiplications decreases as the lag increases, and for $l = N-1$ there is only one multiplication. It should be noted that Table 2.2 described previously is triangulated with respect to the lag.

To summarize these results, the method of computing the spectrum based on the time series defined in the domain of $-\infty < t < \infty$ is either;

1. Estimating from the direct Fourier transform using equation (2.31) (**direct method** or **periodogram method**).

2. Estimating by Fourier transform after calculating the covariance function and using formula (2.34) (**covariance method**).

3

Fourier Transform and its Properties

The Fourier transform, defined in Chapter 2, played the role of changing the perspective and observing time series in frequency domain, which were physically observed in time domain. In this chapter, the Fourier transform is redefined and its properties and relation to the Laplace transform are summarized. See reference books for details[28][43].

3.1 Definition of Fourier Transform

Let us define the Fourier transform of a function $x(t)$ that varies with time t as

$$X(f) = \int_{-\infty}^{\infty} x(t)e^{-i2\pi ft}dt, \tag{3.1}$$

where $i = \sqrt{-1}$ and f is the frequency. From now on, the Fourier transform will be also written as

$$\mathfrak{F}[x(t)] = \int_{-\infty}^{\infty} x(t)e^{-i2\pi ft}dt. \tag{3.2}$$

Using this notation, the Fourier inverse transform is given by

$$x(t) = \mathfrak{F}^{-1}[X(f)] = \int_{-\infty}^{\infty} X(f)e^{i2\pi ft}df. \tag{3.3}$$

If we put $\omega = 2\pi f$, then equation (3.3) becomes

$$x(t) = \frac{1}{2\pi} \int_{-\infty}^{\infty} X(f)e^{i\omega t}d\omega. \tag{3.4}$$

Equations (3.2) and (3.3) are called **Fourier transform pair**. The condition for the existence of a Fourier transformation is the absolute integrability condition

$$\int_{-\infty}^{\infty} |x(t)|dt < \infty. \tag{3.5}$$

That is, $x(t)$ must converge to 0 as $|t|$ increases.

The Fourier transform $X(f)$ is generally complex and can be written as

$$X(f) = R(f) + iQ(f) = |X(f)|e^{i\phi(f)}. \tag{3.6}$$

$|X(f)|$ is called the **magnitude spectrum** and $\phi(f)$ is called the **phase spectrum** of $x(t)$.

When $x(t)$ is a real number, we have

$$R(f) = \int_{-\infty}^{\infty} x(t)\cos(2\pi ft)df, \qquad Q(f) = \int_{-\infty}^{\infty} x(t)\sin(2\pi ft)df. \tag{3.7}$$

DOI: 10.1201/9781003428565-3

The former is called **Fourier cosine transform** and the latter is called **Fourier sine transform**. When $x(t)$ is a real number, the following properties are available:

$$R(f) = R(-f), \qquad Q(f) = -Q(-f). \tag{3.8}$$

3.2 Fourier Transform of Commonly Used Functions

Here, the Fourier transform of functions of time, which are often used in systems analysis[28], are shown.

(1) Boxcar Function

The boxcar function shown in Figure 3.1 is defined by

$$x(t) = \begin{cases} 1, & |t| \leq d \\ 0, & |t| > d. \end{cases}$$

The Fourier transform of the boxcar function is given by (Figure 3.2)

$$\begin{aligned} \mathfrak{F}[x(t)] &= \int_{-\infty}^{\infty} e^{-i2\pi ft} dt = \int_{-d}^{d} e^{-i2\pi ft} dt \\ &= \left[\frac{e^{-i2\pi ft}}{-i2\pi f} \right]_{-d}^{d} = 2d \frac{\sin(2\pi fd)}{(2\pi fd)}. \end{aligned} \tag{3.9}$$

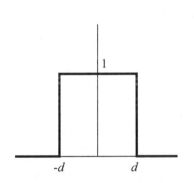

FIGURE 3.1
Boxcar function

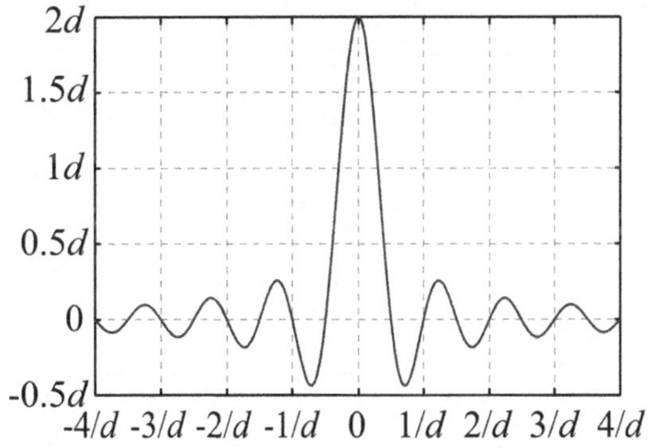

FIGURE 3.2
Fourier transform of a boxcar function

(2) Triangular function

The triangular function shown in Figure 3.3 is given by

$$x(t) = \begin{cases} 1 - |t|/d, & |t| \leq d \\ 0, & |t| > d. \end{cases} \tag{3.10}$$

Its Fourier transform is given by

$$d \left(\frac{\sin \pi fd}{\pi fd} \right)^2, \tag{3.11}$$

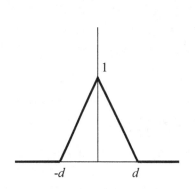

FIGURE 3.3
Triangular function

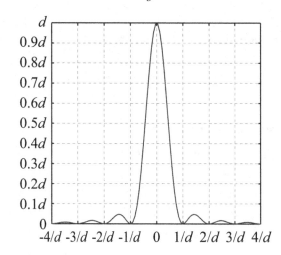

FIGURE 3.4
Fourier transform of a triangular function

and Figure 3.4 shows its shape. Compared with Figure 3.2, there are no drops to the negative side.

(3) Exponential function

Exponential function

$$x(t) = e^{-a|t|} \tag{3.12}$$

converges to zero as $|t| \to \infty$ when $a > 0$. Fourier transform of the exponential function is given by

$$
\begin{aligned}
X(f) &= \int_{-\infty}^{\infty} e^{-a|t|} e^{-i2\pi ft} dt \\
&= \int_{0}^{\infty} \left(e^{-(a+i2\pi f)t} + e^{-(a-i2\pi f)t} \right) dt \\
&= \left[\frac{1}{-(a+i2\pi f)} e^{-(a+i2\pi f)t} + \frac{1}{-(a-i2\pi f)} e^{-(a-i2\pi f)t} \right]_{0}^{\infty} \\
&= \frac{1}{a+i2\pi f} + \frac{1}{a-i2\pi f} = \frac{2a}{a^2 + (2\pi f)^2}. \tag{3.13}
\end{aligned}
$$

If this exponential function is defined on one side, i.e.,

$$x(t) = \begin{cases} e^{-at}, & t \geq 0 \\ 0, & t < 0, \end{cases} \tag{3.14}$$

(Figure 3.5), then its Fourier transform is given by

$$
\begin{aligned}
X(f) &= \int_{-\infty}^{\infty} x(t) e^{-i2\pi ft} dt \\
&= \frac{1}{a+i2\pi f}. \tag{3.15}
\end{aligned}
$$

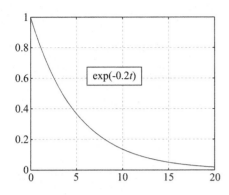

FIGURE 3.5
Exponential function

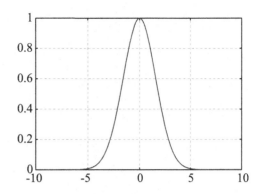

FIGURE 3.6
Normal distribution type exponential function ($\lambda = 0.2$)

(4) Function $x(t) = e^{-\lambda t^2}$

As shown in Figure 3.6 ($\lambda = 0.2$), this function has a normal distribution function form (Section 17.3.5) and it is known that

$$\int_{-\infty}^{\infty} e^{-\lambda t^2} dt = \sqrt{\frac{\pi}{\lambda}}. \tag{3.16}$$

Its Fourier transform is given by

$$
\begin{aligned}
X(f) &= \int_{-\infty}^{\infty} e^{-\lambda t^2} e^{-2\pi ft} dt = \int_{-\infty}^{\infty} e^{-\lambda \left(t^2 + i\frac{1}{\lambda} 2\pi ft\right)} dt \\
&= \int_{-\infty}^{\infty} e^{-\lambda \left(t + \frac{2\pi f}{2\lambda} i\right)^2} dt \, e^{-(2\pi f)^2/4\lambda} = \sqrt{\frac{\pi}{\lambda}} e^{-(2\pi f)^2/4\lambda}.
\end{aligned} \tag{3.17}
$$

It can be seen that the function is of the same form as the original function $x(t)$.

(5) Dirac δ function

For equation (3.16), we consider the function[43]

$$\delta_n(t) = \sqrt{\frac{n}{\pi}} e^{-nt^2} dt. \tag{3.18}$$

Figure 3.7 shows δ_n with n varying from $n = 1$ to 10. As n is increased, the value near 0 becomes higher and also gradually sharper ($\lim_{n\to\infty} \delta_n(t) = 0$ for $t \neq 0$).

However, if we integrate, it always satisfies

$$\int_{-\infty}^{\infty} \delta_n(t) dt = 1. \tag{3.19}$$

From these facts, and since we have from the mean value theorem in calculus that

$$\phi(t) = \phi(0) + \phi'(0)t,$$

for any function $\phi(t)$ that is continuously differentiable and whose derivatives are bounded, it can be shown that the function (3.18) satisfies

$$\lim_{n\to\infty} \int_{-\infty}^{\infty} \delta_n(t)\phi(t) dt = \phi(0). \tag{3.20}$$

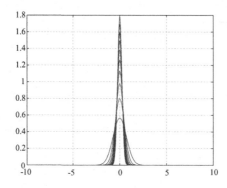

FIGURE 3.7
δ_n function ($n = 1, 2, \ldots, 10$)

Therefore, when M is the maximum value of $|\phi'(t)|$, from the assumed property of ϕ and equation (3.19), it holds that

$$
\begin{aligned}
\left| \int_{-\infty}^{\infty} \delta_n(t)\phi(t)dt - \phi(0) \right| &= \left| \int_{-\infty}^{\infty} \delta_n(t)(\phi(t) - \phi(0))dt \right| \\
&= \int_{-\infty}^{\infty} \delta_n(t)t\phi'(t)dt \leq M\sqrt{\frac{n}{\pi}} \int_{-\infty}^{\infty} |t|e^{-nt^2} dt \\
&= M\sqrt{\frac{1}{\pi n}} \to 0, \quad n \to \infty.
\end{aligned}
$$

Not only equation (3.18), but also, if we narrow the definition interval of

$$
f_n = \frac{\sin n\pi t}{\pi t}, \tag{3.21}
$$

or the square function, and the triangle function while keeping the area constant, we obtain the δ function. Here, we define the δ function as the function that satisfies

$$
\int_{-\infty}^{\infty} \delta(t)dt = 1, \qquad \delta(t) = 0, \quad t \neq 0 \tag{3.22}
$$

and

$$
\int_{-\infty}^{\infty} \delta(t)\phi(t) = \phi(0), \tag{3.23}
$$

for any continuously differentiable function. Dirac named such a function, which only makes sense when integrating, a generalized function (commonly known as a delta (δ) function). From equation (3.23), the Fourier transform of the delta function is given by

$$
\int_{-\infty}^{\infty} \delta(t)e^{-i2\pi ft}dt = 1. \tag{3.24}
$$

(6) Step function
    Figure 3.8 shows the step function defined by

$$
u(t) = \begin{cases} 1, & t > 0 \\ 0, & \text{otherwise.} \end{cases} \tag{3.25}
$$

The Fourier transform of equation (3.25) looks like a δ function, but considering the discontinuity at $t = 0$, as shown in Figure 3.9, we obtain

$$U(f) = \mathfrak{F}(u(t)) = \delta(f) + \frac{1}{i2\pi f}. \tag{3.26}$$

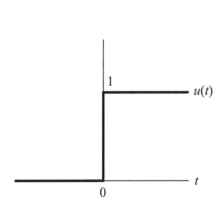

FIGURE 3.8
Step function

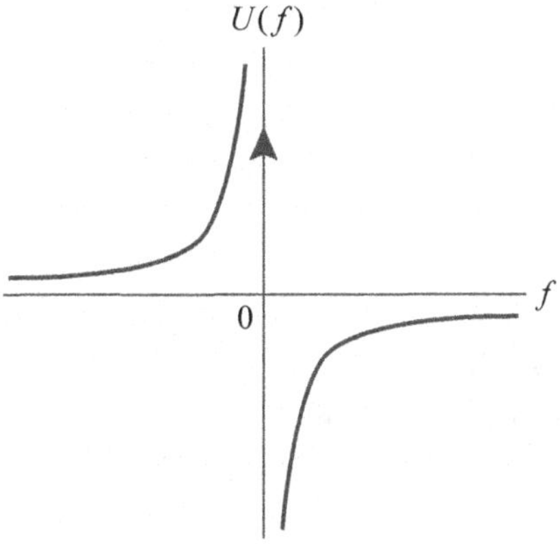

FIGURE 3.9
Fourier transform of a step function

3.3 Properties of the Fourier Transform

Let us summarize some important properties of the Fourier transform.

(1) Linearity
    Let $G_1(f)$ and $G_2(f)$ be the Fourier transforms of $g_1(t)$ and $g_2(t)$, respectively, then the following holds

$$\mathfrak{F}[a_1 g_1(t) + a_2 g_2(t)] = a_1 G_1(f) + a_2 G_2(f). \tag{3.27}$$

(2) Scalability
    If $G(f) = \mathfrak{F}[g(t)]$, then placing $x = at$ and using $dx = a\,dt$, we get

$$\mathfrak{F}[g(at)] = \frac{1}{|a|} G\left(\frac{g}{a}\right). \tag{3.28}$$

By this property, an elongation on the time axis corresponds to a contraction on the frequency axis, and a contraction on the time axis corresponds to an elongation on the frequency axis. That is, if time doubles a times in the time domain, the frequency domain becomes a times higher in the frequency domain. If $a = -1$, then it can be shown that

$$\mathfrak{F}[g(-t)] = G(-g). \tag{3.29}$$

(3) Time shift
    Assume that $G(f) = \mathfrak{F}[g(t)]$, then we have

$$\mathfrak{F}[g(t - L)] = G(f)e^{-i2\pi fL}. \tag{3.30}$$

This relation shows the case where the original function $g(t)$ arises with a delay of time L, as shown in Figure 3.10. In this case, L is called **delay time** or **dead time**. This relationship is relevant to the z-transformation defined later.

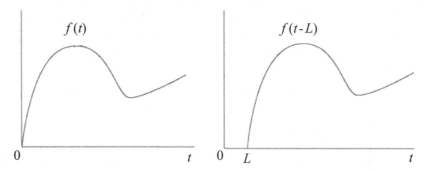

FIGURE 3.10
Delay time (dead time)

In the frequency domain, the following frequency shift holds,

$$\mathfrak{F}[g(t)e^{i2\pi f_0 t}] = G(2\pi(f - f_0)). \tag{3.31}$$

Now setting $g(t) = 1$ and replacing f_0 by $-f_0$ in equation (3.31) and using equation (2.18), the Fourier transforms of $\sin(2\pi f t_0)$ and $\cos(2\pi f t_0)$ are obtained by

$$\begin{cases} \mathfrak{F}\left(\sin(2\pi f t_0)\right) = -i/2\left(\delta(f - f_0) - \delta(f + f_0)\right) \\ \mathfrak{F}\left(\cos(2\pi f t_0)\right) = 1/2\left(\delta(f - f_0) + \delta(f + f_0)\right). \end{cases} \tag{3.32}$$

(4) Symmetry
In the Fourier inverse transform of $G(f)$,

$$g(t) = \int_{-\infty}^{\infty} G(f)e^{i2\pi f t} df,$$

by replacing t by $-t$, we obtain

$$g(-t) = \int_{-\infty}^{\infty} G(f)e^{-i2\pi f t} df. \tag{3.33}$$

Therefore, we have the relation that

$$\mathfrak{F}[G(f)] = \mathfrak{F}[\mathfrak{F}[g(t)]] = g(-t).$$

(5) Convolution integral
Given two functions $g_1(t)$ and $g_2(t)$, the integral defined by

$$g(t) = \int_{-\infty}^{\infty} g_1(x)g_2(t - x)dx \tag{3.34}$$

is called the **convolution integral** which is sometimes written as

$$g(t) = g_1(t) * g_2(t).$$

In the convolution integral, define

$$G_1(f) = \mathfrak{F}[g_1(t)], \qquad G_2(f) = \mathfrak{F}[g_2(t)],$$

then it holds

$$
\begin{aligned}
\mathfrak{F}[g_1(t) * g_2(t)] &= \int_{-\infty}^{\infty} \left\{ \int_{-\infty}^{\infty} g_1(x) g_2(t-x) dx \right\} e^{-i2\pi ft} dt \\
&= \int_{-\infty}^{\infty} g_1(x) e^{-i2\pi fx} dx \int_{-\infty}^{\infty} g_2(t-x) e^{-i2\pi f(t-x)} dt \\
&= G_1(f) G_2(f).
\end{aligned}
\tag{3.35}
$$

That is, the convolution integral of the two functions in the time domain is the product of the Fourier transforms of the two functions in the frequency domain. Equation (3.35) is called the **convolution theorem**.

(6) Differential formula

Assume that $\mathfrak{F}[g(t)] = G(f)$. If $g(t) = 0$ for $t \to \pm\infty$, then from the partial integration formula, it holds that

$$
\begin{aligned}
\mathfrak{F}[g'(t)] &= \int_{-\infty}^{\infty} g'(t) e^{-i2\pi ft} df \\
&= g(t) e^{-i2\pi ft} \Big|_{-\infty}^{\infty} + i2\pi f \int_{-\infty}^{\infty} g(t) e^{-i2\pi ft} dt \\
&= i2\pi f \int_{-\infty}^{\infty} g(t) e^{-i2\pi ft} dt = i2\pi f \mathfrak{F}[g(t)].
\end{aligned}
\tag{3.36}
$$

In general, for n-th order derivative, it holds that

$$\mathfrak{F}[g'^{\cdots\prime}(t)] = (i2\pi f)^n \mathfrak{F}[g(t)]. \tag{3.37}$$

That is, if the Fourier transform $\mathfrak{F}[g(t)]$ of the function $g(t)$ can be obtained, its n-th order derivative can be obtained by multiplying it by $(i2\pi f)$ n times. This property is useful when solving differential equations, as it allows them to be solved algebraically.

3.4   Laplace Transform

The Laplace transform is defined from the Fourier transform

$$\mathfrak{F}(f) = \int_{-\infty}^{\infty} f(t) e^{-i2\pi ft} dt,$$

by replacing $e^{-i2\pi ft}$ by e^{-st} and restricting the range as $t \geq 0$,

$$F(s) = \mathfrak{L}[f(t)] = \int_{0}^{\infty} f(t) e^{-st} dt. \tag{3.38}$$

The s may in general be a complex number $s = \sigma + i\omega$ ($\omega = 2\pi f$). If $f(t)$ is a function of time, this transformation transforms it into a function of s. The original domain is called **time domain** or simply t-domain, and the transformed domain is called **Laplace domain** or s-domain. While the functions that can be Fourier transformed are limited to absolutely integrable functions, a wider range of functions $f(t)$ can be transformed by multiplying the function $f(t)$ by e^{-st}, which has good convergence. While the Fourier transform is a powerful tool for solving boundary value problems, the Laplace transform is a powerful tool for initial value problems. Table 3.1 shows the Laplace transforms of some commonly used functions.

In connection with equation (3.37), it holds that

$$\mathfrak{L}[g'^{\cdots'}(t)] = s^n \mathfrak{L}[g(t)]. \tag{3.39}$$

TABLE 3.1
Commonly used Laplace transformations

$f(t)$	$F(s) = \mathfrak{L}[f]$	Condition
1	$\dfrac{1}{s}$	
t^n	$\dfrac{n!}{s^{n+1}}$	
e^{at}	$\dfrac{1}{s-a}$	$s > 0$
$\sin at$	$\dfrac{a}{s^2 + a^2}$	$s > 0$
$\cos at$	$\dfrac{s}{s^2 + a^2}$	$s > 0$
$\sinh at$	$\dfrac{a}{s^2 - a^2}$	$s > 0$
$\cosh at$	$\dfrac{s}{s^2 - a^2}$	$s > 0$
$\delta(t)$	1	

4

Statistical Estimation of Spectrum

In Chapter 2, we presented two methods for obtaining the power spectrum. However, this theory alone does not provide a reliable spectrum. In this chapter, we discretize an infinitely long time series and clarify the statistical properties when the spectrum is calculated via the covariance function obtained from the finite number of time series. Then, based on the results, we will learn how to use the Fourier transform to reliably estimate the spectrum.

4.1 Revisiting from Stochastic Process Theory

4.1.1 Raw Spectrum

The left panel of Figure 4.1 shows the spectrum obtained by the periodogram method described in Chapter 2 from the data of $N = 400$ points observed from the container ship shown in Figure 1.3. The right panel of Figure 4.1 shows the spectrum for the same data using the covariance method. Since the lag is set to 400, we can see that the results are almost the same as those obtained by the periodogram method. Both spectra are highly variable.

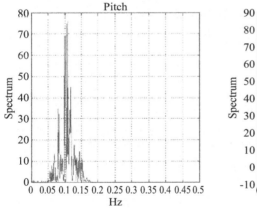 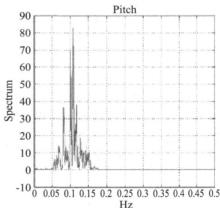

FIGURE 4.1
Raw spectra obtained by periodogram method (left) and covariance method (right).

To investigate the cause of this, it is necessary to introduce the idea that the observed time series is a sampled subset of a long irregular time series called **stochastic process**. That is, the spectrum calculated from the currently obtained sample time series (called the sample process) is the **random variable** obtained from a population with a certain

DOI: 10.1201/9781003428565-4

distribution. What we want to know is the expected value when the spectrum is obtained many times from a population that has the same structure as the spectrum obtained by chance. In this section, we will discuss some statistical devices to obtain a reliable spectrum from the obtained sample.

4.1.2 View from Stochastic Process Theory

The basis for dealing with time series in a stochastic manner is **stochastic process theory**. In stochastic process theory, the source of a time series, such as ship motions, is considered to be a continuous time series $\{x(t), -\infty < t < \infty\}$ that moves stochastically in continuous time[17][35]. Keeping in mind that the time series is cut out of time interval T, $x_T(t)$, and sampled at the sampling interval Δt to be viewed as sampled discrete data $x(1), x(2), \cdots, x(N)$. We will examine how this finiteness and discretization distort the properties of the original time series. For this purpose, we change the notation in the original time series and in the sample time series. For example, the covariance function $R(\tau)$ of the original time series becomes the sample covariance function $C(\tau)$ when it is sampled, and the spectrum $P(f)$ is denoted as the sample spectrum $S(f)$.

Let $p(x(t_1), x(t_2), \cdots, x(t_n))$ be a simultaneous probability density function that takes the values $\{x(t_1), x(t_2), \cdots, x(t_n)\}$ at time $\{t_1 < t_2 < \cdots < t_n\}$. If this probability distribution is independent of time, i.e., for any τ

$$p(x(t_1 + \tau), x(t_2 + \tau), \cdots, x(t_n + \tau)) = p(x(t_1), x(t_2), \cdots, x(t_n))$$

holds, $x(t)$ is said to be a **strongly stationary process**. Also, for the mean and the second-order moment (covariance function) of the time series, if it holds that

$$m_x(t) = E[x(t)] = m_x \text{ (constant)}$$
$$R_{xx}(t_1, t_2) = E[(x(t_1) - m_x)(x(t_2) - m_x)] = R_{xx}(0, t_2 - t_1),$$

then it is called a **weakly stationary process** (second-order stationary processes). $E[\cdot]$ is the expected value (see Chapter 17, and can be thought of as an average) of events occurring within [ ], also called **ensemble mean** or spatial mean. The second equation above shows that the covariance function depends only on time differences. If the mean and the covariance exist, a strongly stationary process is a weakly stationary process, but the converse does not hold. Hereafter, unless otherwise stated, we will treat weakly stationary processes.

Since the mean value of a weakly stationary process is constant, we do not lose generality by converting the observed data into a time series around the mean value m_x and treating it as $x(t) - m_x$. Hereafter, unless otherwise noted, we will treat **mean removed time series** in this manner. In this case, the autocovariance function of the second-order weakly stationary process is expressed by the ensemble mean

$$R_{xx}(\tau) = E[x(t + \tau)\, x(t)]. \tag{4.1}$$

The autocovariance function has the properties (see Section 17.6.1)

$$|R_{xx}(\tau)| \le R_{xx}(0), \quad R_{xx}(\tau) = R_{xx}(-\tau). \tag{4.2}$$

In Chapter 6, we deal with two different second-order weakly stationary time series $\{x(t), y(t)\}$, and define the **cross-covariance function**

$$R_{xx}(\tau) = E[x(t + \tau)\, y(t)]. \tag{4.3}$$

Although the cross-covariance function has no symmetry with respect to time, the following properties are obtained

$$R_{xy}(\tau) = R_{yx}(-\tau), \qquad |R_{xy}(\tau)|^2 \le R_{xx}(0)\, R_{yy}(0). \tag{4.4}$$

4.1.3   Ergodicity

The definitions of mean and covariance function in the previous section are concepts in the sense of ensemble mean. In reality, however, the structure of the original stochastic process is examined from a single sample process. Therefore, we average one sample process in the time direction and define $\frac{1}{T}\int_{-\frac{T}{2}}^{\frac{T}{2}} R_{xx}(\tau)dt$ as the covariance function per unit time (time average of covariance function). Such an average method is called **time mean**. Let the following assumption hold for the covariance function defined as a spatial mean and for the limit

$$E[x(t+\tau)\,x(t)] = \lim_{T\to\infty}\frac{1}{T}\int_{-\frac{T}{2}}^{\frac{T}{2}} x(t+\tau)\,x(t)\,dt. \tag{4.5}$$

For this to hold, $\lim_{T\to\infty}\frac{1}{T}\int_{-\frac{T}{2}}^{\frac{T}{2}}|x(t+\tau)\,x(t)|\,dt < \infty$ must hold. When the time mean and the ensemble mean are equal in this way, we say that **ergodic property** is valid. Hereafter, we will deal with stationary processes for which the ergodic property holds. By this assumption, it is guaranteed that the following is true;

$$\begin{cases} R_{xx}(\tau) = E[x(t+\tau)\,x(t)] = \lim_{T\to\infty}\frac{1}{T}\int_{-\frac{T}{2}}^{\frac{T}{2}} x(t+\tau)\,x(t)\,dt \\[4mm] R_{xy}(\tau) = E[x(t+\tau)\,y(t)] = \lim_{T\to\infty}\frac{1}{T}\int_{-\frac{T}{2}}^{\frac{T}{2}} x(t+\tau)\,y(t)\,dt. \end{cases} \tag{4.6}$$

4.1.4   Autocovariance Function and Spectrum

Let $X(t)$ be a stochastic process defined on $-\infty < t < \infty$. The Fourier transform $X(f)$ of $x(t)$, as described in Chapter 3, is defined as

$$X(f) = \int_{-\infty}^{\infty} x(t)\,e^{-i2\pi ft}dt. \tag{4.7}$$

Conversely, $x(t)$ is obtained by

$$x(t) = \int_{-\infty}^{\infty} X(f)\,e^{i2\pi ft}df. \tag{4.8}$$

Hereafter $\overline{X}(f)$ denotes the complex conjugate of $X(f)$, and $|X(f)|$ denotes the amplitude of $X(f)$ at frequency f. The square of its absolute value

$$|X(f)|^2 = X(f)\,\overline{X(f)}$$

is the square of the strength of the sine and cosine waves at frequency f of the time series, so when considered in the time domain, it can be thought of as the energy at frequency f.

In general, the total energy over $T \to \infty$ by this definition is not bounded, but for a weakly stationary time series such as the one assumed here, **average energy per unit time** (time average)

$$\bar{x}(t)^2 = \frac{1}{T}\lim_{T\to\infty}\int_{-\frac{T}{2}}^{\frac{T}{2}} x(t)^2 dt$$

is finite. Next, we define the spectrum according to this idea. Suppose that a sample process in a finite interval T is obtained from the original continuous time series, and denote the time series as

$$x_T(t) = \begin{cases} x(t), & |t| \leq T/2 \\ 0, & |t| > T/2. \end{cases}$$

The Fourier transform of this $x_T(t)$ is given by

$$X_T(f) = \int_{-\infty}^{\infty} x_T(t)\, e^{-i2\pi ft} dt = \int_{-\frac{T}{2}}^{\frac{T}{2}} x_T(t)\, e^{-i2\pi ft} dt.$$

Using this $X_T(f)$, for a very narrow frequency difference df, we can consider

$$\lim_{T\to\infty} \frac{|X_T(f)|^2}{T} df = \lim_{T\to\infty} \frac{X_T(f)\,\overline{X_T(f)}}{T} df.$$

Since the original $x(t)$ is a stochastic process, $X_T(f)$ is also a stochastic variable. Therefore, when looking at a time series as a stochastic process, it is natural to define the spectrum as the ensemble mean

$$P_{xx}(f) = \lim_{T\to\infty} E\left[\frac{X_T(f)\,\overline{X_T(f)}}{T} \right], \tag{4.9}$$

where

$$I_T(f) = \frac{1}{T}\left| \int_{-\frac{T}{2}}^{\frac{T}{2}} x_T(t)\, e^{-i2\pi ft} dt \right|^2 = \frac{1}{T}\left(X_T(f)\,\overline{X_T(f)} \right)$$

is the periodogram described in Section 2.3.3.

Even using this definition, the following relationship between the autocovariance function and the spectrum in $T \to \infty$ is analogous from equation (2.34)

$$P_{xx}(f) = \int_{-\infty}^{\infty} R_{xx}(\tau)\, e^{-i2\pi ft} d\tau \tag{4.10}$$

$$R_{xx}(\tau) = \int_{-\infty}^{\infty} P_{xx}(f)\, e^{i2\pi ft} df. \tag{4.11}$$

This is the **Wiener-Khinchin theorem** from the viewpoint of stochastic process theory.

4.1.5   Aliasing Caused by Discretization

Next, let us consider the unique quantization error introduced by discretizing an originally continuous phenomenon. When the original time series $x(t)$ is a continuous function, sampling this function at interval Δt, the covariance function is given by

$$R_{xx}(l) = \lim_{N\to\infty} \frac{1}{N} \sum_{s=1}^{N} x(s+l)\, x(s).$$

In addition, the spectrum is given by

$$P_{xx}(f; \Delta t) = \Delta t \sum_{n=\infty}^{\infty} R_{xx}(n\Delta t)\, e^{-i2\pi fn\Delta t}.$$

This agrees in frequency f with the spectrum given in equation (4.10). However, when discretized, the inverse transform becomes

$$R_{xx}(l\Delta t) = \int_{-\infty}^{\infty} e^{i2\pi fl\Delta t} P_{xx}(f)\, df$$

$$= \int_{-\frac{\Delta t}{2}}^{\frac{\Delta t}{2}} e^{i2\pi fl\Delta t} \left\{ \sum_{k=-\infty}^{\infty} P_{xx}\left(f + \frac{k}{\Delta t} \right) \right\} df.$$

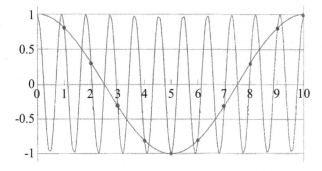

FIGURE 4.2
Aliasing phenomenon.

This is a phenomenon called **aliasing**, where we think we are looking at a spectrum at a frequency f, but in fact we are looking at a statistics that is a superposition of not only f but also a spectrum in which the contributions from the spectrum of integer multiples of $\frac{1}{\Delta t}$ are folded in. In other words, the spectrum of frequencies that are integer multiples of f is folded, and we are looking at a piled-up spectrum. This is the same reason why the wheels of a car appear to rotate in the opposite direction in movies, etc. This is an unavoidable phenomenon when quantization is performed with a large sampling period. The reason is that trigonometric functions have the following characteristics due to their periodicity, as shown in Figure 4.2,

$$\sin \frac{2\pi k}{N} n = \sin \frac{2\pi(k+N)}{N} n$$
$$\cos \frac{2\pi k}{N} n = \cos \frac{2\pi(k+N)}{N} n, \quad n = 0, 1, 2, \cdots.$$

4.2 Statistical Properties of Periodograms, Covariance Functions and Spectra

4.2.1 Variation of Periodograms and Windowing Effects

As mentioned in the previous section, the covariance function $R_{xx}(\tau)$ or spectrum $P_{xx}(f)$, which is defined in infinite time, must be estimated using the sample covariance function $C_{xx}(\tau)$ or the sample spectrum $S_{xx}(f)$ obtained from the sample time series at finite time. This is the role of statistics. In the following, we investigate the statistical properties of covariance functions and spectra from this perspective.

First, let us consider the effect of Fourier transforming such an actually long time series after cutting it at a certain length of time.

Such a sampling method would result in looking at the actual time series through a window of frame of height 1 at some finite time, as shown in Figure 4.3. That is, we can say that we are looking at a function obtained as the product of the function $x(t)$ defined in $-\infty < t < \infty$ and

$$w_T(t) = \begin{cases} 1, & a < t < b \\ 0, & \text{otherwise,} \end{cases}$$

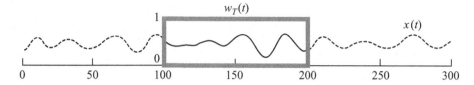

FIGURE 4.3
Effect of sampling on finite interval with length T.

i.e., $x_T(t)$ defined by

$$x_T(t) = w_T(t)\, x(t).$$

A function like $w_T(t)$ is called a **window function** in the time domain. In this case, it is called **do nothing window** because it does nothing but sampling operations. Let us consider how this box window function in the time domain is affected in the frequency domain when it is transformed into the frequency domain by the Fourier transformation. Assume that the Fourier transform of functions $w_T(t)$, $x(t)$ and $x_T(t)$ are given by $W_T(f)$, $X(f)$ and $X_T(f)$, respectively, then from the convolution integral theorem, the Fourier transform of the product of two functions in the time domain is given by

$$X_T(f) = \int_{-\infty}^{\infty} w_T(t)\, x(t)\, e^{-i2\pi ft} dt = \int_{-\infty}^{\infty} W_T(f - f')\, X(f)\, e^{-i2\pi ft} df.$$

The last equation shows that in the frequency domain, when we compute the spectrum of frequency f, the spectrum of frequencies in the neighborhood of the desired f is also taken in with weights. Therefore, depending on what form $W_T(f)$ takes, it is expected to have a significant influence on the result of the Fourier transform.

We will now examine the behavior of this $W_T(f)$. As shown in Chapter 3, a function like $W_T(t)$ is called a boxcar function. The Fourier transform of a boxcar function of width d in the time domain has a peak at frequency 0 (called the main lobe) and decays with alternating positive and negative values for each $1/d$ of frequency width (called the sidelobes), as shown in Figure 3.2. The reason why the periodogram in Figure 4.1 fluctuates so much can be attributed to the effect of the boxcar window, especially the large effect of the edge of the boxcar window. The periodogram in Figure 4.1 fluctuates violently because of the effects of the boxcar window, especially the edge effects.

4.2.2   Statistical Properties of the Sample Covariance Function

Suppose that a covariance function, $C_{xx}(l)$, is computed from a sample process $x_T(s)$ of length T according to the definition. The covariance function thus obtained is called the sample covariance function. In this case, as is clear from Table 2.2, as l increases, the number of counterparts of the product pair decreases by one and finally becomes one. As a result, the number of elements is a triangle with lag at the top. Therefore, the expected value of the sample covariance function becomes

$$E[C_{xx}(l)] = \left(1 - \frac{|l|}{N}\right) R_{xx}(l), \qquad |l| \leq N - 1. \tag{4.12}$$

That is, it is as if the triangular window shown in Figure 3.3 in Chapter 3 were multiplied by the covariance function of the original process. This phenomenon indicates that $C_{xx}(l)$

is not a consistent estimator (see Section 17.6.1) of the true covariance function $R_{xx}(l)$. In order to make it a consistent estimator, we need to modify

$$C_{xx}(l) = \frac{1}{N - |l|} \sum_{r=1}^{n-|l|} x(r + |l|)\, x(r), \quad |l| \le N - 1.$$

However, this modification is not made when there is a large amount of data.

If we examine the variance of the sample autocovariance function, it is approximately given by

$$\mathrm{Var}[C_{xx}(l)] \triangleq E\left[(C_{xx}(l) - E[C_{xx}(l)])^2\right]$$

$$\simeq \frac{1}{N} \sum_{r=-n+|l|}^{n-|l|} \left\{ R_{xx}^2(r) + R_{xx}(r + l)\, R_{xx}(r - l) \right\},$$

where $\mathrm{Var}[\,\cdot\,]$ denotes the variance of the variable in parentheses. For $l = 0$, i.e., the variance of the sample autocovariance function at the origin, can be obtained from the above equation as

$$\mathrm{Var}[C_{xx}(0)] \simeq \frac{2}{N} \sum_{r=-n}^{n} R_{xx}^2(r).$$

According to the assumption of a second-order stationary time series, the sample covariance function is close to zero at points far enough from the origin. Hence, when l is large, we obtain

$$\mathrm{Var}\left[C_{xx}(l)\right] \simeq \frac{1}{N} \sum_{r=-n+|l|}^{n-|l|} R_{xx}^2(r) \simeq \frac{1}{2}\mathrm{Var}[C_{xx}(0)].$$

It follows that the variance of the sample autocovariance function is the largest at the origin, and as lag l increases, it settles to about half of the variance at the origin, but not to zero. On the other hand, from the assumption of a second-order stationary time series, the true autocovariance function should approach zero as l increases. Therefore, the relative error increases as l increases. To reduce this error, the above equation indicates that N must be increased.

4.2.3 Statistical Properties of the Sample Spectrum

Next, we investigate the effect of finite data appearing in the covariance method, i.e., the sample spectrum as the expected value of the Fourier transform of the sample covariance function[34].

From equation (4.12), in the continuous domain, we obtain

$$E[C_{xx}(u)] = \frac{1}{T} \left[\int_{-\frac{T}{2}}^{\frac{T}{2} - |u|} x(t + |u|)\, x(t)\, dt \right] = \left(1 - \frac{|u|}{T}\right) R_{xx}(u)dt. \qquad (4.13)$$

The expected value of the spectrum can be therefore expressed as

$$E[S_{xx}(f)] = \int_{-\frac{T}{2}}^{\frac{T}{2} - |u|} \left(1 - \frac{|u|}{T}\right) R_{xx}(u)\, e^{-i2\pi f u} du. \qquad (4.14)$$

Here, by putting

$$w_0(u) = \begin{cases} \left(1 - \dfrac{|u|}{T}\right), & 0 \le |u| \le T \\ 0, & |u| > T, \end{cases}$$

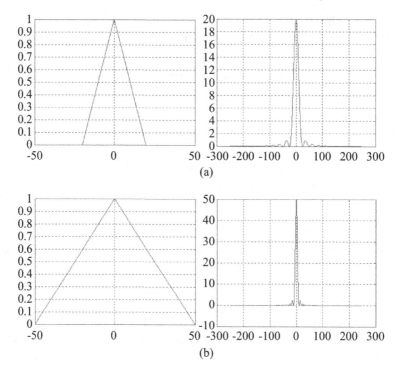

FIGURE 4.4
Truncation effect of triangular window. Left: lag window, Right: Spectral window.

the window over the lag u of a covariance function such as $w_0(u)$ is, in general, called the **lag window**. The triangular window appearing here is called the **Bartlett lag window**. The Fourier transform of the triangular window is given by equation (3.11) in Chapter 3. From this, by the convolution theorem, equation (4.14) becomes

$$E[S_{xx}(f)] = \int_{-\infty}^{\infty} T \left(\frac{\sin \pi T g}{\pi T g} \right)^2 P_{xx}(f - g) \, dg,$$

where by putting

$$W_0(f) = \int_{-\infty}^{\infty} w(u) \, e^{-i2\pi f u} du = T \left(\frac{\sin \pi T g}{\pi T g} \right)^2, \tag{4.15}$$

we obtain

$$E[S_{xx}(f)] = \int_{-\infty}^{\infty} W_0(g) \, P_{xx}(f - g) \, dg.$$

That is, the estimated value of the sample spectrum is a view of the true spectrum $P_{xx}(f)$ in the frequency domain through $W_0(f)$. This $W_0(f)$ is called the **spectral window**, since it acts in the frequency domain. As can be seen from Figure 3.4, $W_0(f)$ is characterized by the fact that it does not cut into the negative side and that the frequency interval between neighboring frequencies that are zero is $2/d$, twice as wide as that of a box window.

Here, in this $W_0(f)$, if T is made smaller as shown in Figure 4.4 (a), the shape of the spectrum has a wider main lobe as shown in the figure on the right. Conversely, as T is increased, the main lobe becomes narrower and higher, as shown in plot (b), and it is as if the spectrum is viewed from inside a slit. Therefore, the spectrum of the target frequency is no longer affected by the surrounding spectrum. In other words, it is asymptotically an unbiased estimator of the spectrum. In this case,

$$B(f) = E[S_{xx}(f)] - P_{xx}(f) \tag{4.16}$$

is called the **spectral bias**.

When the time series is white noise, the sample spectrum obtained has the following statistical properties;

1. Periodograms with frequencies differ by $1/T$ (for continuous case) or $1/N\Delta t$ (for discrete case) are statistically independent.

2. Twice the ratio of the true spectrum $P_{xx}(f)$ and the sample spectrum $S_{xx}(f)$;

$$\frac{2S_{xx}(f)}{P_{xx}(f)}$$

is approximately χ^2-distributed with 2 degrees of freedom (see equation (17.30) in Section 17.5). From this, the properties of the χ^2 distribution (property 1. in Section 17.5), it is approximately given by

$$\text{Var}\left[\frac{2S_{xx}(f)}{P_{xx}(f)}\right] \simeq 4.$$

3. Therefore, it holds approximately that

$$\text{Var}[S_{xx}(f)] \simeq P_{xx}(f)^2. \tag{4.17}$$

From property 1, we know that the coefficient of variation (see equation (17.31)) is 2, i.e., the relative error is 100%. Also, property 2 shows that the variation in the sample spectrum is approximately the square of the true spectrum. Since it is independent of T, we know that no matter how much we increase T, the variation will not be suppressed.

4.2.4 Smoothing of Spectrum

Based on the previous explanations, we now consider how to obtain a spectrum with low variability and high reliability in the covariance method. As mentioned earlier, fixing T and increasing the lag reduces the variance of the covariance function to some extent, but the relative error increases. In other words, the variability will be relatively large. Therefore, in order to cut off the part of the covariance function that fluctuates greatly, it would be better to cut off the covariance function from the origin to a certain range. If the lag u of the covariance function is truncated at τ_M, which is smaller than the largest lag T, the actual covariance function is given by

$$\hat{C}_{xx}(u) = w_\tau(u)\, C_{xx}(u),$$

where

$$w_\tau(u) = \begin{cases} 1, & |u| \le \tau_M \\ 0, & |u| > \tau_M. \end{cases}$$

Figure 4.5 shows the spectrum obtained from the Fourier transform of the covariance function with the lag cut short in this way. Although the spectrum is smoother than in Figure 4.1, there are still some negative spectra.

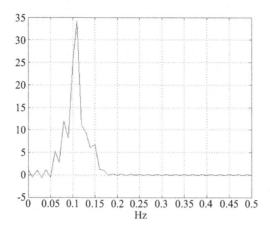

FIGURE 4.5
Spectrum of pitching when covariance function was truncated to 1/10.

The effect of such a censoring effect on the spectrum can be theoretically evaluated from equation (4.15) as

$$
\begin{aligned}
\hat{S}_{xx}(f) &= \int_{-\infty}^{\infty} T \left(\frac{\sin \pi \tau_M g}{\pi \tau_M g} \right)^2 P_{xx}(f - g) \, dg \\
&= \int_{-\infty}^{\infty} W_\tau(g) \, P_{xx}(f - g) \, dg,
\end{aligned}
$$

where

$$
W_\tau(f) = \int_{-\infty}^{\infty} w_\tau(u) \, e^{-i2\pi f u} du = T \left(\frac{\sin \pi \tau_M g}{\pi \tau_M g} \right)^2.
$$

In other words, if the covariance function is terminated at lag τ_M in order to suppress the fluctuation of the spectrum, the spectrum appears as a smoothed value of the spectrum centered at the frequency f to be obtained and separated by g by the weight $W_\tau(g)$. By analogy with $W_\tau(g)$ shown in the right-hand side of Figure 4.4 and other figures, the main lobe at the center will help stabilize the spectrum. However, the sidelobes distort the spectrum, and in some cases, may even take in negative values. However, it is clear that $W_\tau(g)$ helps stabilize the spectrum, by actively devising the lag window $w(u)$, it is possible to obtain a spectrum with high reliability. From this perspective, various spectral windows have been proposed.

4.2.5 Various Window Functions

Instead of a window function that does nothing, such as $w_\tau(u)$ described above, we consider designing effective window function and stabilizing spectrum: the window function proposed by Hann, **Hanning window**, which cut off at τ_M is, in lag window form, given by

$$
w_{Han}(u) = \begin{cases} \dfrac{1}{2} \left(1 + \cos \dfrac{\pi u}{\tau_M} \right), & u \leq \tau_M \\ 0, & u > \tau_M, \end{cases} \tag{4.18}
$$

and gradually decreases to 0 as the lag increases. If this function is multiplied by the covariance function in advance and then Fourier transformed, a smooth spectrum is obtained.

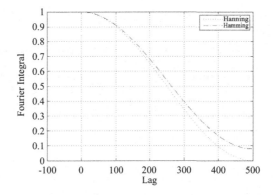

FIGURE 4.6
Hanning and Hamming lag windows.

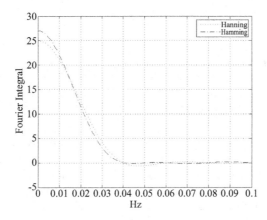

FIGURE 4.7
Hanning and Hamming spectral windows.

By the way, the Fourier transform of $w_{Han}(u)$ can be obtained from Chapter 3 (equation (3.32)) as

$$W_{Han}(f) = \frac{1}{2}W_\tau(f) + \frac{1}{4}\left[W_\tau\left(f + \frac{1}{2\tau_M}\right) + W_\tau\left(f - \frac{1}{2\tau_M}\right)\right].$$

In other words, the operation of equation (4.18) in the lag window indicates that in the spectrum, the frequency f is multiplied by 0.5 and the frequencies on both sides $f \pm \frac{1}{2\tau_M}$ are multiplied by 0.25 for the moving average. The $W_{Han}(g)$ shows the form of the spectral window. It can be seen that a window with no negative dip in the frequency field and with relatively low sidelobe height is obtained. In contrast, Hamming considered the window of the form

$$w_{Ham}(u) = 0.54 + 0.46\cos\frac{\pi u}{\tau_M}. \tag{4.19}$$

Figures 4.6 and 4.7 show the lag and spectral windows of Hanning and Hamming.

Table 4.1 shows such typical spectral windows. W_1, W_2 and W_3 are the windows proposed by Akaike. In order of W_1, W_2 and W_3, the spectral variance increases but the bias decreases.

TABLE 4.1

Typical spectral windows

	Hanning	Hamming	W_1	W_2	W_3
a_0	0.50	0.54	0.5132	0.6398	0.7029
$a_1 = a_{-1}$	0.25	0.23	0.2434	0.2401	0.2228
$a_2 = a_{-2}$	0.0	0.0	0.0	−0.0600	−0.0891
$a_3 = a_{-3}$	0.0	0.0	0.0	0.0	0.0149

In general, if the sample spectrum $S_{xx}(f)$ is smoothed by a spectral window $W(g)$, for data of length $-\frac{T}{2} \leq t < \frac{T}{2}$, the variance of the estimated spectrum $\hat{S}_{xx}(f)$ is given by

$$
\begin{aligned}
\mathrm{Var}[\hat{S}_{xx}(f)] &\simeq \frac{P_{xx}(f)^2}{T} \int_{-\infty}^{\infty} W(g)^2 \, dg \\
&= \frac{P_{xx}(f)^2}{T} \int_{-\infty}^{\infty} w(u)^2 \, du = P_{xx}(f)^2 \frac{I}{T},
\end{aligned}
\tag{4.20}
$$

where $I = \int_{-\infty}^{\infty} w(u)^2 \, du$. Thus, from equation (4.17), we have

$$
\frac{\mathrm{Var}[\hat{S}_{xx}(f)]}{\mathrm{Var}[S_{xx}(f)]} \simeq \frac{\mathrm{Var}[\hat{S}_{xx}(f)]}{P_{xx}(f)^2} = \frac{I}{T}.
\tag{4.21}
$$

The quantity I/T indicates how much the relative error is reduced by the use of the window compared to the case where the raw spectrum $S_{xx}(f)$ is used as the estimate of the spectrum.

For example, in the case of Bartlett's triangular lag window described earlier, we obtain $I = \int_{-\infty}^{\infty} (1 - u/\tau_M)^2 du = 2/3 \, \tau_M$, where τ_M is the maximum lag length of the covariance function. Since the maximum lag is usually about $1/10$ of the number of samples, it is about $2/3 \, (1/10) = 0.067$ and the variance of the spectrum is suppressed to about 6.7%. Thus, according to equation (4.21), the wider the bandwidth of the window (or the smaller the number of censoring lags), the wider the bandwidth of the spectral window of $W(g)$ will be multiplied and smoothed over the sample spectrum in the frequency field, resulting in a more stable spectrum.

On the other hand, however, as discussed in Section 4.2.3, if τ_M is made too small, the estimated spectrum has bias because it incorporates too much of the neighborhood spectrum used for smoothing. To resolve this, using the relationship between variance and bias

$$
E[(\hat{S}_{xx}(f) - P_{xx}(f))^2] = \mathrm{Var}(\hat{S}_{xx}f) + B(f)^2,
$$

the design policy of the spectrum may be to make this value as small as possible, where $B(f)$ is the spectrum bias shown in equation (4.16).

Akaike designed windows W_1, W_2 and W_3 to minimize the variance as much as possible while keeping in mind even higher-order biases in order to achieve the above goal as closely as possible. The larger the trough or peak, the more W_2 should be used rather than W_1 and W_3 rather than W_2.

Note that the fundamental solution to the problem regarding the selection of the length of the covariance function τ_M is solved by the successful modeling of time series in the time domain, which will be discussed later.

4.3 Spectrum Estimation Methods in Practice

4.3.1 Spectrum Estimation by Covariance Method

So far, we have described the spectrum estimation by the covariance method. Here we show specific estimation methods[34][26].

1. Sampling period

   If Δt is too small, the computation becomes both excessive and responsive to high-frequency noise. On the other hand, if Δt is too large, it will cause aliasing, and the spectrum from the high-frequency side will be folded into the low-frequency side. Therefore, the sampling interval should be selected such that the spectrum is sufficiently small outside the Nyquist frequency $\frac{1}{2\Delta t}$.

2. Maximum lag length h.

   If h is too large, the variance of the estimate will be large. Conversely, if it is too small, the spectrum will be highly biased. It is said that about $1/10$ of the number of data N is a good rule of thumb. Alternatively, as shown in Figure 4.8, when determining the bandwidth, one can use

   $$h\Delta t \geq \frac{2}{B}$$

   as a rough guide.

3. Pre-whitening, trend and numerical filter

   As mentioned in the statistical properties of the covariance function, the covariance function has relatively little variation in the estimated value near the origin. Therefore, it is desirable for the majority of the covariance function to be concentrated near the origin. Such a signal is white noise. Therefore, it is desirable to whiten the data in advance through a filter of some nature. This operation is called **pre-whitening**. The presence of long-period signals (trends) that would violate the stationarity assumption is also undesirable and should be removed in advance. These operations are performed using a **numerical filter**. First, to

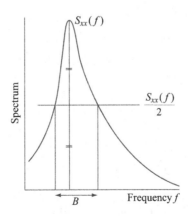

FIGURE 4.8
Definition of the band width B.

remove extremely long-period component data such as trends, we need to make a new time series from $x(s)$ and $x(s-1)$ as

$$y(s) = \frac{1}{2}(x(s) - x(s-1)),$$

and estimate the spectrum of $y(s)$. The above equation can be obtained from its Fourier transform by using

$$|X(f)|^2 = 0.5\,(1 - \cos 2\pi f)$$

and so it plays the role of a **high pass filter**. Conversely, **low pass filter** is defined by

$$y(s) = \frac{1}{2}(x(s) + x(s+1)).$$

To cancel out extremely high-frequency spectrum, for example, we can pre-whiten it by the following second-order autoregressive process

$$x(s) + a(1)\,x(s-1) + a(2)\,x(s-2) = 0.$$

In this case, $a(1)$ and $a(2)$ can be obtained by solving the Yule-Walker equation described later. The method collectively referred to as the **Blackman-Tukey method** is shown below that implements this idea.

4. Sample N pieces of data at an appropriate time interval Δt, and obtain $\{x(1), x(2), \cdots, x(N)\}$. Compute the average of N data, $\bar{x}$, and subtract the average $\bar{x}$ from all the data. Then redefine $\{x(1), x(2), \cdots, x(N)\}$ as follows

$$x(s) \equiv x(s) - \bar{x}, \quad s = 1, 2, \cdots, N, \quad \text{where } \bar{x} = \frac{1}{N}\sum_{s=1}^{N} x(s).$$

5. Set the maximum number of lags h to be about 10% of the number of data N, and calculate the covariance function up to lag h by the following formula

$$C_{xx}(l) = \frac{1}{N}\sum_{s=1}^{N-l} x(s)\,x(s+l), \quad (l = 0, 1, 2, \cdots, h).$$

6. Calculate the raw spectrum using the following formula

$$S_{xx}\left(\frac{r}{h}\frac{1}{2\Delta t}\right) = \Delta t \left\{ C_{xx}(0) + 2\sum_{l=1}^{h-1} C_{xx}(l)\cos 2\pi \frac{r}{2h} l + (-1)^r C_{xx}(h)\right\}.$$

7. The smoothed spectrum at frequency $2\pi \frac{m}{2h}$ is obtained by

$$\hat{S}_{xx}(m) = \sum_{r=-3}^{3} a(r)\,S_{xx}(m-r), \quad (m = 0, 1, 2, \cdots, h),$$

where for $a(r)$, use one of the spectral windows in Table 4.1. For both ends, we use

$$\hat{S}_{xx}(-m) = S_{xx}(m), \quad S_{xx}(h+m) = S_{xx}(h-m).$$

4.3.2 Periodogram Method

In the periodogram method for estimating the spectrum, the observed time series $x(1), x(2), \cdots, x(N)$ is first subjected to Fourier transform and begin by computing

$$d(f) = \frac{1}{N} \sum_{r=-n}^{n-1} x(r) \, e^{-i2\pi f r \Delta t}.$$

Then, the spectrum is obtained as the square of the absolute value

$$I(f) = N \Delta t \, d(f) \, \overline{d(f)}, \quad |f| < \frac{1}{2\Delta t}.$$

The periodogram thus obtained is not an expected value, but a random variable. When the number of samples of this random variable is increased, the expected value agrees with the true spectrum. In other words, we have

$$\lim_{N \to \infty} E[I(f)] = S(f).$$

However, the computed periodograms are independent of each other and do not correlate even between neighboring frequencies. Thus, the added effect of the finite-length window over which the data in the previous section was cut makes the variance large and unusable.

Bartlett proposed a method to reduce the variance by dividing the original time series into M partitions, obtaining a periodogram for each and finally averaging the periodograms at the same frequency f. In this method, however, if N is fixed and the number of divisions M is large, the variance is reduced, but statistical bias occurs. A good rule of thumb is to use about $M = N/10$. In the periodogram method, the **fast Fourier transform** can be used.

4.4 Example of the Spectrum of Ship Motions

Figure 4.9 shows the spectrum of the time series shown in Figure 1.3 obtained by the covariance method. The number of data N is 600 points, the maximum lag is 60 and a Hanning window is used for the spectral window. Figure 4.9 shows the spectra of the pitch (upper left), roll (upper right), yaw (middle left), Zacc (vertical acceleration; middle right), Yacc (lateral acceleration; lower left) and rudder angle (lower right) time series obtained in this way.

In such rough weather conditions, human operators steer the ship with a longer period than the autopilot. Therefore, high peaks are seen on the long-period side. The rolling has a peak at about 15 to 16 seconds. Such a high and sharp peak also appears in the rolling time series of other time periods of the same vessel, with a sharp peak around the same frequency. On the other hand, the period of pitching shows a peak in the frequency band of about 10 seconds in this case. This period may shift and the bandwidth may become wider in other time series. The Yacc (lateral acceleration) is similar to the spectrum of the roll.

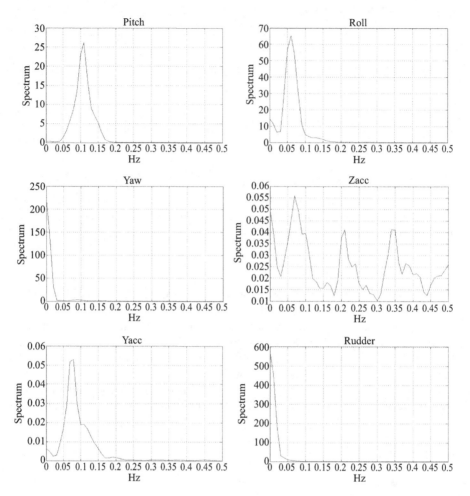

FIGURE 4.9
Spectra estimation by Blackman-Tukey method. Top left: pitch, top right: roll, middle left: yaw, middle right: vertical acceleration, bottom left: lateral acceleration, bottom right: rudder angle.

5

Analysis of Linear Systems

An analytical method that examines the response of a large structure to an external force applied to it, such as a wave striking the hull of a ship, by means of the relationship between the input and output signals, is known as **system analysis**[93]. If the system being treated in this system analysis has a response of proportional magnitude to the input, it is called a **linear system**. The motion of a ship navigating in waves can also be treated as such a linear system. In this chapter, we show how the Fourier and Laplace transformations are used as tools to solve the linear differential equations representing such a linear system, and define the derived frequency response function and impulse response function.

5.1 Input Models Commonly Used in System Analysis

First, we discuss the various inputs used in system analysis. See Chapter 3 for the Fourier transforms of these inputs.

Impulse input

As shown in Figure 5.1, the Dirac's δ function in Chapter 3 is used to model the external forces when a large external force is suddenly applied to the system, such as the yawing motion of a ship when the rudder is quickly moved and returned to its original position, the rolling motion when a ship is suddenly hit by a freak wave and the current change when an electrical short occurs. Representation of impulse input by δ function is given by[93]

$$\delta(t) = \begin{cases} \infty, & t = 0 \\ 0, & t \neq 0. \end{cases} \tag{5.1}$$

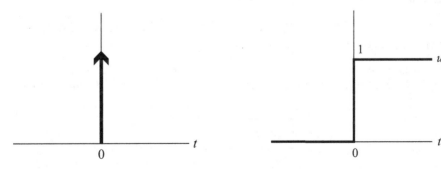

FIGURE 5.1
Impulse input

FIGURE 5.2
Step input

DOI: 10.1201/9781003428565-5

Here, it is assumed that for any positive number τ, the integral of the delta function satisfies

$$\int_{-\tau}^{\tau} \delta(t)\, dt = 1.$$

The Fourier transform of such an impulse input is approximated by, for example (equation (3.21)),

$$\frac{\sin(2\pi f t)}{2\pi f t}, \quad t \to 0. \tag{5.2}$$

Step input

The input that keeps the rudder constant in a certain direction and makes the ship turn, as shown in Figure 5.2, is called **step input** in automatic control theory. In particular, when its height is 1, it is called a unit step function. The figure 5.2 shows a step function with height 1 at $t \geq 0$,

$$u(t) = \begin{cases} 0, & t < 0 \\ 1, & t \geq 0. \end{cases}$$

Its Fourier transform is given by equation (3.26).

Complex amplitude input

The complex amplitude input $u(t)$ with frequency f is defined by

$$u(t) = \cos(2\pi f t) + i \sin(2\pi f t) = e^{i2\pi f t}.$$

The Fourier transform of the output to this input is the frequency response function described in the next section.

Gaussian white noise input

The Gaussian white noise input $Z(t)$ is a Gaussian random variable with

$$E[Z(t)] = 0, \qquad E[Z(t)\, Z(t - t_0)] = \sigma_z^2\, \delta(t_0),$$

where σ_z^2 is the variance of $Z(t)$.

5.2 Fourier Transform, Frequency Response Function and Impulse Response Function

A linear system is defined as a dynamic system as shown in Figure 5.3 where the output is also multiplied by a when the input is multiplied by a constant a in the input-output

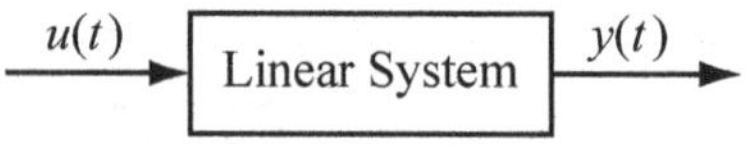

FIGURE 5.3
Linear input-output system

system. That is, suppose a system is given two inputs $u_1(t)$ and $u_2(t)$ with responses $y_1(t)$ and $y_2(t)$, respectively. Then this system is said to be linear[93], if the input to the system is given as the sum of two scalar multiples, $\alpha_1 u_1(t)$ and $\alpha_2 u_2(t)$:

$$u(t) = \alpha_1 u_1(t) + \alpha_2 u_2(t), \tag{5.3}$$

then it responds with a waveform represented by the linear combination of two inputs with the same coefficients

$$y(t) = \alpha_1 y_1(t) + \alpha_2 y_2(t). \tag{5.4}$$

In a linear system, by definition, if the input signal can be further decomposed into some signal, such as a trigonometric function, and the responses to the individual elements can be obtained, as shown in Chapter 2, then the output is obtained by simply adding those individual responses together. This principle is called **principle of linear superposition**. Using this principle, for example, if a wave with various periods of trigonometric period is generated in a tank experiment and the ship motion responses to the wave are determined individually, the actual ship response can be calculated by summing these responses.

Mathematically, it is clear from the definition that a **dynamical system** in which a linear system varies in time is generally represented by a linear differential equation as follows;

$$\frac{d^n y}{dt^n} + a_1 \frac{d^{n-1}y}{dt^{n-1}} + \cdots + a_{n-1}\frac{dy}{dt} + a_n y \tag{5.5}$$
$$= b_0 \frac{d^m u}{dt^m} + b_1 \frac{d^{m-1}u}{dt^{m-1}} + \cdots + b_{m-1}\frac{du}{dt} + b_m u, \quad n \geq m.$$

As shown in equation (3.36) in Chapter 3, Fourier and Laplace transformations have the property that if the results of those transformations of a function are obtained, their derivatives can be obtained by multiplying them by a power of $i2\pi f$ or s by the order of their derivative[28]. Using this property, in finding a solution to the differential equation (5.5) by the **operator method**, one can first perform a Fourier or Laplace transformation of the equation, change the field of consideration to find the solution algebraically in the frequency domain and then apply the inverse transformation to obtain the solution in the time domain. Usually, the Laplace transformation is convenient for solving initial value problems, but here we will consider solving the problem using the Fourier transformation, which will be used more frequently later. First, define the time derivative operator p as follows

$$p = \frac{d}{dt}. \tag{5.6}$$

Multiplying from the left of a function $f(t)$ by this operator p, $f(t)$ is differentiated. This operation is written as $p[f(t)]$. The higher-order derivative operator is written as

$$p^2[f(t)] = \frac{d^2 f(t)}{dt^2}, \cdots, p^n[f(t)] = \frac{d^n f(t)}{dt^n}.$$

Conversely, p^{-1} is an integral operator. From the properties of the Fourier transformation (5) shown in Section 3.3, we know that the Fourier transformation is also such an operator. In the following, $[\cdot]$ is omitted.

Using this operator, equation (5.5) can be written as

$$(p^n + a_1 p^{n-1} + \cdots + a_{n-1} p + a_n) y(t)$$
$$= (b_0 p^m + b_1 p^{m-1} + \cdots + b_{m-1} p + b_m) u(t). \tag{5.7}$$

To simplify the expression, if we define

$$
\begin{aligned}
A(p) &= p^n + a_1\, p^{n-1} + \cdots + a_{n-1}\, p + a_n & (5.8) \\
B(p) &= b_0\, p^m + b_1\, p^{m-1} + \cdots + b_{m-1}\, p + b_m, & (5.9)
\end{aligned}
$$

the equation can also be expressed as follows;

$$
A(p)\, y(t) = B(p)\, u(t). \tag{5.10}
$$

From equation (5.10), the solution $y(t)$ can be obtained algebraically as

$$
y(t) = \frac{B(p)}{A(p)} u(t) = G(p)\, u(t). \tag{5.11}
$$

Therefore, by applying the operator defined by $G(p)$ to $u(t)$, we can obtain the output $y(t)$. In other words, by referring to the results of the Fourier transformation and the table of the Laplace transformation described in Chapter 3 and knowing the relationship between the time domain and frequency domain of the basic function, the differential equation can be solved by returning the answer obtained algebraically in the frequency domain to the time domain.

Referring to the other books[28][95] for the detailed explanations, the solution of equation (5.10) can be obtained as follows:

1. Obtain the solutions of the homogeneous equations $A(p)\, y = 0$, and

2. obtain one special solution.

Then it is well known that the solution is given as the sum of these two solutions.

If all of n roots, μ_i, of the **characteristic equation** defined from $A(p)$,

$$
A(p) = 0, \tag{5.12}
$$

are single roots, the above homogeneous solution is obtained by

$$
\sum_{i=1}^{n} k_i\, e^{\mu_i t}. \tag{5.13}
$$

The special solution is obtained as follows. We know that the usual functions are represented by sums of trigonometric functions. Next, considering an input $u(t) = e^{i2\pi f t}$ with a single frequency to this system, and finding a linear solution of the type $y(t) = G(f)\, e^{i2\pi f t}$, from the differential formula of Fourier analysis, equation (5.5) becomes

$$
\left[\sum_{l=0}^{n} a_l (i2\pi f)^l \right] G(f)\, e^{i2\pi f t} = \left[\sum_{l=0}^{m} b_l (i2\pi f)^l \right] e^{i2\pi f t}.
$$

Here if we put

$$
A(f) = \sum_{l=0}^{n} a_l (i2\pi f)^l \tag{5.14}
$$

$$
B(f) = \sum_{l=0}^{m} b_l (i2\pi f)^l, \tag{5.15}
$$

then $G(f)$ is given by

$$
G(f) = \frac{B(f)}{A(f)}. \tag{5.16}
$$

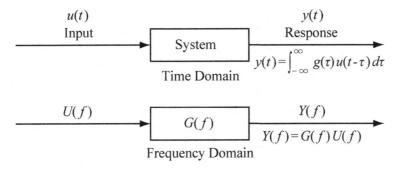

FIGURE 5.4
Linear input-output system

In other words, the response of the system in equation (5.10) with a function with a certain frequency as input is obtained by multiplying the input function $e^{i2\pi ft}$ by the function $G(f)$ defined in equation (5.16). This is the most important relation in linear systems analysis theory, and $G(f)$ is called **frequency response function**.

In the case of a general input function, $u(t)$ is obtained from the inverse transformation of the input $U(f)$ by

$$u(t) = \int_{-\infty}^{\infty} U(f)\, e^{i2\pi ft} df.$$

While equation (5.13) is the solution for a complex amplitude input with a single frequency, we now consider the case where $u(t)$ is a complex amplitude sum over an infinite interval, defined as $U(f)e^{i2\pi ft}\, df$ in the miniscule amplitude interval df. From the principle of linear superposition, considering the integral to be the limit of the sum of function values over an infinitesimal interval, the solution $y(t)$ is obtained by

$$y(t) = \int_{-\infty}^{\infty} G(f)\, U(f)\, e^{i2\pi ft} df. \tag{5.17}$$

On the other hand, if $Y(f)$ is the Fourier transform of $y(t)$, the inverse transform of $Y(f)$ is

$$y(t) = \int_{-\infty}^{\infty} Y(f)\, e^{i2\pi ft} df.$$

Therefore, in the end, we obtain the relation

$$Y(f) = G(f)\, U(f).$$

Using the convolution property of the Fourier transform (equation (3.34)), if the Fourier inverse transform of $G(f)$ is

$$g(t) = \int_{-\infty}^{\infty} G(f)\, e^{i2\pi ft} df,$$

then we obtain (Figure 5.4)

$$y(t) = \int_{-\infty}^{\infty} g(\tau)\, u(t - \tau)\, d\tau. \tag{5.18}$$

Although the above procedure is a mathematical procedure, from the physical point of view of the causality of the system, if t is the present time, the external force $u(t)$ is a "fact" given to the system in the present and past, and the resultant information about

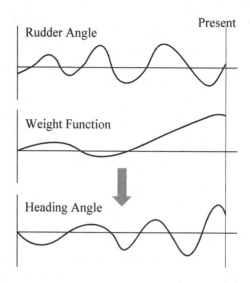

FIGURE 5.5

Implications of convolution in steering systems

y that we can obtain is only up to the "information" at the current time t. This is called the **physical realization condition**. From this physical realization condition requirement, equation (5.18) becomes

$$y(t) = \int_0^\infty g(\tau)\, u(t - \tau)\, d\tau, \tag{5.19}$$

and can be written by one-sided integration. Figure 5.5 shows that if we consider a steering system as a specific system, when the system is linear, the past values of the rudder motion can be multiplied by a weight function and added together to obtain the resulting change in the heading angle.

Furthermore, if the input $u(t)$ is an impulse function, i.e.,

$$u(t) = \delta(t), \tag{5.20}$$

then, according to Chapter 3, $\delta(t)$ is zero outside the origin, and we can see that

$$y(t) = g(t). \tag{5.21}$$

Therefore, the weight function $g(t)$ is called the **impulse response function**.

5.3   Stationarity of Linear Systems

The solution to equation (5.10) is the sum of the solution to equation (5.12) and equation (5.19), as discussed in the previous section. Now consider this in time using the solution of the kinematic system. When a stable external force $u(t)$ is applied to a certain kinematic system, initially the effect of the input $u(t)$ does not appear, so the kinematic system is strongly inertial and governed by the same order equation (5.12) without the effect of external force. Therefore, this solution (equation (5.13)) is called the **transient solution**. As time passes, the influence of the external force gradually increases and a waveform similar to that of equation (5.18), called the **stationary solution**, appears. In this case, however, the real part of the value of the characteristic root, which is the solution of the characteristic equation $A(p) = 0$, holds whether the effect of the transient solution disappears or not.

That is, the characteristic root μ_i is generally given by $\mu_i = \sigma_i + i\,\omega_i$, where $(i = \sqrt{-1})$, substituting this expression for the element $k_i\,e^{\lambda_i t}$ in equation (5.18), we get $k_i\,e^{(\sigma_i + i\,\omega_i)t} = k_i\,e^{\sigma_i t}e^{i\,\omega_i t}$. The absolute value of $e^{i\,\omega_i t}$ is 1, but $e^{\sigma_i t}$ either converges to zero over time, becomes a finite value $(= 1)$, or diverges infinitely, depending on the positive, zero or negative of σ_i. This indicates that the dynamic system under consideration exhibits the following behavior. That is, when a stable external force is applied to the system, at the characteristic root of $A(p) = 0$,

1. If the real part of the characteristic root λ_i is negative, the system response is strongly influenced by the input as the transient solution disappears over time. In this case, the system is said to be **stationary** or **stable**.

2. If the real part of the characteristic root λ_i is positive, the system response will diverge to either positive or negative values as time passes without the transient solution disappearing over time. In this case, the system is said to be either **nonstationary** or **instable**.

3. If the real part of the characteristic root λ_i is 0, we say that the system is **marginal**.

Therefore, for a linear system to be stable, we can say that the **condition for the system to be stationary** is that the real part of all roots μ_i $(i = 1, 2, \cdots, n)$ of $A(p) = 0$ is negative, that is, *the roots of the polynomial equation are located in the left half region* as in Figure 5.6.

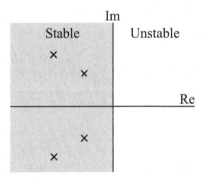

FIGURE 5.6
p-plane and stability of a system

5.4   Maneuvering Equation of Motion for First-Order System

Next, using Nomoto's maneuvering kinematic equation used in the maneuverability study shown in Chapter 15 and the second-order vibration equation used in the sea-keeping ability study, let us calculate the impulse response function, frequency response function, etc. of those dynamic systems[95].

5.4.1   Yaw Angular Velocity Response When Steering

First, consider the equation of maneuvering kinematic motion for the angular velocity of the yaw when steering is given:

$$T\dot{r} + r = K\delta. \tag{5.22}$$

Frequency response function

The frequency response function is then given by

$$G(f) = \frac{K}{1 + i2\pi fT} = K\left(\frac{1}{1 + (2\pi fT)^2} - i\frac{2\pi fT}{1 + (2\pi fT)^2}\right).$$

Thus, we have

$$\text{Amplitude ratio } |G(f)| \quad = \quad \frac{K}{\sqrt{1 + (2\pi fT)^2}}$$

$$\text{Phase difference} \quad = \quad -\tan^{-1}(2\pi fT),$$

and are shown in Figure 5.7. These figures are called the Bode diagrams of the first-order systems, and the x-axis in both figures is shown in logarithmic frequency. The upper figure shows 10 times the logarithm of the amplitude (dB), and the lower figure shows the phase angle (degrees).

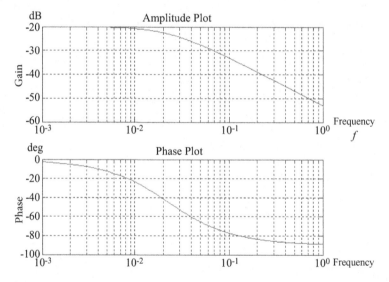

FIGURE 5.7
Frequency response function of a first order system

Impulse Response Function

From the inverse transformation of the frequency response function, the impulse response function of a first-order system is given by

$$g(t) = \begin{cases} \dfrac{K}{T}e^{-\frac{t}{T}}, & t \geq 0 \\ 0, & t < 0. \end{cases} \tag{5.23}$$

5.4.2   Yaw Angle Response When Steer a Ship

In the above example, suppose that the yaw response angle is observed when a rudder angle input is applied in the steering system. From equation (5.22), the characteristics at that time can be expressed as

$$\dot{r} = -\frac{r}{T} + \frac{K}{T}\delta$$
$$\dot{\psi} = r,$$

where ψ is the bow response angle. (In this way, the expression in which the first-order derivative without coefficients is placed on the left-hand side and the expression without a derivative term is placed on the right-hand side is called **normal system expression**. It is closely related to the **state-space representation** described later.)

Eliminating r from the above equation yields

$$T\ddot{\psi} + \dot{\psi} = K\delta.$$

Frequency response function

In this case, putting $\omega = 2\pi f$, the frequency response function is obtained by

$$G(\omega) = \frac{1}{i\omega}\frac{K}{Ti\omega + 1}, \tag{5.24}$$

where it can be expressed as

$$G(\omega) = \frac{K}{i\omega}\left(\frac{1}{Ti\omega + 1}\right) = K\left(\frac{1}{-T(\omega)^2 + i\omega}\right) = K\frac{T\omega^2 + i\omega}{(T\omega^2)^2 + \omega^2}.$$

That is, the amplitude characteristic is given by

$$|G(\omega)| = \frac{K}{\sqrt{(T\omega^2)^2 + \omega^2}}$$

and the phase characteristic is

$$\tan\varepsilon = \frac{K}{T\omega}.$$

The impulse response function can be obtained from equation (5.24) as

$$\frac{1}{i\omega}\frac{K}{Ti\omega + 1} = K\left(\frac{1}{i\omega} - \frac{1}{i\omega + 1/T}\right).$$

Therefore, from equations (3.15) and (3.26), we obtain

$$g(t) = K\left(1 - e^{-\frac{t}{T}}\right).$$

5.4.3   Linear Stochastic Differential Equation of First Order System

The model of the linear homogeneous equation of first order system with white noise forcing input on the right-hand side is given by[88]

$$\frac{dX(t)}{dt} + \alpha_0 X(t) = Z(t). \tag{5.25}$$

Using the linear differential operator p, equation (5.25) becomes

$$(p + \alpha_0) X(t) = Z(t), \tag{5.26}$$

where

$$\begin{cases} E[Z(t)] = 0 \\ E[Z(t) \, Z(t - t_0)] = \sigma_z^2 \, \delta(t_0). \end{cases} \tag{5.27}$$

This equation is called the first-order linear stochastic differential equation in the continuous domain. We write this equation as A(1). Since the inputs $Z(t)$ is a white noise, i.e., values at different times are uncorrelated random variables with mean 0, and the left-hand side is linear, for $\alpha_0 > 0$, the random variable $X(t)$ is also a stationary and stable process with mean 0. To obtain the "solution" of this equation $X(t)$ as a function of $Z(t)$, we formally write

$$X(t) = \frac{Z(t)}{p + \alpha_0}. \tag{5.28}$$

Just as the solution of the linear system (equation (5.5)) is given by the convolution integral of the external force $u(t)$ (equation (5.19)), equation (5.28) is analogously given by the function corresponding to the impulse response function $g(t)$, as follows;

$$X(t) = \int_0^\infty g(\nu) \, Z(t - \nu) \, d\nu = \int_{-\infty}^t g(t - \nu) \, Z(\nu) \, d\nu. \tag{5.29}$$

In this case, from equation (5.28) and the Fourier transformation formula (3.26), the impulse response function $g(t)$ is obtained as

$$g(t) = \begin{cases} e^{-\alpha_0 t} & t \geq 0 \\ 0 & t < 0. \end{cases} \tag{5.30}$$

Thus, $X(t)$ has a integral representatiom as follows;

$$X(t) = \int_0^\infty e^{-\alpha_0 t} Z(t - \nu) \, d\nu = \int_{-\infty}^t e^{-\alpha_0(t-\nu)} Z(\nu) \, d\nu. \tag{5.31}$$

5.4.4   Covariance Function of a First-Order Stochastic Linear System

Stochastic processes are described by covariance functions. Therefore, before finding the covariance function $R(\tau)$ of lag τ of a first-order system given by stochastic linear differential equations, it is necessary to know that the covariance function $R(\tau)$ and equation (5.29) generally have the following relationship;

$$\begin{aligned} R(\tau) &= E[X(t) \, X(t - \tau)] \\ &= E\left[\int_0^\infty g(\nu') \, Z(t - \nu') \, d\nu' \int_0^\infty g(\nu) \, Z(t - \tau - \nu) \, d\nu \right] \\ &= \int_0^\infty \int_0^\infty g(\nu') \, g(\nu) \, E[Z(t - \nu') \, Z(t - \tau - \nu)] d\nu \, d\nu' \\ &= \sigma_z^2 \int_0^\infty \left[\int_0^\infty g(\nu') \, g(\nu) \, \delta(\nu + \tau - \nu') \, d\nu' \right] d\nu \\ &= \sigma_z^2 \int_0^\infty g(\nu) \, g(\nu + \tau) \, d\nu. \end{aligned} \tag{5.32}$$

Using this relationship, the covariance function for a first-order system given by the stochastic linear differential equation is given by

$$
\begin{aligned}
R(\tau) &= \sigma_z^2 \int_0^\infty e^{-\alpha_0 \nu} e^{-\alpha_0(\nu+\tau)} d\nu \\
&= \sigma_z^2 e^{-\alpha_0 \tau} \int_0^\infty e^{-2\alpha_0 \nu} d\nu \\
&= \sigma_z^2 e^{-\alpha_0 \tau} \left[\frac{-e^{-2\alpha_0 \nu}}{2\alpha_0} \right]_0^\infty \\
&= \frac{\sigma_z^2}{2\alpha_0} e^{-\alpha_0 \tau}, \quad \tau \geq 0, \quad \alpha_0 > 0.
\end{aligned}
\tag{5.33}
$$

In particular, the variance of $X(t)$ is given by

$$
R(0) = \frac{\sigma_z^2}{2\alpha_0}.
\tag{5.34}
$$

Since the autocovariance function is an even function, we obtain $R(\tau) = R(-\tau)$. Therefore, if we define $R(\tau)$ over $-\infty < \tau < \infty$, then it becomes

$$
R(\tau) = \frac{\sigma_\nu^2}{2\alpha_0} e^{-\alpha_0 |\tau|}.
\tag{5.35}
$$

Thus, the exponential function has the same form as in equation (5.30), indicating that *observing the covariance function reveals the characteristics of the first order system.*

Putting $\omega = 2\pi f$, the spectrum is obtained as the Fourier transform of equation (5.35),

$$
\begin{aligned}
P(\omega) &= \frac{\sigma_z^2}{|i\omega + \alpha_0|^2} \\
&= \frac{\sigma_z^2}{\omega^2 + \alpha_0^2}, \quad -\infty < \omega < \infty.
\end{aligned}
\tag{5.36}
$$

This result shows that in the case of a rational-type spectrum, it can be obtained by multiplying the variance of the noise Z by the square of the absolute value of the result obtained by substituting $i\omega$ for the denominator (and also the numerator if present) p in equation (5.28).

5.5 Quadratic System and Ship's Kinematic Equation of Motion

5.5.1 Response Function of the Quadratic System

The motion of a ship in waves is expressed by the oscillation equation with damping term (see Chapter 15);

$$
\ddot{\phi} + 2\varsigma\omega_n\dot{\phi} + \omega_n^2\phi = f(t).
\tag{5.37}
$$

The frequency response function is expressed as

$$
\begin{aligned}
G(i\omega) &= \frac{1}{(i\omega)^2 + 2\varsigma\omega_n(i\omega) + \omega_n^2} \\
&= \frac{1}{(\omega_n^2 - \omega^2) + i(2\varsigma\omega_n\omega)} = |G(i\omega)|\angle\theta,
\end{aligned}
$$

where the amplitude characteristic $|G(i\omega)|$ is

$$|G(i\omega)| = \frac{1}{\sqrt{(\omega_n^2 - \omega^2)^2 + (2\varsigma\omega_n\omega)^2}},$$

and the phase characteristic $\angle\theta$ is given by

$$\tan\varepsilon = -\tan^{-1}\frac{2\varsigma\omega_n\omega}{\omega_n^2 - \omega^2}.$$

If we assume that

$$\mu_1, \mu_2 = -\varsigma\omega_n \pm \omega_n\sqrt{1 - \varsigma^2},$$

the impulse response function is given by

$$
\begin{aligned}
g(t) &= \mathfrak{F}^{-1}\left[\frac{1}{(i\omega - \mu_1)(i\omega - \mu_2)}\right] \\
&= \begin{cases} \dfrac{e^{\mu_1 t} - e^{\mu_2 t}}{\mu_1 - \mu_2}, & t \geq 0 \\ 0, & t < 0. \end{cases}
\end{aligned}
\tag{5.38}
$$

Therefore, we have

$$
\begin{aligned}
g(t) &= \frac{1}{2\omega_n\sqrt{\varsigma^2 - 1}}e^{-\varsigma\omega_n t}\left(e^{-\omega_n\sqrt{\varsigma^2-1}\,t} - e^{\omega_n\sqrt{\varsigma^2-1}\,t}\right) \\
&= \frac{1}{\omega_n\sqrt{\varsigma^2 - 1}}\sinh\left(\omega_n\sqrt{\varsigma^2 - 1}\,t\right), \quad \varsigma > 1, t \geq 0 \tag{5.39} \\
&= \frac{1}{\omega_n\sqrt{1 - \varsigma^2}}\sin\left(\omega_n\sqrt{1 - \varsigma^2}\,t\right), \quad \varsigma < 1, t \geq 0, \tag{5.40}
\end{aligned}
$$

where $\sinh x$ is a hyperbolic function given by $\sinh x = (e^x - e^{-x})/2$. Figure 5.8 depicts the frequency response function of the second-order system for $\alpha = \varsigma\omega$.

5.5.2   Impulse Response Function, Covariance Function and Spectrum of a Quadratic Linear Stochastic System

Linear differential equation (5.37) for ship motion with white noise $Z(t)$ added to the model

$$\frac{d^2 X(t)}{dt^2} + 2\varsigma\omega_n\frac{dX(t)}{dt} + \omega_n^2 = Z(t), \tag{5.41}$$

is called a linear stochastic differential equation of the second-order system and is written as A(2)[88].

It is clear from equation (5.38) that the impulse response function of the system given by the linear stochastic differential equation of the second-order system is given by

$$g(t) = \frac{e^{\mu_1 t} - e^{\mu_2 t}}{\mu_1 - \mu_2}. \tag{5.42}$$

As a result, since $X(t)$ is expressed as

$$
\begin{aligned}
X(t) &= \int_0^\infty \frac{e^{\mu_1 v} - e^{\mu_2 v}}{\mu_1 - \mu_2}Z(t - v)\,dv \\
&= \int_{-\infty}^t \frac{e^{\mu_1(t-v)} - e^{\mu_2(t-v)}}{\mu_1 - \mu_2}Z(v)\,dv,
\end{aligned}
$$

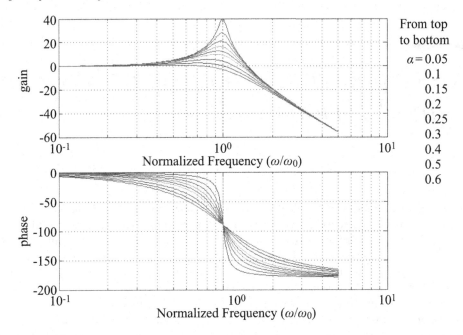

FIGURE 5.8
Frequency response function of a second-order system

the covariance function for lag τ can be expressed as

$$R(\tau) \;=\; \sigma_z^2 \int_0^\infty g(\nu)\, g(\nu + \tau)\, d\nu$$

$$=\; \frac{\sigma_z^2}{2\mu_1\mu_2\,(\mu_1^2 - \mu_2^2)}\,\left(\mu_2\, c^{\mu_1\tau} - \mu_1\, e^{\mu_2\tau}\right). \tag{5.43}$$

Thus, the exponential function is of the same form as equation (5.39), showing that the characteristics of the quadratic system can be determined by observing the **covariance function**.

The variance of $X(t)$ is given by

$$R(0) = -\frac{\sigma_z^2}{2\mu_1\mu_2(\mu_1 + \mu_2)} = \frac{\sigma_z^2}{4\varsigma\omega_n^3}. \tag{5.44}$$

Applying the rules stated in equation (5.36), it is easy to see that the spectrum is given by

$$P(\omega) \;=\; \frac{\sigma_z^2}{|(i\omega)^2 + 2\varsigma\omega_n(i2\pi f) + \omega_n^2|^2}$$

$$=\; \frac{\sigma_z^2}{|(\omega_n^2 - \omega^2)^2 + 4\varsigma^2\omega_n^2\varsigma^2|}, \quad -\infty < \omega < \infty. \tag{5.45}$$

6

Cross-Spectrum Analysis of Ship and Engine Motions in Waves

In the previous chapters, we have dealt with time series of a single variable that fluctuates irregularly. However, when we observe the response of rolling motion in irregular ocean waves, it is thought that irregular disturbances to the system, such as waves, affect the ship motion time series with a certain relationship, sometimes with time delay. The ship motions are also considered to mutually influence each other with a certain causal relationship. In this section, we will first discuss the **cross-covariance function**, which expresses the statistical relationship between time series of two variables in the domain of time, and the **cross-spectrum**, which expresses the relationship in the frequency domain. Frequency response functions and impulse response functions used in system analysis are defined and their estimation methods are described. We also extend the obtained methods to the analysis of multi-input systems that show the relationship between two or more variables.

6.1 Cross-Covariance Function and Cross-Spectrum

First, let us define the **cross-spectrum** in time series theory[26][95]. Suppose now that there are two ergodic weakly stationary time series in the continuous time domain $\{x(t), y(t); -\infty \leq t \leq \infty\}$, measured at the same time. It is further assumed that these two time series are transformed into quantities around the mean as a preprocessing step. In other words, suppose that we replace

$$x(t) \Leftarrow x(t) - m_x, \quad y(t) \Leftarrow y(t) - m_y,$$

where m_x and m_y are the mean values of the time series $x(t)$ and $y(t)$, respectively. For these two time series, the same concept as the autocovariance function can be applied. For example, the cross-covariance function can be defined as a quantity that represents the average relationship between the product of the value $x(t)$ at time t of the wave time series $\{x(t), -\infty \leq t \leq \infty\}$ shown in Figure 6.1 and the value $y(t + \tau)$ at time $t + \tau$, a lag τ away, of the rolling time series $\{y(t), -\infty \leq t \leq \infty\}$,

$$R_{xy}(\tau) = E\left[x(t + \tau)\, y(t)\right] = \lim_{T \to \infty} \frac{1}{T} \int_{-\frac{T}{2}}^{\frac{T}{2}} x(t + \tau)\, y(t)\, dt. \tag{6.1}$$

The cross-covariance function defined in this way is *not symmetric with respect to time lag 0* like the autocovariance function. In this case, the results are generally different when summing over a time series $y(t + \tau)$ in the forward direction at regular intervals from the time series $x(t)$, and when summing over a time series $y(t - \tau)$ in the backward direction. However, in the case of stationary time series, it holds that

$$R_{xy}(\tau) = R_{yx}(-\tau). \tag{6.2}$$

DOI: 10.1201/9781003428565-6

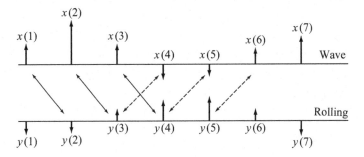

FIGURE 6.1
Sum of products in cross-covariance function. (Solid line: lag=1, dotted line: lag=−1)

Also, it does *not necessarily take the maximum value at lag 0*. This can be understood from the fact that there is generally a lag between the input of the system and the output as its response.

Let $x_T(t)$ and $y_T(t)$ be the time series measured in the time interval with length T from the population time series. Let $X_T(f)$ and $Y_T(f)$ be the discrete Fourier transforms of those time series (equation (7.17)). The **cross-spectral density function** of the two time series is defined as

$$P_{xy}(f) = \lim_{T \to \infty} E\left[\frac{X_T(f)\,\overline{Y_T(f)}}{T}\right], \tag{6.3}$$

where $\overline{Y_T(f)}$ is the complex conjugate of $Y_T(f)$. Corresponding to the Wiener-Khinchin theorem, which states the relationship between the autocovariance function and the spectrum, there is a relationship between the cross-covariance function and the cross-spectrum as

$$P_{xy}(f) = \int_{-\infty}^{\infty} R_{xy}(\tau) \exp(-i2\pi f\tau)\, d\tau \tag{6.4}$$

$$R_{xy}(\tau) = \int_{-\infty}^{\infty} P_{xy}(f) \exp(i2\pi f\tau)\, df. \tag{6.5}$$

The cross-spectrum has the following properties from equation (6.2)

$$P_{xy}(f) = \overline{P_{yx}(f)} = P_{yx}(-f), \tag{6.6}$$

where $P_{yx}(-f)$ denotes the Fourier transform of the covariance function for $y(t)$ and $x(t)$ at frequency $-f$.

6.2   Linear System Analysis and Cross-Spectrum Analysis

Let us consider a one-input and one-output dynamic system in the discrete-time domain. Let $x(1), x(2), \cdots, x(N)$ be the input signal time series and $y(1), y(2), \cdots, y(N)$ be the response signals, and suppose that there is a linear relationship

$$y(s) = \sum_{m=0}^{\infty} g_{yx}(m)\, x(s - m), \quad s = 1, 2, \cdots, N. \tag{6.7}$$

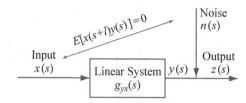

FIGURE 6.2
Linear systems considered here

This model describes the current output signal as the sum of the current and past input signals multiplied by some weight $g_{yx}(m)$, and is called the **linear single-input single-output model**[95]. Here $g_{yx}(m)$, $m = 0, 1, 2, \cdots, \infty$ is the impulse response function defined in Chapter 5. However, if $z(s)$ is the signal on the output side that is actually observed, the observed signal may be disturbed by noise that is uncorrelated with the input. Therefore, we consider that the output signal is disturbed by the additive observation noise $n(s)$ and can be expressed as

$$z(s) = y(s) + n(s). \tag{6.8}$$

However, if there is a feedback loop, even if the disturbance appears to be uncorrelated with the output, there will be a resulting correlation, as shown in Chapter 12. In this section, we first consider an **open loop** system in which there is no feedback from the output side, such as the relationship between ocean waves and rolling motions shown in Figure 6.2. From this assumption, the following holds

$$E\left[x(s + l)\,n(s)\right] = 0. \tag{6.9}$$

In this model, the disturbance is considered to contain a component that could not be represented by the linear response and observed noise. From equations (6.7) and (6.8), $z(s)$ at time s can be expressed using $g_{yx}(m)$ as follows

$$z(s) = y(s) + n(s) = \sum_{m=0}^{\infty} g_{yx}(m)\,x(s - m) + n(s), \tag{6.10}$$

where $n(s)$ is assumed to be noise, not necessarily white noise, and equation (6.9) holds. In this case, multiplying $y(s)$ by $x(s + l)$ and taking the expected value, we obtain

$$R_{yx}(l) = \sum_{m=0}^{\infty} g_{yx}(m)\,R_{xx}(l - m).$$

Using the convolution theorem (equation (3.34)) and Fourier transforming both sides yields

$$P_{yx}(f) = G_{yx}(f)\,P_{xx}(f). \tag{6.11}$$

From this, we obtain

$$G_{yx}(f) = \frac{P_{yx}(f)}{P_{xx}(f)}. \tag{6.12}$$

This $G_{yx}(f)$ is the frequency response function from input $x(s)$ to output $y(s)$, which is also the Fourier transform of the impulse response function $g_{yx}(m)$.

On the other hand, multiplying $y(s)$ by $y(s+l)$ from the left and taking the expectation, we obtain

$$R_{yy}(l) = \sum_{m=0}^{\infty} \sum_{m'=0}^{\infty} g_{yx}(m)\, g_{yx}(m')\, R_{xx}(l+m-m').$$

Then discrete Fourier transforming both sides (equation (7.17)) yields

$$
\begin{aligned}
P_{yy}(f) &= \sum_{l=0}^{\infty} R_{yy}(l)\, e^{-i2\pi fl} \\
&= \sum_{l=0}^{\infty} e^{-i2\pi fl} \sum_{m=0}^{\infty} \sum_{m'=0}^{\infty} g_{yx}(m)\, g_{yx}(m')\, R_{xx}(l+m-m') \\
&= \sum_{m=0}^{\infty} g_{yx}(m)\, e^{-i2\pi fm} \sum_{m'=0}^{\infty} g_{yx}(m')\, e^{i2\pi fm'} \sum_{k=0}^{\infty} R_{xx}(k)\, e^{-i2\pi fk},
\end{aligned}
$$

where $k = l + m - m'$. Therefore, we have

$$P_{yy}(f) = G(f)_{yx}\, G(-f)_{yx}\, P_{xx}(f) = |G_{yx}(f)|^2\, P_{xx}(f). \tag{6.13}$$

Equations (6.11) and (6.13) show two ways to obtain the frequency response function. However, in equation (6.13), the phase information is lost.

For a system with such a linear relationship, consider a model (equation (6.10)) that assumes that the noise $n(s)$ is uncorrelated with the input. By assumption, since we have

$$R_{nx}(l) = 0, \quad l = 0, \pm 1, \pm 2, \cdots, \tag{6.14}$$

the covariance of the observed value $z(s)$ satisfies

$$R_{zz}(l) = R_{yy}(l) + R_{nn}(l).$$

Therefore, for the spectrum, we have

$$P_{zz}(f) = P_{yy}(f) + P_{nn}(f).$$

Then, for each frequency, we can introduce the following quantity as the *proportion of the linearly related signal in the spectrum*,

$$\gamma^2(f) = \frac{P_{yy}(f)}{P_{zz}(f)}, \tag{6.15}$$

and referred to as the **coherency function**[95]. From equation (6.14), since the following holds

$$P_{zx}(f) = P_{yx}(f) + P_{nx}(f) = P_{yx}(f),$$

the numerator of equation (6.12) can be replaced with $P_{zx}(f)$. This allows the frequency response function to be expressed as

$$G_{yx}(f) = \frac{P_{zx}(f)}{P_{xx}(f)}. \tag{6.16}$$

This result implies that the coherency function is also computed using[103][104]

$$\gamma^2(f) = \frac{|P_{zx}(f)|^2}{P_{xx}(f)\, P_{zz}(f)}. \tag{6.17}$$

6.3   Spectrum Analysis of Multi-Input Systems

We next extend the results of the previous section to a spectrum analysis of a multi-input, one-output system. This case can be discussed in the same way as the multivariate analysis in the usual multivariate regression analysis[8][105].

Assume that the q-variate time series

$$x(s) = (x_1(s), x_2(s), \cdots, x_q(s))^t$$

is given as input signal. Further, suppose that the resulting one-dimensional observation $z(s)$ of the output is obtained. Here, assume that $y(s)$ is the signal that has a linear relationship with the input $x(s)$, i.e.

$$y(s) = \sum_{i=1}^{q} y_i(s) = \sum_{i=1}^{q} \sum_{m=0}^{\infty} g_{yi}(m)\, x_i(s-m), \tag{6.18}$$

and the output $z(s)$ can be expressed as the sum of $y(s)$ and a one-dimensional noise $n(s)$ uncorrelated with $x(s)$ (which includes nonlinear relationships that cannot be represented by the above model). In other words

$$z(s) = y(s) + n(s), \quad E\left[x_i(s+l)\, n(s)\right] = 0.$$

For convenience, let $y(s)$ be $x_{q+1}(s)$ and the impulse response vector be

$$g_{q+1}(m) = (g_{q+1\,1}(m), g_{q+1\,2}(m), \cdots, g_{q+1\,q}(m)).$$

Then, equation (6.18) can be written as

$$x_{q+1}(s) = \sum_{m=0}^{\infty} g_{q+1}(m)\, x(s-m).$$

Multiplying both sides of the above equation by $x_{q+1}(s)$ and taking the expectation, it can be seen that the autocovariance function of variable $x_{q+1}(s)$ ($= y(s)$, hereafter omitted) satisfies

$$
\begin{aligned}
R_{q+1\,q+1}(l) &= E\left[\sum_{m=0}^{\infty} g_{q+1}(m)\, x(s+l-m) \sum_{m'=0}^{\infty} g_{q+1}(m')\, x(s-m')\right] \\
&= \sum_{i=1}^{k} \sum_{j=1}^{k} \sum_{m=0}^{\infty} \sum_{m'=0}^{\infty} g_{q+1\,i}(m)\, g_{q+1\,j}(m')\, R_{ij}(l-m+m').
\end{aligned}
$$

As a result, by discrete Fourier transforming both sides (equation (7.17)), the spectrum of $y(s)$ is given by

$$P_{q+1\,q+1}(f) = \sum_{i=1}^{q} \sum_{j=1}^{q} G_{q+1\,i}(f)\, G_{q+1\,j}(f) P_{ij}(f). \tag{6.19}$$

Next, let us calculate the cross-spectrum of $y(s)$ and $x_j(s)$. From equation (6.18), the cross-covariance function between $y(s+l)$ and $x_j(s)$ is given by

$$
\begin{aligned}
R_{q+1\,j}(l) &= E\left[\sum_{m=0}^{\infty} g_{q+1}(m)\,x(s+l-m)\,x_j(s)\right] \\
&= \sum_{m=0}^{\infty} g_{q+1}(m)\,E\left[x(s+l-m)\,x_j(s)\right] \\
&= \sum_{m=0}^{\infty} g_{q+1}(m)\begin{pmatrix} R_{1j}(l-m) \\ R_{2j}(l-m) \\ \vdots \\ R_{qj}(l-m) \end{pmatrix} \\
&= \sum_{i=1}^{q}\sum_{m=0}^{\infty} g_{q+1\,i}(m)\,R_{ij}(l-m), \quad j=1,2,\cdots,q.
\end{aligned}
$$

Therefore, by the Fourier transformation of both sides, the cross-spectrum between the output $y(s)$ and the input $x_j(s)$ is given by

$$
P_{q+1\,j}(f) = \sum_{i=1}^{q} G_{q+1\,i}(f)\,P_{ij}(f), \quad j=1,2,\cdots,q. \tag{6.20}
$$

Note that, using this result, equation (6.19) can be written as

$$
P_{q+1\,q+1}(f) = \sum_{j=1}^{q} P_{q+1\,j}(f)\,\overline{G_{q+1\,j}(f)}. \tag{6.21}
$$

Equation (6.21) is the power spectrum for the noise-free part of the model. Here the ratio of $P_{q+1\,q+1}(f)$ and the power spectrum of the output observation z, $P_{zz}(f)$, expresses the extent to which the power spectrum of the observation is represented by the model. This quantity is therefore a coherency function, and in this case, since there are multiple inputs, it is called a **multiple coherency function**, and can be computed by[104]

$$
\gamma^2_{y,x} = \frac{P_{yy}(f)}{P_{zz}(f)} = \frac{\displaystyle\sum_{j=1}^{q} P_{q+1\,j}(f)\,\overline{G_{q+1\,j}(f)}}{P_{zz}(f)}. \tag{6.22}
$$

6.4 Estimation Methods for Linear Multiple Input Systems

6.4.1 Multiple Regression Models and Linear Multiple Input Systems

The frequency response function and coherency function (equation (6.22)) described in the previous section are applications of multiple regression analysis techniques in multivariate analysis in the frequency domain. In this section, we investigate how to estimate a frequency-based model by applying the least squares method and other methods of multiple regression analysis in the frequency domain. Consider a q-variate time series $x(s)$ and let these variables be the q explanatory variables in multiple regression analysis $(x_1(s), x_2(s), \cdots, x_q(s))$ and the objective variable be the output $y(s)$ $(= x_{q+1}(s))$ (Figure 6.3).

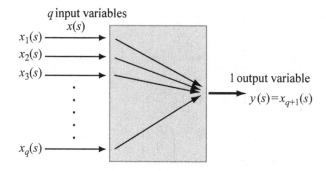

FIGURE 6.3
Multiple input system

The covariance matrix, which is the statistics in the time domain between those variables, and its Fourier transform, the spectrum, are regarded as the variance-covariance matrix in the multiple regression analysis. Note that the spectrum matrix is a complex matrix.

In this case, the spectrum matrix $P_{xx}(f)$ whose elements are the spectra of the multivariate input variable $x(s)$ is defined as

$$P_{xx}(f) = \begin{bmatrix} P_{11}(f) & P_{12}(f) & \cdots & P_{1q}(f) \\ P_{21}(f) & P_{22}(f) & \cdots & P_{2q}(f) \\ \vdots & \vdots & \ddots & \vdots \\ P_{q1}(f) & P_{q2}(f) & \cdots & P_{qq}(f) \end{bmatrix},$$

where $P_{ij}(f) = \overline{P_{ji}(f)}$.

By the way, in equation (6.20), we consider

$$P_{j\,q+1}(f) = \sum_{i=1}^{q} P_{ji}(f)\,G_{i\,q+1}(f), \quad j = 1, 2, \cdots, q. \tag{6.23}$$

In the matrix representation, it is given by

$$\begin{bmatrix} P_{1\,q+1}(f) \\ P_{2\,q+1}(f) \\ \vdots \\ P_{q\,q+1}(f) \end{bmatrix} = \begin{bmatrix} P_{11}(f) & P_{12}(f) & \cdots & P_{1q}(f) \\ P_{21}(f) & P_{22}(f) & \cdots & P_{2q}(f) \\ \vdots & \vdots & \ddots & \vdots \\ P_{q1}(f) & P_{q2}(f) & \cdots & P_{qq}(f) \end{bmatrix} \begin{bmatrix} G_{1\,q+1}(f) \\ G_{2\,q+1}(f) \\ \vdots \\ G_{q\,q+1}(f) \end{bmatrix}.$$

Here, if we define

$$P_{q+1}(f) = \begin{bmatrix} P_{1\,q+1}(f) \\ P_{2\,q+1}(f) \\ \vdots \\ P_{q\,q+1}(f) \end{bmatrix}, \quad G_{q+1}(f) = \begin{bmatrix} G_{1\,q+1}(f) \\ G_{2\,q+1}(f) \\ \vdots \\ G_{q\,q+1}(f) \end{bmatrix}$$

it can be expressed as

$$P_{q+1}(f) = P_{xx}(f)\,G_{q+1}(f). \tag{6.24}$$

This equation is a system of linear equations for $G_{q+1}(f)$ and can be easily solved. Note that, however, the complex conjugate of the obtained result is the required frequency response function $G_{q+1\,j}(f)$.

6.4.2 Relation Between Frequency Response Function and Cross-Spectrum

In this section, we will leave the cross-spectrum estimation method for later and show how to calculate the frequency response function from equation (6.24). The first-order linear system (6.24) can be solved by the well-known method called the **sweep-out method**. The sweep-out method proceeds as follows;

1. Make the following matrix by augmenting the cross-spectrum matrix:

$$\begin{pmatrix} P_{11}(f) & P_{12}(f) & \cdots & P_{1q}(f) & P_{1\,q+1}(f) & 1 & 0 & \cdots & 0 \\ P_{21}(f) & P_{22}(f) & \cdots & P_{2q}(f) & P_{2\,q+1}(f) & 0 & 1 & \cdots & 0 \\ \vdots & \vdots & \vdots & \vdots & \vdots & \vdots & \vdots & \ddots & \vdots \\ P_{q1}(f) & P_{q2}(f) & \cdots & P_{qq}(f) & P_{q\,q+1}(f) & 0 & 0 & \cdots & 1 \\ P_{q+1\,1}(f) & P_{q+1\,2}(f) & \cdots & P_{q+1\,q}(f) & P_{q+1\,q+1}(f) & 0 & 0 & \cdots & 0 \end{pmatrix}. \tag{6.25}$$

2. Focusing on the diagonal element $P_{11}(f)$, divide the first row by $P_{11}(f)$.

3. Then subtract from the second row the result of multiplying each element of the first row by $P_{21}(f)$, so that the second row of the first column is 0 and the second row of the other column becomes

$$\left(P_{22}(f) - P_{21}(f)\frac{P_{12}(f)}{P_{11}(f)}\right), \left(P_{23}(f) - P_{21}(f)\frac{P_{13}(f)}{P_{11}(f)}\right), \cdots,$$

$$\left(P_{2q}(f) - P_{21}(f)\frac{P_{1q}(f)}{P_{11}(f)}\right), \left(P_{2\,q+1}(f) - P_{21}(f)\frac{P_{1\,q+1}(f)}{P_{11}(f)}\right).$$

4. Similarly, performing the same operation below the third line, the left half of equation (6.25) becomes

$$\begin{pmatrix} 1 & P_{12}^{(1)}(f) & \cdots & P_{1q}^{(1)}(f) & P_{1\,q+1}^{(1)}(f) \\ 0 & P_{22}^{(1)}(f) & \cdots & P_{2q}^{(1)}(f) & P_{2\,q+1}^{(1)}(f) \\ \vdots & \vdots & \vdots & \vdots & \vdots \\ 0 & P_{q2}^{(1)}(f) & \cdots & P_{qq}^{(1)}(f) & P_{q\,q+1}^{(1)}(f) \\ 0 & P_{q+1\,2}^{(1)}(f) & \cdots & P_{q+1\,q}^{(1)}(f) & P_{q+1\,q+1}^{(1)}(f) \end{pmatrix}. \tag{6.26}$$

(The same operation is performed for the right half, but omitted here.)

5. Divide the second row by the element $P_{22}^{(1)}(f)$ in the second column of the second row of equation (6.26) obtained in step 4..

6. Perform the same operation as steps 3 and 4 for $j = 1, 3, 4, \cdots, q+1$ rows yields $(2, 2)$ element being 1 and the other elements in the second column being 0.

7. Performing this sequence of operations up to the qth row, we obtain

$$\begin{pmatrix} 1 & 0 & \cdots & 0 & \alpha_1 & \gamma_{11} & \gamma_{12} & \cdots & \gamma_{1q} \\ 0 & 1 & \cdots & 0 & \alpha_2 & \gamma_{21} & \gamma_{22} & \cdots & \gamma_{2q} \\ \vdots & \vdots & \cdots & \vdots & \vdots & \vdots & \vdots & \cdots & \vdots \\ 0 & 0 & \cdots & 1 & \alpha_q & \gamma_{q1} & \gamma_{q2} & \cdots & \gamma_{qq} \\ 0 & 0 & \cdots & 0 & \varepsilon & \beta_1 & \beta_2 & \cdots & \beta_q \end{pmatrix}. \tag{6.27}$$

The $q \times q$ matrix that is formed on the right of equation (6.27) gives the inverse $P(f)^{-1}$ of the original matrix $P(f)$.

Since $\alpha_j(f)$ is a solution of equation (6.24), its complex conjugate $\overline{\alpha_j(f)}$ gives the frequency response function from the input variable $x_j(s)$, $j = 1, 2, \cdots, q$, to the output variable $x_{q+1}(s) = y(s)$. Furthermore, it is equal to $-\beta_j(f)$. Because the last line of equation (6.27) is the residual defined by subtracting linear contributions from the input component,

$$\varepsilon = P_{q+1\,q+1}(f) - G_{q+1\,1}(f)\,P_{q+1\,1}(f) - G_{q+1\,2}(f)\,P_{q+1\,2}(f) - \cdots - G_{q+1\,q}(f)\,P_{q+1\,q}(f).$$

(see remark below). As a result, the multiple coherency can be calculated from the real part of ε,

$$\gamma^2(f) = 1 - \frac{\varepsilon}{P_{zz}(f)}. \tag{6.28}$$

Akaike's definition of **partial coherency function** is given by

$$\gamma^2_{q+1,j\cdot x(j)}(f) = \frac{|\alpha_j|^2}{|\alpha_j|^2 + \varepsilon\,\gamma_{jj}}. \tag{6.29}$$

The meaning of this quantity can be understood, since from equation (6.13), $|\alpha_j|^2/\gamma_{jj}$ represents the effect of $x_{q+1}(s)$ on the power spectrum after excluding linear effects from input variables other than $x_j(s)$, and ε is the estimate of the square of the residuals.

Remark:

The operation to be performed in the above sweep-out method,

$$P_{ij}^{(k)} = P_{ij}^{(k-1)} - P_{ik}^{(k-1)}\frac{P_{kj}^{(k-1)}}{P_{kk}^{(k-1)}}, \tag{6.30}$$

can be interpreted as an operation to further remove the influence of X_k from the influence of random variables $X_1, X_2, \cdots, X_{k-1}$ on the regression coefficients up to the random variable $P_{ij}^{(k-1)}$, which was removed in the previous stage $k - 1$.

6.4.3   Estimation of Cross-Spectrum

The cross-spectrum estimation takes the following steps[34].

Preprocessing of Data

Compute the mean of the data and remove the mean from each data. After this preprocessing, let $x_{q+1}(s)$ be the variables considered as output and $x_1(s), x_2(s), \ldots, x_q(s)$ the other input variables, yielding $(q+1)$-variate data of length N, $\{x_1(s), \ldots, x_q(s), x_{q+1}(s)\}$, $s = 1, 2, \ldots, N$.

Maximum lag of the covariance function h

Determine the maximum lag of the covariance function h with $h \le N/5$ as a guide.

Cross-covariance function

Compute the cross-covariance function for $j, l = 0, 1, 2, \cdots, k$,

$$C_{jl}(m) = \frac{1}{N}\sum_{s=1}^{N-m} x_j(s+m)\,x_l(s), \quad 0 \le m < h.$$

However for $m = h$, modify as follows;

$$C_{jl}(h) = \frac{1}{2}C_{jl}(h).$$

Calculation of Raw Cross-Spectrum

For $j = 1, 2, \cdots, q+1$ and $l = j, j+1, \cdots, q+1$, obtain $\{C_{jl}(-h), C_{jl}(-h+1), \cdots, C_{jl}(-1), C_{jl}(0), C_{jl}(1), \cdots, C_{jl}(h-1), C_{jl}(h)\}$ and then calculate

$$Q_{jl}(r) = \sum_{m=-h}^{h} \exp\left(-i2\pi\frac{r}{2h}m\right) C_{jl}(m)\Delta t, \quad r = 0, 1, 2, \cdots, h.$$

Note that here we use the property

$$C_{jl}(-m) = C_{lj}(m), \quad m = 1, 2, \cdots, h.$$

Spectral Window Operations

For $j = 1, 2, \cdots, q+1$ and $l = j, j+1, \cdots, q+1$, obtain the smoothed spectrum by

$$P_{jl}(r; i) = \sum_{n=-2}^{2} a_i(n) Q_{jl}(r-n).$$

For the spectral window $a_i(n)$, see Chapter 4. Here, it satisfies

$$Q_{jl}(-n) = \overline{Q_{jl}(n)}, \quad Q_{jl}(h+n) = \overline{Q_{jl}(h-n)}.$$

The resulting $P_{jl}(r; 1)$ is the estimate of the smoothed cross-spectrum $P_{jl}(r/2\Delta t)$. The subsequent calculation of the frequency response function and coherency function using the obtained cross-spectrum should follow the method given in the previous subsection.

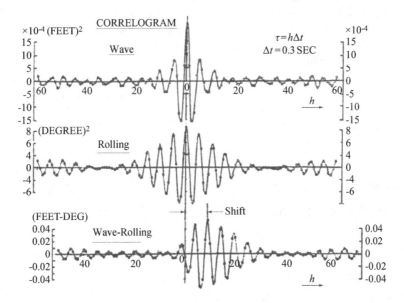

FIGURE 6.4

Auto- and cross-covariance functions of the response of rolling motion to irregular waves generated in the experimental tank

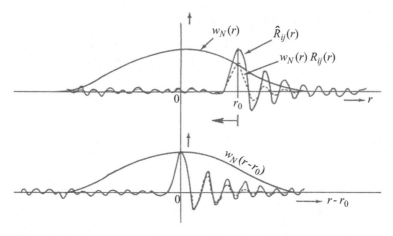

FIGURE 6.5
Variation of information obtained depending on the position of the lag window

6.4.4   Correction of Bias in Cross-Spectrum Estimation

By the way, the estimate of the cross-spectrum obtained in this way has a bias, and to obtain a cross-spectrum without bias, it is better to shift the zero point of the covariance function by K as shown below.

Figure 6.4 shows the auto- and cross-covariance functions of the model ship in irregular waves generated in the experimental tank. As can be seen from the cross-covariance function (bottom panel), the maximum value of the roll response to the given wave is shifted in the positive direction from lag 0. In this case, as shown in the upper panel of Figure 6.5, multiplying by the lag window whose peak is at lag 0 may result in underestimation of the covariance. To correct this, it is necessary to find the lag K of the point that takes the peak value in the cross-covariance function and match the highest value of the window to that point, as shown in the lower panel of Figure 6.5. Dr. Yamanouchi was the first to point this out, and thus the correction by K is called the **Yamanouchi shift**[103][8]. This effect can also be confirmed in the cross-spectrum field, since it is also the point at which the cross-spectrum phase changes abruptly.

6.5   Examples of Frequency Response Functions and Coherency Functions

6.5.1   Frequency Response Analysis of Rolling in Irregular Waves in a Tank

Figure 6.6 shows an example of the frequency response analysis of rolling to wave height obtained from a model test in irregular waves in a tank using the method described in this chapter. From the top, the coherency function and the phase and amplitude characteristics of frequency response function are shown. It is shown that the linear response of the rolling motion to the waves is high at angular frequencies of 3 to 5.

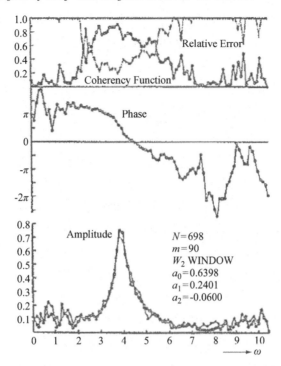

FIGURE 6.6
Rolling response to irregular waves in the experimental tank

6.5.2 Frequency Response Analysis of Actual Ship Data

As shown in Chapter 15, the flow around the stern of a ship is complex, and is called a **wake**. Assuming that the ship is navigating in calm sea and that the fluid is inviscid, the fluid that flows in from the bow does not flow straight away, but returns a little toward the bow (potential flow). The wake flow becomes more complicated when pitching, rolling and heaving motion are added. The propeller then accelerates the flow and it hits the rudder. The propeller in this complex flow is affected by the stern or rudder, and the effect is transmitted to the main engine via the propeller. In this subsection, we investigate the effects of rolling and pitching on the propeller revolution and the effects of rudder motion and fluctuation of propeller revolution on the ship's thrust in the frequency domain using the methods described in this chapter.

(1) Effects of Rolling and Pitching Motion on Propeller Revolution

Figure 6.7 shows the time series of propeller revolution (top), rolling (middle) and pitching (bottom) motion of a large container ship. Figure 6.8 shows their spectra (cutting above 0.25 Hz). As can be seen from the spectra of the rolling and pitching, the main periods of the two time series are almost coincident. The propeller revolution also peaks at the same location. The pitching has a strong damping force and is strongly affected by waves, as described in Chapter 15, and can be regarded as a kind of wave height meter. The damping force is weak for rolling. Considering this, the figure shown here can be regarded as a **synchronization** of the waves and the rolling motion. This can be seen by examining the time series of the pitching and rolling motion.

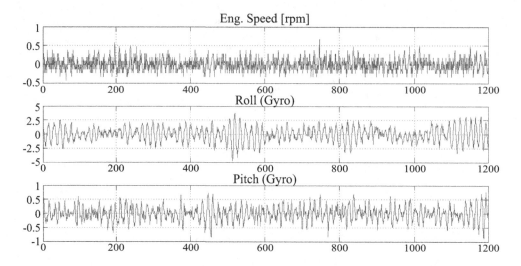

FIGURE 6.7
Time series of propeller revolutions variation, rolling and pitching of a container ship

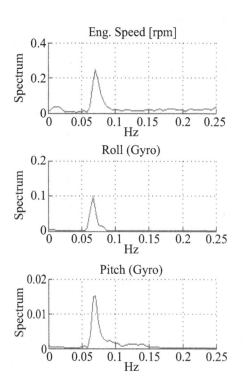

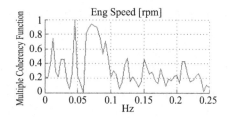

FIGURE 6.9
Multiple coherency to propeller revolution variation

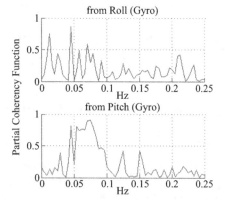

FIGURE 6.8
Spectrum of the time series in Figure 6.7

FIGURE 6.10
Partial coherency to propeller revolution variation

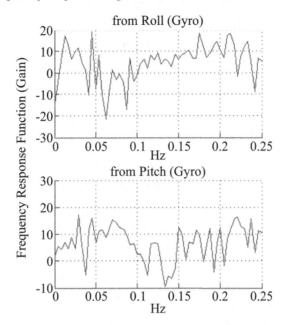

FIGURE 6.11
Frequency response functions of rolling and pitching oscillations to propeller revolution variation (Gain in dB)

Figures 6.9 and 6.10 show the multiple coherency and partial coherency for the goodness of fit of the linear model with propeller revolution as the output and the rolling and pitching as the input. From each figure, it can be seen that the degree of linear influence is strong near the frequency band representing the peak of the spectrum. Figure 6.11 shows the frequency response functions (amplitude characteristics) from the rolling and pitching to the propeller revolution (phase characteristics are omitted). From this figure, it can be seen that in the frequency band of interest, the influence from the pitching to the fluctuations of propeller revolution is high. It can be seen that the vertical motion caused by the stern's pitching motion has a significant effect on the fluctuation of propeller revolution.

(2) Effects of Rudder Angle and Propeller Revolution on Thrust

Next, let us examine the effects of fluctuation of propeller revolution and rudder angle variation on the thrust of the ship measured at the propeller shaft. From a causal point of view, we should assume that the rudder angle variation causes the fluctuation of propeller revolution and that causes the thrust variation; however, we will treat this as a multiple-input analysis. Figure 6.12 shows the time series of rudder angle, propeller revolution and thrust measured on a certain container ship, which is larger than the ship for Figure 6.7. As seen in the spectrum in Figure 6.13, the propeller speed and thrust vary accordingly due to the large long-period rudder angle. Figures 6.14 and 6.15 show the multiple coherency and partial coherency functions that represent the validity of the model when thrust is considered as output, rudder angle and propeller revolution, and a linear model is assumed. It can be seen that the validity of the model and the contribution from other variables to the thrust are higher in the long-period side of the band where the spectrum is higher.

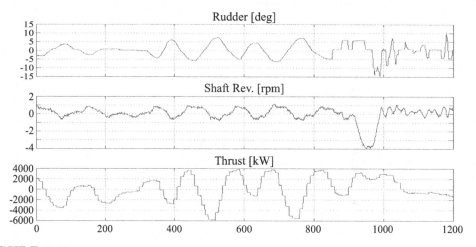

FIGURE 6.12

Time series of thrust, propeller revolution and rudder angle of a large container ship ($N = 1200$, $\Delta t = 1$ sec)

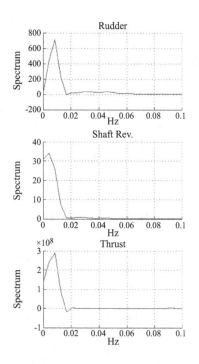

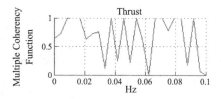

FIGURE 6.14

Multiple coherency of rudder angle and propeller revolution to thrust variation

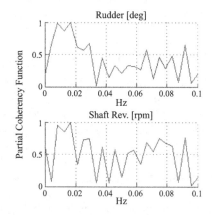

FIGURE 6.13

Spectrum of thrust, propeller revolution and rudder angle of a large container ship

FIGURE 6.15

Partial coherency of rudder angle and propeller revolution to thrust variation

Part II

Applications of Time Series Statistical Models

7

Continuous and Discrete Time Series

When using a computer to analyze ship data or control ship in the time domain, we are dealing with discretized data taken at regular intervals from the ship motions. In this case, it is necessary to discretize the data, which are originally continuously changing. In such a case, the question arises as to how much interval sampling should be done without losing the information in the original time series. In this chapter, we will discuss the relationship between the continuous model and the discrete model, which will be dealt with in Chapter 8 and thereafter.

7.1 Sampling of Time Series

Now assume that the time series $x(t)$ is defined in the frequency domain $|f| \leq f_m$ and is zero outside it[43]. When the Fourier transform of the time series $x(t)$ is given by

$$X(f) = \int_{-\infty}^{\infty} x(t)\, e^{-i2\pi ft} dt, \qquad (7.1)$$

the condition mentioned above is

$$X(f) = 0, \qquad |f| \geq f_m. \qquad (7.2)$$

The Fourier inverse transform is obtained by

$$x(t) = \int_{-\infty}^{\infty} X(f)\, e^{i2\pi ft} df = \int_{-f_m}^{f_m} X(f)\, e^{i2\pi ft} df. \qquad (7.3)$$

Now consider the case where $x(t)$ is sampled every Δt, as shown in Figure 7.1. In this case, expressing $\Delta t = \frac{1}{2f_x}$, we obtain equally spaced sampled value $x(k\Delta t)$ of the time series

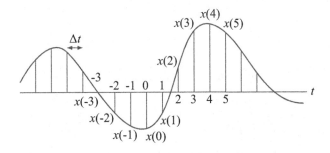

FIGURE 7.1
Sampling a continuous time series at period Δt

DOI: 10.1201/9781003428565-7

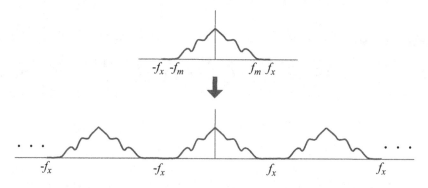

FIGURE 7.2
Periodic connection of the original frequency function

$x(t)$ observed at time

$$t = k\Delta t = k\frac{1}{2f_x}, \quad k = 0, \pm 1, \cdots, \pm\infty,$$

and equation (7.3) becomes

$$x(k\Delta t) = x\left(\frac{k}{2f_x}\right) = \int_{-f_x}^{f_x} X(f)\, e^{i\frac{2\pi k f}{2f_x}}\, df, \quad k = 0, \pm 1, \cdots, \pm\infty. \tag{7.4}$$

From the assumption that $x(t)$ has no value in the frequency domain $|f| \geq f_m$, if f_x satisfies

$$f_x = \frac{1}{2\Delta t} \geq f_m, \tag{7.5}$$

then, even if creates a periodic function $X_0(f)$ with period $2f_x$ on the frequency axis from $X(f)$ by repeatedly connecting the functions at $2f_x$ intervals, as shown in the lower panel of Figure 7.2, $X_0(f)$ is equal to $X(f)$ on $|f| < f_x$.

Then, expanding this periodic function $X_0(f)$ into a Fourier series with respect to frequency in the frequency domain (not time domain), we obtain from the definition of the Fourier series

$$X_0(f) = \sum_{k=-\infty}^{\infty} d_k\, e^{-i\frac{2\pi k f}{2f_x}} \tag{7.6}$$

$$d_k = \frac{1}{2f_x} \int_{-f_x}^{f_x} X_0(f)\, e^{i\frac{2\pi k f}{2f_x}}\, df. \tag{7.7}$$

(see equations (2.21) and (2.22)). Here d_k can be expressed as

$$d_k = \frac{1}{2f_x} \int_{-f_m}^{f_m} X(f)\, e^{i\frac{2\pi k}{2f_m} f}\, df, \quad k = 0, \pm 1, \cdots, \pm\infty. \tag{7.8}$$

Since the integral interval can be considered to be $-\infty$ to $+\infty$, from equation (7.3), the above equation is the Fourier inverse transform at $t = k\Delta t$. Therefore, we obtain

$$d_k = x(k\Delta t). \tag{7.9}$$

Substituting equation (7.9) into equation (7.6), we use equations (7.3) and (7.4) to obtain

$$2f_x X(f) = \frac{X(f)}{\Delta t} = \sum_{k=-\infty}^{\infty} x(k\Delta t)\, e^{-i2\pi f k \Delta t}. \tag{7.10}$$

This equation shows that the Fourier coefficient of $X(f)$ is the sequence $\{x(k\Delta t)\}_{k=-\infty}^{\infty}$ of values sampled every Δt of $x(t)$.

Substituting $x(k\Delta t)$ into equation (7.3), we obtain from equation (3.9) that $x(t)$ is quantized as

$$\begin{aligned}
x(t) &= \int_{-f_x}^{f_x} X(f)\, e^{i2\pi f t} df = \Delta t \int_{-f_x}^{f_x} \left(\sum_{k=-\infty}^{\infty} x(k\Delta t)\, e^{-i2\pi f k \Delta t} \right) e^{i2\pi f t} df \\
&= \Delta t \sum_{k=-\infty}^{\infty} x(k\Delta t) \int_{-f_x}^{f_x} e^{-i2\pi f(k\Delta t - t)} df \\
&= \sum_{k=-\infty}^{\infty} x(k\Delta t) \frac{\Delta t \sin \frac{1}{\Delta t}\pi(k\Delta t - t)}{\pi(k\Delta t - t)} \\
&= \sum_{k=-\infty}^{\infty} x(k\Delta t) \frac{\sin \pi(k - t/\Delta t)}{\pi(k - t/\Delta t)}, \tag{7.11}
\end{aligned}$$

where

$$\frac{\sin \pi(k - t/\Delta t)}{\pi(k - t/\Delta t)}$$

has the property of δ function (see equation (3.22)) that has value 1 only around $t = k\Delta t$ and is 0 otherwise. Using the δ function, the above equation can be written as

$$x(t) \sim \sum_{k=-\infty}^{\infty} x(kT)\, \delta(k - t/T). \tag{7.12}$$

Equation (7.12) represents a function where t has a value for every $k\Delta t$ and the rest are 0, where k is an arbitrary integer.

This result shows that the value $x(t)$ at any time t in the continuous domain can be completely determined from a sampled $\{x(k\Delta t)\}$ sequence, i.e., at a point in time $\Delta t = \frac{1}{2f_x}$ apart from each other, when the spectrum is zero in the outer region of frequency ranges $2f_x$. This is called the **Shannon-Someya sampling theorem**. The spectrum described in Chapter 4 is obtained from a discrete time series, and if the sampling period is Δt, the highest frequency detected (Nyquist frequency) is $1/(2\Delta t)$, so the above results show that *when the highest frequency contained in the continuous time series of interest is smaller than the **Nyquist frequency**, $1/(2\Delta t)$, the discrete time series sampled every Δt from the continuous domain Δt retains the full information in the continuous time series.*

Conversely, what would happen in the case of a function that does not satisfy the sampling theorem, as shown in Figure 7.3?

From equation (7.3), we have

$$x(k\Delta t) = \int_{-\infty}^{\infty} X(f)\, e^{i2\pi f k \Delta t} df. \tag{7.13}$$

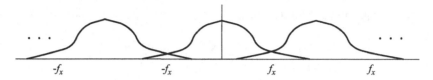

FIGURE 7.3
Case that the sampling theorem is not satisfied

Here note that $e^{i2\pi fk\Delta t} = e^{i2\pi fk\Delta t+2\pi k} = e^{i2\pi(f+\frac{1}{\Delta t})k\Delta t}$, it can be written as

$$x(k\Delta t) = \left(\cdots \int_{-3/2\Delta t}^{-1/2\Delta t} + \int_{-1/2\Delta t}^{1/2\Delta t} + \int_{1/2\Delta t}^{3/2\Delta t} + \cdots\right) e^{i2\pi fk\Delta t} X(f)\, df$$

$$= \int_{-1/2\Delta t}^{1/2\Delta t} \left(\sum_{m=-\infty}^{\infty} e^{2\pi fk\Delta t} X\left(f + \frac{m}{\Delta t}\right)\right) df$$

$$= \int_{-1/2\Delta t}^{1/2\Delta t} e^{2\pi fk\Delta t} \sum_{m=-\infty}^{\infty} X\left(f + \frac{m}{\Delta t}\right) df.$$

From this, if we put

$$X^*(f) = \sum_{m=-\infty}^{\infty} X\left(f + \frac{m}{\Delta t}\right), \tag{7.14}$$

we obtain

$$x(k\Delta t) = \int_{-1/2\Delta t}^{1/2\Delta t} e^{i2\pi fk\Delta t} X^*(f)\, df. \tag{7.15}$$

Multiplying the above equation by $e^{-i2\pi fm\Delta t}$ and adding them together for m, from the orthogonality of $e^{i2\pi fk\Delta t}$, we obtain

$$\sum_{m=-\infty}^{\infty} x(k\Delta t)\, e^{-i2\pi fm\Delta t} = \sum_{m=-\infty}^{\infty} \int_{-1/2\Delta t}^{1/2\Delta t} X^*(f)\, e^{i2\pi f(k-m)\Delta t} df = X^*(f).$$

Rewriting this, we obtain

$$X^*(f) = \sum_{k=-\infty}^{\infty} x(k\Delta t)\, e^{-i2\pi fk\Delta t}. \tag{7.16}$$

Namely, if there is an effective component with an absolute value of frequency outside of $1/2\Delta t$, then instead of $X(f)$, $X^*(f)$ defined in equation (7.14) satisfies equation (7.15). Figure 7.4 illustrates this phenomenon, i.e., the aliasing phenomenon discussed in Chapter 4.

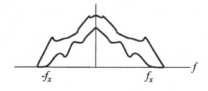

FIGURE 7.4
Eliasing phenomenon

7.2 Discrete Fourier Transform

The $X^*(f)$ shown in equation (7.16) can be thought of as a discretization of the Fourier transform in the continuous domain and is called the **discrete Fourier transform** (DFT). If the range of k is a finite integer L, then it follows that

$$X(f) = \sum_{k=-L}^{L} x(k\Delta t)\, e^{-i2\pi f k \Delta t}. \tag{7.17}$$

The discrete Fourier transform preserves the properties of the Fourier transform in the continuous domain described above. *This definition is used when considering the discrete Fourier transform in the discrete statistical model described below.*

7.3 z-transform

As shown in Section 7.1, the discrete Fourier transform $X(f)$ of the sampled discrete time series $\{x(k\Delta t)\}$ is connected to the Fourier transform of the original function $x(t)$ by equation (7.14) and $X^*(f)$ is expressed by equation (7.16). Here, on the right-hand side of equation (7.16), if we put

$$z = e^{i2\pi f \Delta t}, \tag{7.18}$$

then $X^*(f)$ can be simply expressed as

$$X^*(z) = \sum_{k=-\infty}^{\infty} x(k\Delta t)\, z^{-k}. \tag{7.19}$$

$X^*(z)$ is called **z-transform** of the discrete time series $\{x(k\Delta t)\}$. In the z-transform, multiplying some time function $x(s)$ by z moves time forward one step to $x(s+1)$, and multiplying by z^{-1} moves it back one step to $x(s-1)$. Table 7.1 shows the z-transforms of the typical time-domain functions[96]. Often in time series analysis, z^{-1} is written as B and is referred to as the **Box operator**.

From equation (7.12), $x(k\Delta t)$ can be written as $x(t) = \sum_{k=-\infty}^{\infty} x(k\Delta t)\,\delta(k - t/\Delta t)$. Applying the Laplace transform defined in equation (3.38) to both sides, we obtain

$$\begin{aligned}
\mathcal{L}\{x(t)\} &= \int_{-\infty}^{\infty} \sum_{k=-\infty}^{\infty} x(k\Delta t)\,\delta(k - t/\Delta t)\, e^{-st} dt \\
&= \sum_{k=-\infty}^{\infty} x(k\Delta t) \int_{-\infty}^{\infty} \delta(k - t/\Delta t)\, e^{-st} dt = \sum_{k=-\infty}^{\infty} x(k\Delta t)\, e^{-sk\Delta t}
\end{aligned}$$

and by putting

$$z = e^{s\Delta t}, \tag{7.20}$$

the above equation can be expressed as

$$\mathcal{L}\{x(t)\} = \sum_{k=-\infty}^{\infty} x(k\Delta t)\, z^{-k}.$$

This result is the same as in equation (7.19).

TABLE 7.1

Examples of major z-transforms[43]

$x(n)$	$X(z)$
$\delta(n)$	1
1	$\dfrac{1}{1 - z^{-1}}$
n	$\dfrac{z^{-1}}{(1 - z^{-1})^2}$
$\dfrac{n(n-1)}{2!}$	$\dfrac{z^{-2}}{(1 - z^{-1})^3}$
α^n	$\dfrac{1}{1 - \alpha z^{-1}}$
$n\alpha^{n-1}$	$\dfrac{z^{-1}}{(1 - \alpha z^{-1})^2}$
$\sin \beta\, n\Delta t$	$\dfrac{\sin \beta\Delta t \cdot z^{-1}}{1 - 2\cos \beta\Delta t \cdot z^{-1} + z^{-2}}$
$\cos \beta\, n\Delta t$	$\dfrac{1 - \cos \beta\Delta t \cdot z^{-1}}{1 - 2\cos \beta\Delta t \cdot z^{-1} + z^{-2}}$

7.4   Low-Order Discrete-Time Autoregressive Model

7.4.1   First-Order Autoregressive Model (AR1)

The equations of ship's motions are represented by second-order stochastic linear differential equations in the continuous domain under the condition that the disturbance is white Gaussian noise. The problem of what stochastic discrete-time model can be equivalently fitted to the time series obtained by recording the time series of ship motions in waves[96], which is governed by such equations in a continuous domain, was initially considered by Box, Bartlet and Pandit and Wu[88] et al. as low-order discrete-time models called a first-order autoregressive model, a second-order autoregressive model or a second-order autoregressive moving-average model. These models are valid under very specific disturbances, but are considered here because they are helpful in interpreting the results obtained when treated as statistical models, which we will deal with later[88].

Assuming that s is a parameter of time indicating that the time series is sampled at interval Δt, meaning $s = s\Delta t$, a simple linear discrete model representing the time series in the time domain is a first order autoregressive model

$$y(s) = a(1)\, y(s - 1) + v(s), \qquad (7.21)$$

where $v(s)$ is a Gaussian white noise with mean 0 and variance σ_v^2. This first order model is a special case of the model, generally called the autoregressive model, which will be discussed in a later chapter. From now on, we write this model as **AR(1)**. In order to express AR(1) with respect to $y(s)$, if we expand as

$$
\begin{aligned}
y(s) &= a(1)(a(1)\,y(s-2)+v(s-1))+v(s) \\
&= \cdots \\
&= \sum_{j=0}^{\infty} a(1)^j\, v(s-j),
\end{aligned} \tag{7.22}
$$

we see that $y(s)$ can be expressed in terms of the past and present values of $v(s)$.

Alternatively, the z-transformation introduced earlier for both sides of the model (equation (7.21)) yields

$$
\left(1 - a(1)\, z^{-1}\right) y(s) = v(s).
$$

Using this expression to find $y(s)$ formally, we obtain

$$
y(s) = \left(1 - a(1)\, z^{-1}\right)^{-1} v(s),
$$

and series expansion with respect to z^{-1} yields equation (7.22). Here if we define

$$
G_j = a(1)^j, \tag{7.23}
$$

then $y(s)$ can be expressed only in terms of the past and present of the white noise term $v(s)$ as

$$
y(s) = \sum_{j=0}^{\infty} G_j\, v(s-j) = \sum_{j=-\infty}^{s} G_{s-j}\, v(j).
$$

This representation of $y(s)$ using orthogonal random variables that are uncorrelated with each other is called **Wald decomposition**. Also, G_j is generally called **impulse response sequence**.

From equation (7.23), if

$$
|a(1)| < 1, \tag{7.24}
$$

then G_j approaches 0 when j becomes large. In this case, we say that **AR(1)** is **stable** or **stationary**.

Next, the covariance function of this **AR(1)** is obtained using the impulse response sequence as

$$
\begin{aligned}
R(l) &= E[y(s)\, y(s+l)] \\
&= E\left[\left(\sum_{j=0}^{\infty} G_j\, v(s-j)\right)\left(\sum_{j=0}^{\infty} G_j\, v((s+l)-j)\right)\right] \\
&= \left(\sum_{j=0}^{\infty} G_{j+l}\, G_j\right)\sigma_v^2 = \left(a(1)^l \sum_{j=0}^{\infty} a(1)^{2j}\right)\sigma_v^2 \\
&= \frac{\sigma_v^2}{1 - a(1)^2} a(1)^l.
\end{aligned} \tag{7.25}
$$

In particular, the variance of the **AR(1)** is given by

$$
R(0) = \frac{\sigma_v^2}{1 - a(1)^2}. \tag{7.26}
$$

7.4.2   Second-Order Autoregressive Moving Average Model (ARMA(2,1))

Second-order autoregressive moving average model is defined by

$$y(s) - a(1)\,y(s-1) - a(2)\,y(s-2) = v(s) - b(1)\,v(s-1). \tag{7.27}$$

Applying the z-transformation to both sides yields

$$\left(1 - a(1)\,z^{-1} - a(2)\,z^{-2}\right) y(s) = \left(1 - b(1)\,z^{-1}\right) v(s), \tag{7.28}$$

where it can be factorized as

$$1 - a(1)\,z^{-1} - a(2)\,z^{-2} = \left(1 - \lambda_1\,z^{-1}\right)\left(1 - \lambda_2\,z^{-1}\right). \tag{7.29}$$

Here $\lambda_1 C \lambda_2$ are roots of the characteristic equation of **ARMA(2,1)**:

$$1 - a(1)\,\lambda - a(2)\,\lambda^2 = 0,$$

and are given by

$$\lambda_1, \lambda_2 = \frac{1}{2}\left(a(1) \pm \sqrt{a(1)^2 + 4a(2)}\right). \tag{7.30}$$

Therefore, the Wald decomposition of **ARMA(2,1)** is given by

$$y(s) = \frac{1 - b(1)}{1 - a(1)\,z^{-1} - a(2)\,z^{-2}}v(s) = \frac{1 - b(1)}{\left(1 - \lambda_1\,z^{-1}\right)\left(1 - \lambda_2\,z^{-1}\right)}v(s). \tag{7.31}$$

If λ_1 and λ_2 are not multiple roots, the partial fraction expansion of equation (7.31) yields

$$
\begin{aligned}
y(s) &= \left[\frac{\lambda_1 - b(1)}{\lambda_1 - \lambda_2}\frac{1}{1 - \lambda_1\,z^{-1}} + \frac{\lambda_2 - b(1)}{\lambda_2 - \lambda_1}\frac{1}{1 - \lambda_2\,z^{-1}}\right] v(s) \\
&= \sum_{j=0}^{\infty}\left[\frac{\lambda_1 - b(1)}{\lambda_1 - \lambda_2}\lambda_1^j + \frac{\lambda_2 - b(1)}{\lambda_2 - \lambda_1}\lambda_2^j\right] v(s).
\end{aligned}
\tag{7.32}
$$

Therefore, the **impulse response function of ARMA(2,1)** is given by

$$
\begin{aligned}
G_j &= \frac{\lambda_1 - b(1)}{\lambda_1 - \lambda_2}\lambda_1^j + \frac{\lambda_2 - b(1)}{\lambda_2 - \lambda_1}\lambda_2^j \\
&= g_1\,\lambda_1^j + g_2\,\lambda_2^j,
\end{aligned}
\tag{7.33}
$$

where

$$g_1 = \frac{\lambda_1 - b(1)}{\lambda_1 - \lambda_2}, \quad g_2 = \frac{\lambda_2 - b(1)}{\lambda_2 - \lambda_1}. \tag{7.34}$$

Equation (7.32) converges to zero over time when it satisfies

$$|\lambda_1| < 1, \quad |\lambda_2| < 1.$$

In this case, we say that **ARMA(2,1)** is **stable** or **stationary**. This condition is equivalent to

$$a(1) + a(2) < 1, \quad a(2) - a(1) < 1, \quad |a(2)| < 1. \tag{7.35}$$

The covariance function is given by

$$
\begin{aligned}
R(l) &= E\left[y(s)\,y(s-l)\right] \\
&= \left(\sum_{j=0}^{\infty} G_{j+l}\,G_j\,v\right)\sigma_v^2 \\
&= \sigma_v^2 \sum_{j=0}^{\infty}\left(g_1\,\lambda_1^{l+j} + g_2\,\lambda_2^{l+j}\right)\left(g_1\,\lambda_1^{j} + g_2\,\lambda_2^{j}\right) \\
&= \sigma_v^2\left[\frac{g_1^2}{1-\lambda_1^2}\lambda_1^l + \frac{g_2^2}{1-\lambda_2^2}\lambda_2^l + \frac{g_1 g_2}{1-\lambda_1\lambda_2}(\lambda_1^l + \lambda_2^l)\right].
\end{aligned}
\tag{7.36}
$$

In particular, the variance is given by

$$
R(0) = \sigma_v^2\left[\frac{g_1^2}{1-\lambda_1^2} + \frac{g_2^2}{1-\lambda_2^2} + \frac{2g_1 g_2}{1-\lambda_1\lambda_2}\right].
\tag{7.37}
$$

If we define

$$
\begin{aligned}
d_1 &= \sigma_v^2\,g_1\left[\frac{g_1}{1-\lambda_1^2} + \frac{g_2}{1-\lambda_1\lambda_2}\right] \\
&= \frac{\sigma_v^2(\lambda_1 - b(1))}{(\lambda_1-\lambda_2)^2}\left[\frac{\lambda_1 - b(1)}{1-\lambda_1^2} - \frac{\lambda_2 - b(1)}{1-\lambda_1\lambda_2}\right]
\end{aligned}
\tag{7.38}
$$

$$
\begin{aligned}
d_2 &= \sigma_v^2\,g_2\left[\frac{g_2}{1-\lambda_1^2} + \frac{g_1}{1-\lambda_1\lambda_2}\right] \\
&= \frac{\sigma_v^2(\lambda_2 - b(1))}{(\lambda_1-\lambda_2)^2}\left[\frac{\lambda_2 - b(1)}{1-\lambda_2^2} - \frac{\lambda_1 - b(1)}{1-\lambda_1\lambda_2}\right]
\end{aligned}
\tag{7.39}
$$

then the covariance function is given by

$$
\begin{aligned}
R(l) &= d_1\lambda_1^l + d_2\lambda_2^l & (7.40) \\
R(0) &= d_1 + d_2, & (7.41)
\end{aligned}
$$

and it can be seen that the covariance function can be expressed as a linear combination of λ_i^l. Note that in the above results, the results for **AR(2)** are obtained when the moving average term $b(1)$ is zero.

7.5 Discretization of the State-Space Model

Now consider the following quadratic system

$$
\ddot{x}_2(t) + a(1)\,\dot{x}_2(t) + a(2)\,x_2(t) = b(1)\,u(t).
\tag{7.42}
$$

Here, we introduce a new variable and putting

$$
\dot{x}_2(t) = x_1(t),
$$

equation (7.42) can be expressed as

$$
\begin{pmatrix} \dot{x}_1(t) \\ \dot{x}_2(t) \end{pmatrix} = \begin{pmatrix} -a(1) & -a(2) \\ 1 & 0 \end{pmatrix}\begin{pmatrix} x_1(t) \\ x_2(t) \end{pmatrix} + \begin{pmatrix} b(1) \\ 0 \end{pmatrix} u(t).
\tag{7.43}
$$

Such a representation is called a state-space representation in systems engineering. In general, an n-order linear differential equation can be represented by

$$\dot{x}(t) = Ax(t) + Bu(t),$$
(7.44)

where A is an $n \times n$ matrix and B is an $n \times 1$ matrix. Next, consider the case where such a continuous-type state-space representation is sampled with a sampling period Δt. In this case, the state at some time t between time t_k and the next sampling time t_{k+1} ($t_k < t \leq t_{k+1}$) is given by

$$x(t) = e^{A(t-t_k)} x(t_k) + \int_{t_k}^{t} e^{A(t'-t_k)} Bu(t') dt'.$$

Note here, that this operation is a matrix operation. Assuming that the input $u(t)$ is held (this sampling method is called **zero-order hold**), the state at the next sampling time t_{k+1} is obtained as

$$
\begin{aligned}
x(t_{k+1}) &= e^{A(t_{k+1}-t_k)} x(t_k) + \int_{t_k}^{t_{k+1}} e^{A(t'-t_k)} Bu(t') dt' \\
&= e^{A(t_{k+1}-t_k)} x(t_k) + \int_{t_k}^{t_{k+1}} e^{A(t'-t_k)} dt' \, Bu(t_k) \\
&= \Phi(t_{k+1}, t_k) x(t_k) + \Gamma(t_{k+1}, t_k) u(t_k).
\end{aligned}
$$
(7.45)

Here we define

$$\Phi(t_{k+1}, t_k) = e^{A\Delta t}$$
(7.46)

$$\Gamma(t_{k+1}, t_k) = \int_{0}^{\Delta t} e^{As} ds \, B,$$
(7.47)

where $\Delta t = t_{k+1} - t_k$. These Φ and Γ can be computed numerically. A simple way to do this is to expand the matrix exponential function

$$e^{A\Delta t} = I + A\Delta t + \frac{1}{2!}(A\Delta t)^2 + \frac{1}{3!}(A\Delta t)^3 + \cdots,$$
(7.48)

and cut in an appropriate order. In addition to this method, we can also use

1. Laplace transform $(sI - A)^{-1}$ of $e^{A\Delta t}$.

2. Using Cayley-Hamilton's theorem.

3. Solving by matrix transformation to Jordan form.

For the inverse problem described here, i.e., how to obtain a continuous model from a discrete model, there is no general method because of the appearance of multivalued functions[14].

8

Autoregressive Modeling of Ship and Engine Motions

The analysis of time series described in Chapters 2 and 4 has been performed by transforming the observed time series into the frequency domain using the Fourier transformation as is or via the covariance function, and analyzing them in the frequency domain by auto spectrum and cross-spectrum. In contrast, in the following chapters, we will construct a statistical model that can effectively represent the observed time series obtained in the time domain, such as ship and engine motions, and consider how to proceed with the analysis, considering that the time series changes according to this constructed model.

8.1 Statistical Model of Time Series in Discrete Time Domain

8.1.1 Observation Process and Statistical Model of Time Series

In general, the objective of the analysis with a statistical model is to express the time series as

$$\text{Observed value} = \text{Statistical model} + \text{Random error}, \tag{8.1}$$

and clarify the characteristics of the population under study by the obtained statistical model. The analysis with the statistical model in the time series is also performed by creating a function that assumes that the values of the time series at each time point obtained in the discrete domain vary with time, which is called the time series model. In other words, we assume that

$$\begin{aligned} &\text{Observed process of the time series} \\ &= \text{Statistical model of the time series} + \text{Noise process.} \end{aligned} \tag{8.2}$$

In this case, what kind of statistical model is a time series model? When a time series is obtained, it is natural to assume that the observed values at a certain point in time are more or less influenced by the past values of the time series itself. The statistical model of a time series that we will discuss hereafter is also based on this idea. In this case, an important question in selecting a statistical model to represent the desired time series is how far in the past the time series data should be used and with what weights to be most effective.

8.1.2 Gaussian White Noise Process

Figure 2.3 in Chapter 2 shows a time series in which the past is assumed to have no effect at all, and the values of the next time series at each stage are assumed to be dominated by coincidental movements that have nothing to do with the past. The noise process on

DOI: 10.1201/9781003428565-8

the right-hand side of equation (8.2) is also such noise, and is called a white noise process because of its flat spectrum. If further, the frequency distribution of this waveform has a distribution that can be regarded as a Gaussian distribution, as shown in the lower plot in Figure 2.3, it is called **Gaussian white noise**.

8.1.3   Stationary Time Series Model

The time series of rolling and pitching are not irregular like Gaussian white noise, but appear to be affected by their own past motion and follow certain rules. The spectrum is also not white, but has characteristic peaks for each motion. Unlike the white noise process, the statistical model of time series described below is a model that incorporates the effect of past values of the time series with weights. Let us begin our consideration with the simple univariate stationary time series model.

A time series is said to be stationary if the parameters representing the statistical model of the time series under consideration do not change over time and remain around the mean values. When the surrounding ocean environment is considered stationary, ship motions such as rolling and pitching can also be considered stationary. However, as time passes, the wind waves generated by the wind gradually change and the marine environment varies. As a result, the characteristics of ship motions navigating over the environment will also change. A time series whose structure changes with time is called nonstationary. When the obtained time series is stationary, the following three models are often used.

1. **Autoregressive (AR) model**

2. **Moving average (MA) model**

3. **Autoregressive moving average (ARMA) model**

Among these models, the **autoregressive model** is a statistical model in the time domain introduced by Yule[108] in his analysis of the periodicity of the appearance of sunspots on the sun. Later, G. E. P. Box and G. M. Jenkins[18] vigorously examined such models in the time domain. Given an observed time series $y(1), y(2), \cdots, y(N)$, an autoregressive model expresses the variable of interest $y(s)$ at one point as a linearly weighted sum of its own values at an earlier time as (Figure 8.1)

$$
\begin{aligned}
y(s) &= a(1)\,y(s-1) + a(2)\,y(s-2) + \cdots + a(M)\,y(s-M) + v(s) \\
&= \sum_{m=1}^{M} a(m)\,y(s-m) + v(s).
\end{aligned}
\tag{8.3}
$$

From the standpoint of regression analysis often used in statistics, this model can be considered as a regression model for the objective variable $y(s)$ represented by the explanatory variables $y(s-1), \cdots, y(s-M)$. Here M is the number of past observations used in the model and is called the **order** of the model.

The noise $v(s)$ is uncorrelated with its own past values, i.e., it is assumed to hold

$$
E[v(s)\,v(s-l)] = \left\{ \begin{array}{ll} \sigma_v^2, & l = 0 \\ 0, & l \neq 0. \end{array} \right.
\tag{8.4}
$$

Therefore, if the spectrum of equation (8.4) is $P_{vv}(f)$, then by the Fourier transformation of equation (8.4), we obtain

$$
P_{vv}(f) = \sigma_v^2.
$$

In the autoregressive model, one spectrum peak can generally be represented by two autoregressive orders.

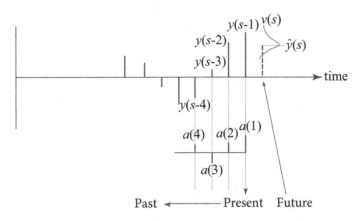

FIGURE 8.1
Concept of AR model

In contrast, the **moving average model** represents the value of the variable of interest $y(s)$ at a point in time by the weighted sum (moving average) of past Gaussian white noise $v(s-1), v(s-2), \cdots, v(s-L)$, i.e.,

$$
\begin{aligned}
y(s) &= v(s) + c(1)\,v(s-1) + c(2)\,v(s-2) + \cdots + c(L)\,v(s-L) \\
&= \sum_{l=1}^{L} c(l)\,v(s-l) + v(s).
\end{aligned}
\tag{8.5}
$$

Here L is the order of the moving average model. This model is suitable for representing the trough of the spectrum.

An **autoregressive moving average model** is a mixture of the above two models, and is given by

$$
y(s) = \sum_{m=1}^{M} a(m)\,y(s-m) + \sum_{l=1}^{L} c(l)\,v(s-l) + v(s),
\tag{8.6}
$$

where (M, L) is the order, and from physical feasibility it is assumed that $M \geq L$. Using the autoregressive moving average model, a time series model can be represented by the smallest order (in other words, the smallest number of parameters). Among the above three models, the autoregressive model has very good properties in terms of parameter estimation and statistical properties. We will examine it in more detail in the next section.

8.2 Autoregressive Model

8.2.1 Definition and Properties of Autoregressive Model

Let $\{y(s), s = 1, 2, \cdots, N\}$ be a univariate discrete time series. The **univariate autoregressive model** of order M is then defined as a linearly weighted sum of the M values of $y(s)$ in the past as follows

$$
y(s) = \sum_{m=1}^{M} a(m)\,y(s-m) + v(s).
\tag{8.7}
$$

The $v(s)$ is a Gaussian white noise whose distribution follows a normal distribution with mean 0 and variance σ_v^2. Applying the z transformation introduced in Chapter 7 to both sides, we obtain

$$\varphi(z^{-1})\,y(s) = v(s),\qquad(8.8)$$

where

$$\varphi(z^{-1}) = 1 - a(1)\,z^{-1} - a(2)\,z^{-2} - \cdots - a(M)\,z^{-M}.$$

Now define an equation

$$\varphi(z^{-1}) = 1 - a(1)\,z^{-1} - a(2)\,z^{-2} - \cdots - a(M)\,z^{-M} = 0.\qquad(8.9)$$

This equation is called **characteristic equation** and the M roots of this equation are called **characteristic roots**. In general, the characteristic roots contain complex roots. Let $\lambda_1, \lambda_2, \cdots, \lambda_M$ be the roots for z^{-1} in equation (8.9). Then the equation (8.9) can be factorized as

$$\varphi(z^{-1}) = (1 - \lambda_1 z^{-1})(1 - \lambda_2 z^{-1})\cdots(1 - \lambda_M z^{-1}) = 0.$$

For simplicity, let $\lambda_1, \lambda_2, \cdots, \lambda_M$ be roots with respect to z^{-1} that differ from each other, and from equation (8.8), formally $y(s)$ can be expanded into partial fractions,

$$y(s) = \frac{1}{\varphi(z^{-1})}\,v(s) = \sum_{m=1}^{M} \frac{c_m}{1 - \lambda_m z^{-1}} v(s),$$

where c_m is a constant that does not contain z^{-1}. Since

$$\frac{1}{1 - \lambda_m z^{-1}} = \sum_{k=0}^{\infty} \lambda_m^k z^{-k},$$

equation (8.7) is formally expressed as

$$y(s) = \sum_{m=1}^{M} c_m \sum_{k=0}^{\infty} \lambda_m^k z^{-k}\, v(s).\qquad(8.10)$$

In this case, for equation (8.10) to converge, it is necessary to satisfy

$$|\lambda_m| < 1.\qquad(8.11)$$

That is, for all m, the absolute value of the root λ_m must be less than 1. Thus, we can say that for the time series model to be **stationary** or **stable**, all the roots of equation (8.9) must be in the **unit circle** as shown on the right side of Figure 8.2.

8.2.2   Relationship Between Stationary Regions in Continuous and Discrete Domains

Let us investigate the relationship between stationary regions in continuous and discrete domains in relation to the Laplace transform. As considered in Section 5.3, in the stable region of a linear system, the real part of all roots of the equation

$$A(p) = 0,$$

are negative. For a given root $p = \sigma + i\omega$ in the p-plane, $e^{(\sigma+i\omega)\Delta t}$ corresponds in the z-plane sampled by the period Δt. As is clear from Figure 8.2, the imaginary axis of p in the p-plane

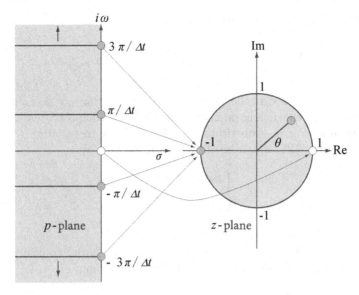

FIGURE 8.2
Correspondence between p-plane and z-plane

is mapped to a circle of radius 1 in the z-plane. In the z-plane, however, $e^{i\omega\Delta t}$ returns to its original position with period $\pi/\Delta t$, so there remains an ambiguity in the period $\pi/\Delta t$. This is due to the aliasing effect of sampling. As described in Chapter 5, the real part α of the p-plane is negative and is called the stable region, where the solution converges. Any strip of negative area in the p-plane that is on the imaginary axis $[-\pi/\Delta t, \pi/\Delta t]$ of length $2\pi/\Delta t$ in the vertical direction is mapped into a circle of radius 1 in the z-plane. The stationary conditions of the discrete system are that all roots of the corresponding characteristic equations

$$A(z^{-1}) = 0 \qquad (8.12)$$

lie in the **unit circle** of the z-plane (discrete domain).

Also, when ω in the continuous domain (p-plane) is non-zero, it indicates the presence of an oscillating root with frequency ω, which means that the corresponding angle in the z-plane also gives the oscillating frequency. In other words, as shown in the right part of Figure 8.2, there is a relationship between the phase angle θ and the frequency f in the z-plane, as

$$f = \frac{\theta}{2\pi}. \qquad (8.13)$$

8.2.3 The Impulse Response Sequence and Frequency Response Sequence of the Autoregressive Model

Equation (8.10) can be further transformed as follows;

$$
\begin{aligned}
y(s) &= \sum_{m=1}^{M} c_m \sum_{k=0}^{\infty} \lambda_m^k z^{-k} v(s) \\
&= \sum_{k=0}^{\infty} \sum_{m=1}^{M} c_m \lambda_m^k z^{-k} v(s) = \sum_{k=0}^{\infty} g_k v(s-k), \qquad (8.14)
\end{aligned}
$$

where g_k satisfies the following relation

$$g_0 = \sum_{m=1}^{M} c_m = 1, \quad g_k = \sum_{m=1}^{M} c_m \lambda_m^k, \quad k = 1, 2, \cdots, \infty. \qquad (8.15)$$

The $g_k, k = 0, 1, 2, \cdots, \infty$ is the representation in the discrete domain of the impulse response function in the continuous domain, as described in Chapter 5, and is called **impulse response sequence**. For g_k, substitute equation (8.14) into equation (8.7), and from a comparison of the two sides, it can be obtained successively by

$$\begin{cases} g_0 = 1 \\ g_i = \sum_{j=1}^{i} a_j g_{i-j}, \end{cases} \qquad (8.16)$$

where $a_j = 0$ when $j > M$. In the continuous domain, the Fourier transform of the impulse response function is a frequency response function. By defining

$$G\left(z^{-1}\right) = \frac{1}{\varphi\left(z^{-1}\right)},$$

it was shown in equation (7.18) in Chapter 7 that the z-transform is related to the discrete Fourier transform by $z = e^{i2\pi f \Delta t}$. Since we have

$$G\left(z^{-1}\right) = \frac{1}{\varphi\left(z^{-1}\right)} = \sum_{k=0}^{\infty} g_k z^{-k},$$

substituting $e^{-i2\pi ft}$ for z^{-1}, we get **frequency response sequence** as

$$G\left(e^{-i2\pi f}\right) = \sum_{k=0}^{\infty} g_k e^{-i2\pi fk}. \qquad (8.17)$$

The above results show that an infinite number of past $v(s - k)$ is needed to represent the AR model in terms of the white noise process $v(s)$.

8.2.4   Autocovariance Function

To obtain the autocovariance function of the autoregressive model, we multiply equation (8.7) by $y(s - l)$ for $l \geq 1$ to obtain

$$\begin{aligned} y(s)\,y(s - l) \;=\; & a(1)\,y(s - 1)\,y(s - l) + a(2)\,y(s - 2)\,y(s - l) + \cdots \\ & + a(M)\,y(s - M)\,y(s - l) + v(s)\,y(s - l), \end{aligned}$$

where since $E[v(s)\,y(s - l)] = 0$ for $l \neq 0$, taking the expected values of both sides yields

$$R_{yy}(l) = a(1)\,R_{yy}(l - 1) + a(2)\,R_{yy}(l - 2) + \cdots + a(M)\,R_{yy}(l - M), \quad l \geq 1. \qquad (8.18)$$

$R_{yy}(l)$ is the autocovariance function for lag l. Also for $l = 0$, using equation (8.7) to compute $E[y(s)\,y(s)]$, we obtain

$$R_{yy}(0) = a(1)\,R_{yy}(1) + a(2)\,R_{yy}(2) + \cdots + a(M)\,R_{yy}(M) + \sigma_v^2. \qquad (8.19)$$

(Note that $E[v(s)\,y(s)] = \sigma_v^2$.)

8.2.5   Rational Spectrum

The autoregressive model can be rewritten as

$$y(s) - \sum_{m=1}^{M} a(m)\, y(s-m) = v(s). \tag{8.20}$$

Considering this equation, we can think of the disturbance $v(s)$ as acting on the (discrete) linear system represented by the left-hand side of the equation for the linear system in the continuous domain. Now, given the input

$$v(s) = e^{i2\pi f s},$$

to obtain the solution of the type $y(s) = A(f)\, e^{i2\pi f s}$, since

$$\left(1 - \sum_{m=1}^{M} a(m)\, e^{-i2\pi f m}\right) A(f)\, e^{i2\pi f s} = e^{i2\pi f s},$$

it holds that

$$A(f) = \frac{1}{\left(1 - \sum\limits_{m=1}^{M} a(m)\, e^{-i2\pi f m}\right)}. \tag{8.21}$$

This $A(f)$ is equal to equation (8.17) by its definition and is a frequency response sequence.

Then, as we learned in Chapter 6, the spectrum $p_{xx}(f)$ of the input $x(s)$ and the spectrum $p_{yy}(f)$ of the output $y(t)$ are connected via the frequency response function

$$p_{yy}(f) = |A(f)|^2\, p_{xx}(f).$$

In this case, if we replace $e^{i2\pi f s}$ with Gaussian white noise $v(s)$ as input $x(s)$, its spectrum $P_{vv}(f)$ is given by

$$P_{vv}(f) = \sigma_v^2.$$

Therefore, it can be seen that the spectrum (rational spectrum) of $y(s)$ is given by

$$P_{yy}(f) = \frac{\sigma_v^2}{\left|1 - \sum\limits_{m=1}^{M} a(m) \exp\{-i2\pi f m\}\right|^2}, \quad |f| \le \frac{1}{2}. \tag{8.22}$$

In general, among the eigen roots of a characteristic polynomial, the complex root represents a periodically oscillating waveform. Since the complex root is paired with the conjugate root, the two roots in effect represent a single peak in the spectrum. Also, the larger the absolute value of λ_m, the sharper the peak.

8.2.6   Arrangement of Characteristic Roots of a First-Order Autoregressive Model

We consider here the first order autoregressive model

$$y(s) = a(1)\, y(s-1) + v(s), \tag{8.23}$$

as an example of autoregressive model and investigate the relationship with the eigen roots. The characteristic equation of the model is given by

$$1 - a(1)\, z^{-1} = 0. \tag{8.24}$$

and the characteristic root is $z^{-1} = 1/a$. Therefore, to satisfy the stationarity condition, the absolute value of the root must be greater than 1, i.e., $a(1)$ must satisfy

$$|a(1)| < 1. \tag{8.25}$$

Example 8.1 *First-order trend model*

Figure 8.3 shows the simulation of the first-order autoregressive model with $a(1) = 1$, and the time series slowly descends with fluctuations. The following model

$$y(s) = y(s-1) + v(s),$$

is used to express the first-order trend.

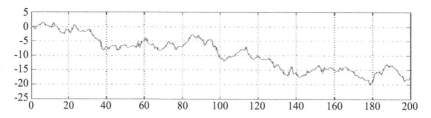

FIGURE 8.3

Simulation of the first-order autoregressive model with $a(1) = 1$

Example 8.2 *Change in time series due to sign of autoregressive coefficient*

Let $|a(1)| < 1$ and consider the cases where a is positive and negative. Figure 8.4 shows the result of simulation when the autoregressive coefficient $a(1)$ is negative and Figure 8.5 shows the case when $a(1)$ is positive. When $a(1)$ is negative, if the value is positive at one point in time, there is a high probability that it will be negative at the next point in time, so the waveform violently moves back and forth between positive and negative values. In contrast, when $a(1)$ is positive, it stays in the positive or negative region for a while.

8.2.7   Arrangement of Characteristic Roots of Second-Order Autoregressive Model

Second order autoregressive model is given by

$$y(s) = a(1)\, y(s-1) + a(2)\, y(s-2) + v(s). \tag{8.26}$$

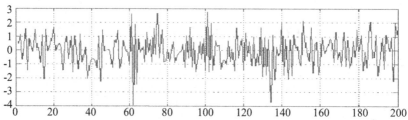

FIGURE 8.4

Case that the coefficient a(1) is negative

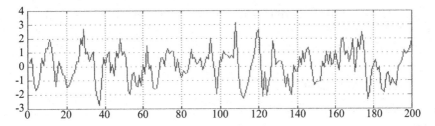

FIGURE 8.5
Case that the coefficient a(1) is positive

Its characteristic equation is given by

$$\varphi(z^{-1}) = 1 - a(1)\, z^{-1} - a(2)\, z^{-2} = 0. \tag{8.27}$$

Letting λ_1 and λ_2 be the roots with respect to z in the characteristic equation, the stationary conditions are given by

$$
\begin{aligned}
|a(2)| &< 1 \\
a(1) + a(2) &< 1 \\
a(2) - a(1) &< 1.
\end{aligned}
\tag{8.28}
$$

Figure 8.6 shows the regions of $a(1)$ and $a(2)$ that satisfy the above conditions. Here it can be seen that;

$$
\begin{aligned}
\text{if } a(1)^2 + 4a(2) < 0, && \lambda_1, \lambda_2 \text{ are complex roots} \\
\text{if } a(1)^2 + 4a(2) = 0, && \lambda_1, \lambda_2 \text{ are multiple roots} \\
\text{if } a(1)^2 + 4a(2) > 0, && \lambda_1, \lambda_2 \text{ are two real roots.}
\end{aligned}
$$

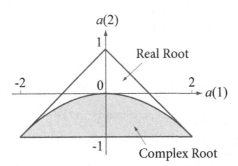

FIGURE 8.6
Stationary region of the AR(2) model and the type of characteristic roots

8.3 Estimation of Autoregressive Models

In the previous section, we have examined the characteristics of waveforms of low-order autoregressive models such as first-order and second-order models. In this section, we deal with the problem of how to obtain a meaningful statistical model given certain data. Such a problem is called the **fitting method** or identification method of a statistical model to an observed time series. The model is not limited to a second-order model, but higher-order models are also considered.

8.3.1 Yule-Walker Method

In equation (8.18), varying l from 1 to M and considering that the autocovariance function has symmetry $R_{yy}(-l) = R_{yy}(l)$ and that the noise term is uncorrelated with its own past term, it can be expressed in matrix form as

$$
\begin{pmatrix}
R_{yy}(0) & R_{yy}(1) & \cdots & R_{yy}(M-1) \\
R_{yy}(1) & R_{yy}(0) & \cdots & R_{yy}(M-2) \\
\vdots & \vdots & \ddots & \vdots \\
R_{yy}(M-1) & R_{yy}(M-2) & \cdots & R_{yy}(0)
\end{pmatrix}
\begin{pmatrix}
a(1) \\
a(2) \\
\vdots \\
a(M)
\end{pmatrix}
=
\begin{pmatrix}
R_{yy}(1) \\
R_{yy}(2) \\
\vdots \\
R_{yy}(M)
\end{pmatrix}.
\tag{8.29}
$$

Also, from equation (8.19) the variance of the residuals can be obtained as

$$
\sigma_v^2 = R_{yy}(0) - a(1)\,R_{yy}(1) - a(2)\,R_{yy}(2) - \cdots - a(M)\,R_{yy}(M).
\tag{8.30}
$$

Equations (8.29) and (8.30) are called the **Yule-Walker equation**. In this equation, all covariance functions R_{yy} on both sides can be computed from observed data. Therefore, solving this equation for $a(i), i = 1, \cdots, M$ will yield the regression coefficients. The method of using the coefficients obtained in this way as estimates of the parameters of the model is called the **Yule-Walker method**. The matrix of the first term on the left-hand side of equation (8.29) is denoted as R_M and is called **Toeplitz matrix**.

8.3.2 Parameter Estimation by Least Squares and Maximum Likelihood Methods

Thus, we find that the parameters can be easily estimated by solving Yule-Walker equation for $a(m)$. In this section, we will show how the coefficients obtained in this way are estimators from a statistical point of view.

Consider the following autoregressive model of order M:

$$
y(s) = a(1)\,y(s-1) + a(2)\,y(s-2) + \cdots + a(M)\,y(s-M) + v(s).
\tag{8.31}
$$

In equation (8.31), the following part corresponds to the statistical model,

$$
\hat{y}(s) = a(1)\,y(s-1) + a(2)\,y(s-2) + \cdots + a(M)\,y(s-M).
$$

To estimate this statistical model, we can use the following method.

First, let us consider the least squares method advocated by Gauss. The least squares method selects parameters so as to minimize the expected value of the square of the residuals that remain when the statistical model is applied to all available data. In other words, in equation (8.31), we choose $a(1), a(2), \cdots, a(M)$ that minimizes

$$
\begin{aligned}
L &= E\left[(y(s) - \hat{y}(s))^2\right] \\
&= E\left[(y(s) - a(1)\,y(s-1) - a(2)\,y(s-2) - \cdots - a(M)\,y(s-M))^2\right] \\
&= E\left[v(s)^2\right].
\end{aligned}
\tag{8.32}
$$

The $\hat{a}(m)$ estimated in this way is called the **least square estimate (LSE)** of $a(m)$.

The second method is the **maximum likelihood method** proposed by Ronald A. Fisher. In the maximum likelihood method, the realization y of a random event Y is considered to be sampled from a population whose distribution is given by $p(y|\theta)$ specified by the parameter θ. That is, we consider that the observed value at a given time was sampled independently from a population of a number of possibilities at that time. From this assumption, it can be immediately understood that the probability of N samples $y(1), \cdots, y(N)$ occurring simultaneously is given by

$$L(y|\theta) = p(y(1)|\theta)\, p(y(2)|\theta) \cdots p(y(N)|\theta), \tag{8.33}$$

where θ is unknown, but the observations $y = (y(1), y(2), \cdots, y(N))^t$ are known. This simultaneous distribution considered as a function of θ is called **likelihood function**. The maximum likelihood method estimates the parameter θ so that this likelihood function is maximized, that is, $\max_\theta L(y|\theta)$. The parameter θ chosen by such a criterion is written $\hat{\theta}$ and is called the **maximum likelihood estimator (MLE)**. The maximum likelihood estimator $\hat{\theta}$ is a function of the data y. For ease of computation, in actual computation, we often use the **log-likelihood function**

$$l(y|\theta) = \log L(y|\theta). \tag{8.34}$$

In this case, the maximum likelihood estimator is obtained by solving[20]

$$\frac{\partial \log L(y|\theta)}{\partial \theta} = 0. \tag{8.35}$$

The unknown parameters of the autoregressive model of order M are $\theta = \big(a(1), a(2), \cdots, a(M), \sigma_v^2\big)^t$. Now, given N data of $y(s)$ and the initial values $y(0), \ldots, y(1-M)$, from the Gaussianity of its distribution, the likelihood of the autoregressive model is given by

$$L\big(a(1), a(2), \cdots, a(M), \sigma_v^2\big) = \frac{1}{\big(\sqrt{2\pi}\,\sigma_v\big)^N} \exp\left[-\frac{1}{2\sigma_v^2} \sum_{s=1}^{N} \left\{ y(s) - \sum_{m=1}^{M} a(m)\, y(s-m) \right\}^2 \right]. \tag{8.36}$$

And the log-likelihood function is given by

$$l\big(a(1), a(2), \cdots, a(M), \sigma_v^2\big) = -\frac{N}{2}\log 2\pi - \frac{N}{2}\log \sigma_v^2 - \frac{1}{2\sigma_v^2}\sum_{s=1}^{N}\left\{ y(s) - \sum_{m=1}^{M} a(m)\, y(s-m) \right\}^2. \tag{8.37}$$

From equation (8.35), by setting the partial derivative of $l(\theta)$ with respect to σ_v^2 to zero, the maximum likelihood estimator for σ_v^2 is given by

$$\hat{\sigma}_v^2 = \frac{1}{N} \sum_{s=1}^{N} \left\{ y(s) - \sum_{m=1}^{M} \hat{a}(m)\, y(s-m) \right\}^2. \tag{8.38}$$

Substituting this result into equation (8.37) yields

$$l(a(1), \cdots, a(M)) = -\frac{N}{2}\log 2\pi\hat{\sigma}_v^2 - \frac{N}{2}. \tag{8.39}$$

Thus, we see that to apply the maximum likelihood principle to parameter estimation, it is necessary to choose $a(1), \cdots, a(M)$ that minimizes equation (8.38). This is consistent with the criterion of the least-squares method, which aims to minimize equation (8.32). In other words, we have obtained the result that the least squares estimator of the coefficients of the *autoregressive model approximately coincides with the maximum likelihood estimator.*

By the way, given N observations $y(1), y(2), \cdots, y(N)$, the autoregressive model of order M for this data is given by

$$
\begin{cases}
y(M+1) = a(1)\,y(M) + a(2)\,y(M-1) + \cdots + a(M)\,y(1) + v(M+1) \\
y(M+2) = a(1)\,y(M+1) + a(2)\,y(M) + \cdots + a(M)\,y(2) + v(M+2) \\
\quad\vdots \\
y(N) = a(1)\,y(N-1) + a(2)\,y(N-2) + \cdots + a(M)\,y(N-M) + v(N).
\end{cases}
$$

Here, if we put

$$
\mathbf{y} = \begin{pmatrix} y(M+1) \\ y(M+2) \\ \vdots \\ y(N) \end{pmatrix}, \quad
\mathbf{X} = \begin{pmatrix}
y(M) & y(M-1) & \cdots & y(1) \\
y(M+1) & y(M) & \cdots & y(2) \\
\vdots & \vdots & \ddots & \vdots \\
y(N-1) & y(N-2) & \cdots & y(N-M)
\end{pmatrix},
$$

$$
\alpha = \begin{pmatrix} a(1) \\ a(2) \\ \vdots \\ a(M) \end{pmatrix}, \quad
\mathbf{e} = \begin{pmatrix} v(M+1) \\ v(M+2) \\ \vdots \\ v(N) \end{pmatrix},
$$

the autoregressive model of order M can be written as

$$
\mathbf{y} = \mathbf{X}\alpha + \mathbf{e}. \tag{8.40}
$$

In this case, the squared error S of the residual is given by

$$
S = \mathbf{e}^t \mathbf{e} = (\mathbf{y} - \mathbf{X}\alpha)^{\mathbf{t}}\,(\mathbf{y} - \mathbf{X}\alpha).
$$

Differentiating S by α to find α that minimizes S, we obtain

$$
(\mathbf{X}^t \mathbf{X})\alpha = \mathbf{X}^{\mathbf{t}}\,\mathbf{y}. \tag{8.41}
$$

Since

$$
\mathbf{X}^t \mathbf{X} = \sum_{s=M+1}^{N} \begin{pmatrix}
y(s)^2 & y(s-1)\,y(s) & \cdots & y(s-M)\,y(s) \\
y(s)\,y(s-1) & y(s-1)^2 & \cdots & y(s-M)\,y(s-1) \\
\vdots & \vdots & \ddots & \vdots \\
y(s)\,y(s-M) & \cdots & \cdots & y(s-M)^2
\end{pmatrix}
$$

$$
\mathbf{X}^t \mathbf{y} = \sum_{s=M+1}^{N} \begin{pmatrix}
y(s)^2 \\
y(s-1)\,y(s) \\
\vdots \\
y(s-M)\,y(s)
\end{pmatrix},
$$

taking the expected value of equation (8.41) (multiply by $1/N$ and let $N \to \infty$), each converges to the covariance function of the corresponding lag, so from the relation below the second line, we obtain the following Yule-Walker equation

$$
\sum_{m=1}^{M} a(m)\,R(l-m) = R(l), \quad l = 1, 2, \cdots, M.
$$

These results indicate that in the *autoregressive modeling, the Yule-Walker and least squares and maximum likelihood estimators of* $\mathbf{a(m)}$ *are approximately equal.*

8.3.3 Autoregressive Models and FPE

In the discussion up to the previous section, the parameters of the model have been estimated based on the a priori assumption of the order M of the autoregressive model. As a result, the determination of the order is left to the subjective judgment of the analyst. In this case, a model with too small order is not reliable because the model error is too large. Conversely, a model with too large order may be sensitive to errors (Figure 8.7). In this situation, Dr. Hirotugu Akaike introduced a statistical criterion called **final prediction error (FPE)**[2][7], which objectively solved the problem of order determination in autoregressive modeling. FPE is an estimator of the expected value of the prediction error variance when predicting the future data obtained from a population with the same probability structure as the one that generated the data used to estimate the model.

As shown in the previous section, given N observations $y(1), y(2), \cdots, y(N)$, fitting an autoregressive model of order M to this data yields

$$\mathbf{y} = \mathbf{X}\alpha + \mathbf{e},$$

where $E[\mathbf{e}] = 0$ and the variance is $\mathrm{Var}(\mathbf{e}) = \sigma_v^2 I_{N \times N}$ with

$$\sigma_v^2 = \frac{1}{N} \sum_{s=M+1}^{N} (y(s) - \hat{y}(s))^2.$$

The least squares estimator of α of this model is

$$\hat{\alpha} = (\mathbf{X}^t \mathbf{X})^{-1} \mathbf{X}^t \mathbf{y}, \tag{8.42}$$

and the unbiased estimator of the variance, σ_e^2, is given by (see equation (17.61))[96]

$$\sigma_e^2 = \frac{1}{N - M} \sigma_v^2. \tag{8.43}$$

On the other hand, the sum of squares of the prediction errors when using $\hat{\alpha}$ obtained here to predict an N-dimensional future observation vector $\mathbf{z}_0$ obtained independently from the same stochastic structure as the observation $\mathbf{y}$ used for estimation, is evaluated as

$$S_M^2 = (\mathbf{z}_0 - \hat{\mathbf{y}})^t (\mathbf{z}_0 - \hat{\mathbf{y}}). \tag{8.44}$$

If we put $H = \mathbf{X}(\mathbf{X}^t \mathbf{X})^{-1} \mathbf{X}^t$, we have $\hat{\mathbf{y}} = H\mathbf{y}$ and $H\mathbf{X} = \mathbf{X}$. Using the fact that $H^2 = H$ and the equality $\alpha^t H \alpha = \mathrm{tr}(H \alpha \alpha^t)$ for the trace of a symmetric matrix in quadratic form

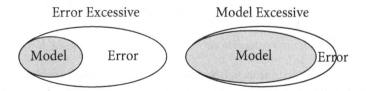

FIGURE 8.7
Error excessive and model excessive cases

(hereafter written as tr), the expected value of equation (8.44) can be obtained as follows

$$
\begin{aligned}
E\left[S_M^2\right] &= E\left[(\mathbf{z}_0 - \hat{\mathbf{y}})^t (\mathbf{z}_0 - \hat{\mathbf{y}})\right] \\
&= E\left[\{\mathbf{z}_0 - \mathbf{X}\alpha - (\hat{\mathbf{y}} - \mathbf{X}\alpha)\}^t \{\mathbf{z}_0 - \mathbf{X}\alpha - (\hat{\mathbf{y}} - \mathbf{X}\alpha)\}\right] \\
&= E\left[(\mathbf{z}_0 - \mathbf{X}\alpha)^t (\mathbf{z}_0 - \mathbf{X}\alpha)\right] + E\left[(\hat{\mathbf{y}} - \mathbf{X}\alpha)^t (\hat{\mathbf{y}} - \mathbf{X}\alpha)\right] \\
&= N\sigma_v^2 + E\left[(H\mathbf{y} - H\mathbf{X}\alpha)^t (H\mathbf{y} - H\mathbf{X}\alpha)\right] \\
&= N\sigma_v^2 + E\left[(\mathbf{y} - \mathbf{X}\alpha)^t H (\mathbf{y} - \mathbf{X}\alpha)\right] \\
&= N\sigma_v^2 + \operatorname{tr}\{H\operatorname{Var}[\mathbf{y}]\} \\
&= N\sigma_v^2 + M\sigma_v^2 = (N + M)\sigma_v^2.
\end{aligned}
\tag{8.45}
$$

Substituting the unbiased estimator of σ_v^2 (equation (8.43)) into equation (8.45), the final prediction error FPE is given by

$$
\text{FPE} = \frac{N + M}{N - M}\sigma_v^2.
\tag{8.46}
$$

This result shows that the estimate of the variance of the error σ_v^2 by the maximum likelihood method, $\hat{\sigma}_v^2$, becomes monotonically smaller as the order M increases from equation (8.38), whereas in FPE, its numerator gives a penalty for increasing the parameter M and that there is an order that gives the smallest overall prediction error variance.

8.3.4   Akaike Information Criterion: AIC

The key point in deriving the final prediction error, FPE, is that the model is evaluated from the standpoint of prediction. Namely, it is not evaluated using the same observed data used to estimate the model by the least squares method, but rather evaluate the expected prediction error variance using data independently obtained from a population with the same probability structure.

While the least squares method is an effective method for estimating the parameters of statistical models, the maximum likelihood method is a more precise estimation method. Based on the same idea in determining the model by the maximum likelihood method, Akaike clarified that when the model is estimated by the maximum likelihood method, the expected log-likelihood has a bias by the number of parameters due to the use of the same observations for model estimation and estimation of the expected log-likelihood[4][5]. Akaike considered this estimate of twice the expected maximum log-likelihood as the quantity that measures the goodness of predictive ability of the model, and defined **Akaike Information Criterion, AIC**:

$$
\begin{aligned}
\text{AIC}(k) &= -2l(\hat{\theta}) + 2k \\
&= -2(\text{maximum log-likelihood}) + 2(\text{number of free parameters in the model}).
\end{aligned}
\tag{8.47}
$$

Then, according to this definition, when there are several candidate models for the same data, the model that minimizes the AIC should be selected as the best model instead of the conventional maximum likelihood method. This method is called the **minimum AIC estimation procedure**, or **MAICE method** for short.

For the autoregressive model of order M, the maximum log-likelihood is approximately given by

$$
-\frac{1}{2}\left\{N\log(2\pi\hat{\sigma}_M^2) + N\right\},
$$

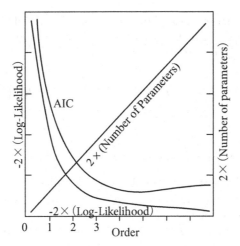

FIGURE 8.8
Meaning of AIC

and since the number of parameters to be estimated is $M + 1$ including the variance of the residuals, AIC is obtained by

$$\text{AIC}(M) = N \log(2\pi\hat{\sigma}_M^2) + N + 2(M + 1). \tag{8.48}$$

Taking the logarithm of the FPE (equation (8.46)), we obtain

$$
\begin{aligned}
\log(\text{FPE}(M)) &= \log\hat{\sigma}_M^2 + \log\left(1 + \frac{M}{N}\right) - \log\left(1 - \frac{M}{N}\right) \\
&\simeq \log\hat{\sigma}_M^2 + \frac{2M}{N},
\end{aligned}
$$

thus it can be seen that FPE and AIC give approximately the same criterion in the case of the autoregressive model.

Figure 8.8 schematically illustrates the changes of the two terms comprising the AIC with order. The first term of the AIC, i.e., -2 times the maximum log-likelihood for each order, decreases with increasing order. In contrast, the number of parameters increases linearly. The trade-off point between the two can be interpreted as the MAICE order.

8.4 Sequential Estimation of the Autoregressive Model

As mentioned so far, autoregressive models have some convenient properties such that the maximum likelihood estimator of the coefficients approximately agrees with the least squares estimator. In this section, we present a practical fitting method using the MAICE method for various autoregressive models[7].

8.4.1 Estimation Method Using the Levinson-Durbin Algorithm

The Toeplitz matrix (equation (8.29)) has a special form, and its properties can be used to obtain the regression coefficients $a_m(i)$ by a recursive formula with respect to the order.

This method is called the **Levinson-Durbin algorithm**, and if the covariance function for lag m is $R_{xx}(m)$, it is given by the following recursive formula with respect to the order. (The proof is given in Appendix B at the end of this chapter).

$$a_{m+1}(m+1) = \frac{1}{\sigma_v^2(m)}\left\{R_{yy}(m+1) - \sum_{j=1}^{m} a_m(j)\, R_{yy}(m+1-j)\right\} \tag{8.49}$$

$$a_{m+1}(i) = a_m(i) - a_{m+1}(m+1)\, a_m(m+1-i), \quad i = 1, 2, \cdots, m \tag{8.50}$$

$$\sigma_v^2(m+1) = \sigma_v^2(m)\left\{1 - a_{m+1}(m+1)^2\right\} \tag{8.51}$$

Equation (8.50) shows that if the final term of each order, which we call **PARCOR** (partial autocorrelation coefficient) $a_m(m), m = 1, \cdots, M$, is estimated, all AR coefficients up to order M can be obtained using equation (8.50) repeatedly. Conversely, if each regression coefficient of the maximum order M is known, then from equation (8.50), it holds that

$$a_M(i) = a_{M-1}(i) - a_M(M)\left\{a_M(M-i) + a_M(M)\, a_{M-1}(i)\right\}.$$

So, solving for this, the autoregressive coefficient of order $M-1$ is given by

$$a_{M-1}(i) = \frac{a_M(i) + a_M(M)\, a_M(M-i)}{1 - (a_M(M))^2}. \tag{8.52}$$

By repeating the above calculations, PARCOR $a_1(1), a_2(2), \cdots, a_M(M)$ can be calculated. Using these relationships as well as AIC for order selection, we obtain the following estimation algorithm for the autoregressive model[7][6][45].

1. Compute the average $\bar{y}$ of the data $y(s), s = 1, 2, \cdots, N$, subtract $\bar{y}$ from the original data and rewrite it as $y(s)$. In other words

$$y(s) \leftarrow y(s) - \bar{y}, \quad s = 1, 2, \cdots, N.$$

2. Determine the maximum order M of the autoregressive model. Limit M to about $\max(M) = N/10$.

3. Calculate the covariance function by;

$$C(l) = \frac{1}{N}\sum_{s=1}^{N-l} y(s+l)\, y(s), \quad l = 0, 1, 2, \cdots, M.$$

4. Set initial values to $\sigma_v^2(0) = C(0)$, $a_0(m) = 0$, $m = 1, 2, \cdots, M$ and $\mathrm{AIC}(0) = N\left(\log 2\pi\, \sigma_v^2(0) + 1\right) + 2$.

5. For $m = 0, 1, 2, \cdots, M-1$, compute

   (a) PARCOR:
   $$a_{m+1}(m+1) = \left(C(m+1) - \sum_{j=1}^{m} a_m(j)\, C(m+1-j)\right)\left(\sigma_v^2(m)\right)^{-1}$$
   (b) $a_{m+1}(i) = a_m(i) - a_{m+1}(m+1)\, a_m(m+1-i), \quad i = 1, 2, \cdots, m$
   (c) $\sigma_v^2(m+1) = \sigma_v^2(m)\left(1 - a_{m+1}(m+1)^2\right)$
   (d) $\mathrm{AIC}(m+1) = N\left(\log 2\pi\, \sigma_v^2(m+1) + 1\right) + 2(m+2).$

   and find the order $\hat{m}$ that minimizes the value of $\mathrm{AIC}(m)$ among $m = 0, 1, \ldots, M$. The autoregressive model of order $\hat{m}$ is selected as the optimal model.

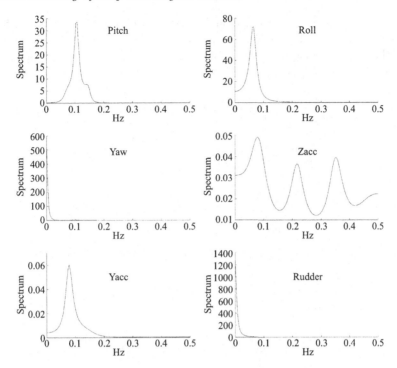

FIGURE 8.9
Spectra of ship motions obtained by AR method

8.5 Analysis of Ship Motion Time Series by Autoregressive Model

The method described in this chapter has been widely applied as a method for analyzing real data in nautical science, naval architecture and many other fields[107][80][106], and its field of application is still expanding. In this section, we apply the autoregressive model introduced in this chapter to the time series of ship motions observed at sea and investigate the characteristics of the time series of ship motions.

8.5.1 Comparison with the Spectra Obtained by the Blackman-Tukey Method

Figure 8.9 shows the spectra obtained by the AR method for the six motions of the medium-sized container ship shown in Figure 1.3. It can be seen that the spectra are smoother than the spectra obtained for the same time series using the Blackman-Tukey method shown in Figure 4.10. The orders of the autoregressive models are 12 for pitching, 4 for rolling, 14 for yawing, 9 for Z_{acc} (vertical acceleration), 7 for Y_{acc} (lateral acceleration) and 4 for rudder angle.

8.5.2 Time Series of Rolling and Pitching Motions

A group of data in which a very long time series is cut to an appropriate length to form a single time series data is called **batch data**.

As shown in Figure 8.10, the batch data with 10 batch interval were obtained from time series of 12,000 points of rolling and pitching measured continuously every second from a certain large container ship. In Figure 8.11, 10 spectra obtained by the estimated autoregressive models for each batch interval are superimposed. As shown in Figure 8.11, the spectrum of the rolling time series (left) is concentrated around 0.055 Hz, while the peak position of the pitching spectrum (right) shifts from 0.09 Hz to around 0.11 Hz, with some batch data showing two peaks.

FIGURE 8.10
Batch data

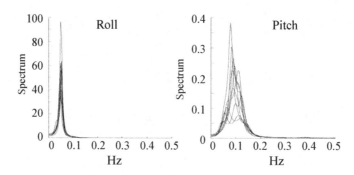

FIGURE 8.11
The spectra of the 10 data of rolling and pitching

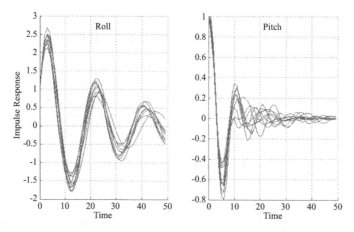

FIGURE 8.12
Impulse response functions of rolling and pitching

Figure 8.12 shows the impulse response sequence obtained by equation (8.16). Considering these figures from a dynamical point of view, it can be inferred that the *rolling motion is weakly damped and therefore has a strong natural period, while the pitching is a strongly damped motion that is more responsive to disturbances.* This is reflected in the fact that the peak position of the spectrum of pitching is not constant. This is the reason why pitching is called a kind of wave height meter. In addition, pitching motion also shows the same tendency as that of heaving motion.

Figures 8.13 and 8.14 show the two complex conjugate characteristic roots with the largest absolute values and positive phase angles (**dominant root**) of the characteristic

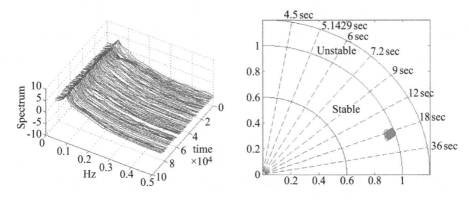

FIGURE 8.13
Spectra and distribution of dominant period of batch data of rolling

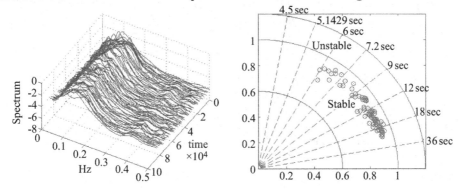

FIGURE 8.14
Spectra and distribution of dominant period of batch data of pitching

roots of rolling and pitching for different consecutive batches (24 hours) of data for the same container ship, respectively. It can be seen that the major roots of the pitching time series are spread between 18 and 5 seconds, while those of the rolling time series are concentrated around 18 seconds.

8.5.3 Relationships Among the Main Cycles of Ship Motion and Engine Motion

Figure 8.15 shows a scatter plot of the main cycles of each of the 800 batch data of the same rolling and pitching time series as in the previous subsection. The straight line in the center of the figure indicates the line where the pitching and rolling periods coincide. If the wave spectrum can be substituted by the pitching, it may be considered that the encounter period of the disturbance wave and the rolling motion coincide. However, the ellipse in the center indicates the area where there is a large amount of data for ship speeds between 10 knots and 15 knots. In the case of this data, it can be said that the pitching and rolling tend to oscillate within the same period when the speed is reduced. The straight line below is the line where twice the period of pitching corresponds to the period of rolling. Data on this line can be said to have the possibility of parametric rolling.

Figure 8.16 shows a scatter plot between the main roots of the main engine propeller revolution and pitching for the same batch time series. It can be seen that the synchronization between pitching and propeller revolution is at the same location as the synchronization period between rolling and pitching described in Figure 8.15. In addition, a positive

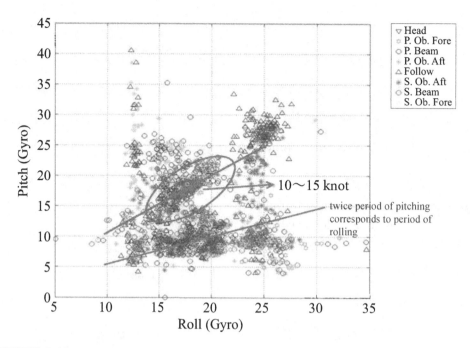

FIGURE 8.15
Scatter plot of dominant periods of rolling and pitching

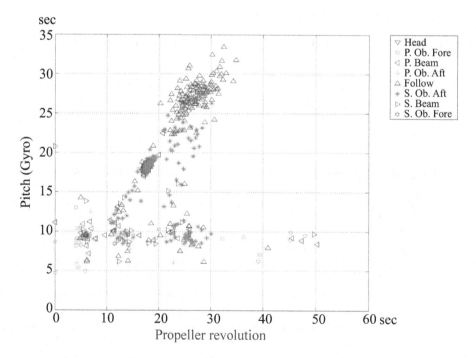

FIGURE 8.16
Scatter plot of dominant periods of pitching and propeller revolution

correlation is observed between the main period of the propeller revolution and the main period of the pitching in the tailwind condition. This indicates that the propeller revolution fluctuates due to the change in the propeller submergence depth caused by the pitching.

8.5.4 Characteristics of Yawing

Figure 8.17 shows the distribution of characteristic roots for a time series of a yawing. A characteristic root with an absolute value of nearly 1 (unit root) can be seen near the unit circle. This indicates that there is an integral term in the yawing. If we take the difference of this time series (which is the yaw rate), this root disappears. This indicates that the time series of the yawing is astatic.

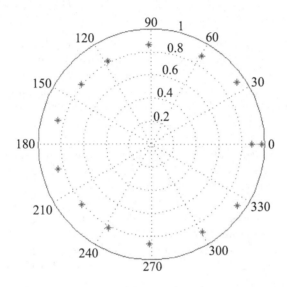

FIGURE 8.17
Distribution of characteristic roots of yawing motion

Appendix A: Information Criterion AIC

This appendix shows the derivation of the **AIC (Akaike Information Criterion)** [91][53] referred in this chapter. Readers who do not need it may skip this appendix.

The maximum likelihood method is an effective method for estimating parameters in statistical model building. Under the assumption that the data under consideration are generated from a true distribution $g(y|\theta)$ with a certain parameter θ, the simultaneous probability that the existing data $\mathbf{y} = \{y_1, y_2, \cdots, y_N\}$ are observed is called **likelihood** and is denoted by $L(\theta)$. Also, $l(\theta)$ is the logarithm of $L(\theta)$. Let $g(y)$ be the true distribution and $f(y)$ be the approximate model in the neighborhood. In this case, as a measure of how close $f(y)$ is to $g(y)$, there is a concept called **Kullback-Leibler information** (henceforth **KL information**) as follows[56]

$$\begin{aligned}
I(g; f) &= E_Y\left[\log \frac{g(Y)}{f(Y)}\right] = \int g(y) \log \frac{g(y)}{f(y)} dy \\
&= \int g(y) \log g(y) dy - \int g(y) \log f(y) dy.
\end{aligned} \tag{8.53}$$

We review the maximum likelihood method from the perspective of this KL information. In this case, the first term in the last equation of $I(g, f)$ cannot be evaluated because the true distribution $g(y)$ is unknown. However, since this term is a constant independent of the model $f(y)$, the larger the second term $E_Y[\log f(y)] = \int g(y) \log f(y) dy$, the closer it is to the true model. Since the second term is the expected value of the log-likelihood of the model, $\log f(x)$, this expected value $E_Y[\log f(Y)]$ with respect to the true distribution $g(y)$ is called the **expected log-likelihood**. This expected log-likelihood also cannot be calculated since $g(y)$ is unknown, but if $g(y)$ is replaced by the empirical distribution, from the law of large numbers, as N grows, it is expected to have

$$\frac{1}{N} \sum_{n=1}^{N} E_Y[\log f(y)] \longrightarrow E_Y[\log f(Y)]. \tag{8.54}$$

Thus, if we let N times of equation(8.54) be the **log-likelihood**, we see that the log-likelihood defined in this way is N times of a natural estimator of the expected log-likelihood.

By the way, what we are considering here is limited to the statistical distribution expressed by a certain parameter θ. Then, according to the maximum likelihood method, which considers that the larger the likelihood, the more reliable the model is, we select θ with the largest likelihood among the parameters of the model. Let us write $\hat{\theta}_{\max}$ for θ selected in this way.

Next, let us consider comparing the goodness of the estimated models based on the maximum likelihood values. Namely, after obtaining the maximum likelihood estimate of tha parameter $\hat{\theta}_{\max}$ that gives the maximum log-likelihood of each model, the resulting maximum log-likelihoods $l(\hat{\theta}_{\max})$ are compared, and the model with the largest maximum log-likelihood may be considered to be the model that best approximates the true model. However, this is not appropriate to determine the best order of the statistical model. This can be understood, for example, for the autoregressive model, the maximized likelihood is the monotone increasing with the order, and the model with the highest order is always selected by this criterion.

The cause of this problem is that when the maximum likelihood estimate $\hat{\theta}_{\max}$ is used as the optimal parameter, $1/N$ of the log-likelihood, $N^{-1}l(\hat{\theta}_{\max})$, has a positive bias as an estimator of the expected log-likelihood $E_Y \log f\left(Y|\hat{\theta}(y)\right)$ of the model specified by $\hat{\theta}(\mathbf{y})$. Namely, the mean value of the log-likelihood does not match the estimate of the expected maximum log-likelihood. That is, the maximum log-likelihood has a bias as an estimate of the log-likelihood when the maximum likelihood estimate is substituted. This discrepancy arises from the fact that the observation $\mathbf{y} = \{y_1, y_2, \cdots, y_N\}$ is used twice in total, once for estimating the parameter that maximizes the log-likelihood and once for evaluating the goodness of the estimated model, that is, the same members of the data used for estimation also conduct the evaluation. (i.e., patronage arises from the fact that the performance is evaluated within the group).

As a result, as can be seen from Figure 8.18, the true evaluation criterion should be $E_Y[\log f(Y|\hat{\theta}_{\max})] \leq E_Y[\log f(Y|\theta_0)]$, but the log-likelihood always evaluates as $l(\hat{\theta}_{\max}) \geq l(\theta_0)$. Therefore, for a fair evaluation, it is necessary to evaluate the bias caused by using the log-likelihood to estimate the expected log-likelihood. Hereafter, we consider independent data $Z = \{z_1, z_2, \cdots, z_N\}$ generated from the same distribution as the observed data, and evaluate the bias $E_Y \log f(Y|\hat{\theta}_{\max}) - \frac{1}{N} \sum \log f(y_n|\hat{\theta}_{\max})$ by the expectation with respect to the data Z as follows

$$D = E_Z\left\{ E_Y \log f(Y|\hat{\theta}_{\max}) - \frac{1}{N} \sum_{n=1}^{N} \log f(y_n|\hat{\theta}_{\max}) \right\}. \tag{8.55}$$

Here, the expectation about the random variable Z is taken with respect to its simultaneous distribution $\prod\limits_{n=1}^{N} g(z_n)$.

The Akaike Information Criterion is proposed by calculating this bias and compensating for this bias to use as "an information criterion" to measure the goodness of the model,

$$\text{AIC} = -2(\text{maximum log-likelihood} - \text{estimated bias}) \tag{8.56}$$

$$= -2 \sum_{n=1}^{N} \log f(Y_N|\hat{\theta}) + 2(\text{estimate of } D).$$

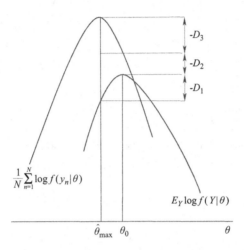

FIGURE 8.18
Expected log-likelihood and log-likelihood

Evaluation of Bias

Figure 8.18 shows the relationship between the expected log-likelihood function and the log-likelihood function of the statistical model specified by one-dimensional parameter θ[91][53].

According to the figure, the estimate of bias consists of three terms D_1, D_2 and D_3. That is, when θ_0 is the true parameter that maximizes the expected log-likelihood function $E_Y \log f(Y|\theta)$, then the bias can be decomposed as

$$
\begin{aligned}
D &= E_Z \left\{ E_Y \log f(Y|\hat{\theta}) - E_Y \log f(Y|\theta_0) \right\} \\
&\quad + E_Z \left\{ E_Y \log f(Y|\theta_0) - \frac{1}{N} \sum_{n=1}^{N} \log f(y_n|\theta_0) \right\} \\
&\quad + E_Z \left\{ \frac{1}{N} \sum_{n=1}^{N} \log f(y_n|\theta_0) - \frac{1}{N} \sum_{n=1}^{N} \log f(y_n|\hat{\theta}) \right\} \tag{8.57} \\
&= D_1 + D_2 + D_3.
\end{aligned}
$$

Note that for θ_0, we have

$$\frac{\partial}{\partial \theta} E_Y \log f(Y|\theta_0) = 0. \tag{8.58}$$

On the other hand, for $\hat{\theta}_{\max}$, since it is the maximizer of the likelihood function $l(\theta) = \sum_{n=1}^{N} \log f(Y_n|\theta)$, it holds that

$$\frac{\partial}{\partial \theta} \sum_{n=1}^{N} \log f\left(y_n|\hat{\theta}_{\max}\right) = 0.$$

Computation of D_2

Among D_1, D_2 and D_3, D_2 is the only one that does not contain an estimator. Therefore, we start with the calculation of D_2. When θ_0 is fixed, the expected value of each term of $\sum_{n=1}^{N} \log f(y_n|\theta_0)$ is identical to the expected log-likelihood, so we have

$$
\begin{aligned}
D_2 &= E_Z \left[\log f(Y|\theta_0) - \frac{1}{N} \sum_{n=1}^{N} \log f(y_n|\theta_0) \right] \\
&= E_Z \log f(Y|\theta_0) - E_Z \left[\frac{1}{N} \sum_{n=1}^{N} \log f(y_n|\theta_0) \right] \\
&= E_Z \log f(Y|\theta_0) - E_Z \log f(Y|\theta_0) = 0.
\end{aligned}
$$

That is, this term fluctuates but on average it always holds that

$$D_2 = 0.$$

Computation of D_1

Taylor expansion of the expected log-likelihood of the maximum likelihood estimator, $E_Y \log f(Y|\hat{\theta}_{\max})$, around θ_0 to a quadratic term yields

$$
\begin{aligned}
E_Y \log f(Y|\hat{\theta}_{\max}) &\approx E_Y \log f(Y|\theta_0) + \left\{ \frac{\partial}{\partial \theta} E_Y \log f(Y|\theta_0) \right\} (\hat{\theta}_{\max} - \theta_0) \\
&\quad + \frac{1}{2}(\hat{\theta}_{\max} - \theta_0)^t \left\{ \frac{\partial^2}{\partial \theta \, \partial \theta'} E_Y \log f(Y|\theta_0) \right\} (\hat{\theta}_{\max} - \theta_0) \\
&= E_Y \log f(Y|\theta_0) + E_Y \left\{ \frac{\partial}{\partial \theta} \log f(Y|\theta_0) \right\} (\hat{\theta}_{\max} - \theta_0) \\
&\quad + \frac{1}{2}(\hat{\theta}_{\max} - \theta_0)^t E_Y \left\{ \frac{\partial^2}{\partial \theta \, \partial \theta'} \log f(Y|\theta_0) \right\} (\hat{\theta}_{\max} - \theta_0) \\
&= E_Y \log f(Y|\theta_0) - \frac{1}{2}(\hat{\theta}_{\max} - \theta_0)^t J(\theta_0)(\hat{\theta}_{\max} - \theta_0).
\end{aligned}
$$

Here

$$J(\theta_0) \equiv -E_Y \left\{ \frac{\partial^2}{\partial \theta \, \partial \theta'} \log f(Y|\theta_0) \right\}$$

and we use the fact that θ_0 is the solution of

$$E_Y \left\{ \frac{\partial}{\partial \theta} \log f(Y|\theta_0) \right\} = \int g(y) \frac{\partial}{\partial \theta} \log f(y|\theta_0) dy = 0.$$

Furthermore, if $\hat{\theta}_{\max}$ is the maximum likelihood estimator of θ_0 and $I(\theta_0)$ is the Fisher information matrix:

$$I(\theta_0) \equiv E_Y \left\{ \frac{\partial}{\partial \theta} \log f(Y|\theta_0) \right\} \left\{ \frac{\partial}{\partial \theta'} \log f(Y|\theta_0) \right\}^t,$$

then, by the central limit theorem and from equation (17.64), $\sqrt{N}\,(\hat{\theta}_{\max} - \theta_0)$ asymptotically follows a normal distribution with mean 0 and the variance-covariance matrix $J(\theta_0)^{-1}\,I(\theta_0)\,J(\theta_0)^{-1}$[20][96]. Therefore, taking expectation with respect to Z, we get

$$
\begin{aligned}
E_Z\left[(\hat{\theta}_{\max} - \theta_0)^t J(\theta_0)(\hat{\theta}_{\max} - \theta_0)\right] &= E_Z\left[\operatorname{tr}\left\{J(\theta_0)(\hat{\theta}_{\max} - \theta_0)(\hat{\theta}_{\max} - \theta_0)^t\right\}\right] \\
&= \operatorname{tr}\left\{J(\theta_0)E_Z\left[(\hat{\theta}_{\max} - \theta_0)(\hat{\theta}_{\max} - \theta_0)^t\right]\right\} \\
&= \frac{1}{N}\operatorname{tr}\left\{J(\theta_0)\,J(\theta_0)^{-1}\,I(\theta_0)\,J(\theta_0)^{-1}\right\} \\
&= \frac{1}{N}\operatorname{tr}\left\{I(\theta_0)\,J(\theta_0)^{-1}\right\},
\end{aligned}
$$

where tr is the trace of the matrix. Further, it is known that, if $g(y) = f(y|\theta_0)$ holds, then $J(\theta_0) = I(\theta_0)$. Therefore, for the last term the following holds approximately

$$
E_Z\left[(\hat{\theta}_{\max} - \theta_0)^t\,J(\theta_0)(\hat{\theta}_{\max} - \theta_0)\right] \approx \frac{k}{N},
$$

where k is the rank of the matrix I. Using this result, we can go back to the beginning and find that it holds asymptotically

$$
E_Z\left[E_Y \log f(Y|\hat{\theta}_{\max}) - E_Y \log f(Y|\theta_0)\right] \approx -\frac{k}{2N}.
$$

Computation of D_3

Taylor expansion of

$$
\frac{1}{N}\sum_{n=1}^{N}\log f(y_n|\theta_0)
$$

around $\hat{\theta}_{\max}$ yields

$$
\begin{aligned}
\frac{1}{N}\sum_{n=1}^{N}\log f(y_n|\theta_0) \approx\ & \frac{1}{N}\sum_{n=1}^{N}\log f(y_n|\hat{\theta}_{\max}) + \frac{1}{N}\sum_{n=1}^{N}\frac{\partial}{\partial\theta}\log f(y_n|\hat{\theta}_{\max})(\theta_0 - \hat{\theta}_{\max}) \\
& + \frac{1}{2}(\theta_0 - \hat{\theta}_{\max})^t\frac{1}{N}\sum_{n=1}^{N}\frac{\partial^2}{\partial\theta\,\partial\theta'}\log f(y_n|\hat{\theta}_{\max})(\theta_0 - \hat{\theta}_{\max}).
\end{aligned}
$$

Here, since $\hat{\theta}_{\max}$ is the maximum likelihood estimator, the second term on the right side is zero.

From the law of large numbers (see Subsection 17.3.5), if $N \to \infty$ then

$$
\frac{1}{N}\sum_{n=1}^{N}\frac{\partial^2}{\partial\theta\,\partial\theta'}\log f(y_n|\hat{\theta}_{\max}) \Longrightarrow E_Y\left\{\frac{\partial^2}{\partial\theta\,\partial\theta'}\log f(Y|\theta_0)\right\} \equiv -J(\theta_0),
$$

therefore we have

$$
\frac{1}{N}\sum_{n=1}^{N}\log f(y_n|\theta_0) \approx \frac{1}{N}\sum_{n=1}^{N}\log f(y_n|\hat{\theta}_{\max}) - \frac{1}{2}(\theta_0 - \hat{\theta}_{\max})^t J(\theta_0)(\theta_0 - \hat{\theta}_{\max}).
$$

Finally, taking the expectation with respect to the random variable Z of both sides, for the same reason as described for D_1, we obtain

$$
E_Z\left[(\theta_0 - \hat{\theta}_{\max})(\theta_0 - \hat{\theta}_{\max})^t\right] \approx \frac{1}{N}\left\{J(\theta_0)^{-1}\,I(\theta_0)\,J(\theta_0)^{-1}\right\}.
$$

Therefore, D_1 is evaluated as

$$D_3 = E_Z \left\{ \frac{1}{N} \sum_{n=1}^{N} \log f(y_n|\theta_0) - \frac{1}{N} \sum_{n=1}^{N} \log f(y_n|\hat{\theta}_{\max}) \right\} \approx -\frac{k}{2N}.$$

Appendix B: Proof of the Levinson-Durbin Algorithm

From the Yule-Walker equation (8.29) of order $M + 1$, we have

$$\sum_{j=1}^{M+1} a_{M+1}(j) R_{yy}(m - j) = R_{yy}(m), \quad m = 1, 2, \cdots, M + 1. \tag{8.59}$$

This equation can be written as

$$\sum_{j=1}^{M} a_{M+1}(j) R_{yy}(m - j) + a_{M+1}(M + 1) R_{yy}(m - M - 1) = R_{yy}(m), \quad m = 1, \cdots, M + 1. \tag{8.60}$$

From this, if R_M is the Teplitz matrix (see Subsection 8.3.1), then

$$R_M \begin{bmatrix} a_{M+1}(1) \\ a_{M+1}(2) \\ \vdots \\ a_{M+1}(M) \end{bmatrix} = \begin{bmatrix} R_{xx}(1) \\ R_{xx}(2) \\ \vdots \\ R_{xx}(M) \end{bmatrix} - a_{M+1}(M+1) \begin{bmatrix} R_{xx}(M) \\ R_{xx}(M-1) \\ \vdots \\ R_{xx}(1) \end{bmatrix}. \tag{8.61}$$

On the other hand, from the Yule-Walker equation (8.29), $R_M^{-1} R_{xx}(m) = a_M(m)$, so we have

$$\begin{bmatrix} a_{M+1}(1) \\ a_{M+1}(2) \\ \vdots \\ a_{M+1}(M) \end{bmatrix} = R_M^{-1} \begin{bmatrix} R_{xx}(1) \\ R_{xx}(2) \\ \vdots \\ R_{xx}(M) \end{bmatrix} - a_{M+1}(M+1) R_M^{-1} \begin{bmatrix} R_{xx}(M) \\ R_{xx}(M-1) \\ \vdots \\ R_{xx}(1) \end{bmatrix}$$

$$= \begin{bmatrix} a_M(1) \\ a_M(2) \\ \vdots \\ a_M(M) \end{bmatrix} - a_{M+1}(M+1) \begin{bmatrix} a_M(M) \\ a_M(M-1) \\ \vdots \\ a_M(1) \end{bmatrix}$$

From this, we first obtain equation (8.50):

$$a_{M+1}(m) = a_M(m) - a_{M+1}(M + 1) a_M(M+1-m), \quad m = 1, 2, \cdots, M. \tag{8.62}$$

In equation (8.60), if we put $m = M + 1$, then from equation (8.62) we have

$$\sum_{j=1}^{M} a_{M+1}(j) \, R_{yy}(M+1-j) + a_{M+1}(M + 1) \, R_{yy}(0)$$

$$= \sum_{j=1}^{M} a_M(j) \, R_{yy}(M+1-j)$$

$$+ a_{M+1}(M + 1) \left\{ R_{yy}(0) - \sum_{j=1}^{M} a_M(M+1-j) \, R_{yy}(M+1-j) \right\}$$

$$= R_{yy}(M + 1). \tag{8.63}$$

Now, substituting equation (8.19) into $\{\ \}$ in equation (8.63), we obtain equation (8.49). Finally, consider equation (8.30). Using equation (8.62) for $a_{M+1}(j)$ in the second line of the following equation, we obtain

$$\begin{aligned}
\sigma_v^2(M + 1) &= R_{yy}(0) - \sum_{j=1}^{M+1} a_{M+1}(j) \, R_{yy}(j) \\
&= R_{yy}(0) - \sum_{j=1}^{M} a_{M+1}(j) \, R_{yy}(j) - a_{M+1}(M + 1) \, R_{yy}(M + 1) \\
&= R_{yy}(0) - \sum_{j=1}^{M} a_M(j) \, R_{yy}(j) \\
&\quad - a_{M+1}(M + 1) \left\{ R_{yy}(M + 1) - \sum_{j=1}^{M} a_M(M+1-j) \, R_{yy}(j) \right\} \\
&= \sigma_v^2(M) - a_{M+1}(M + 1) \, a_{M+1}(M + 1) \, \sigma_v^2(M) \\
&= \sigma_v^2(M) \left\{ 1 - (a_{M+1}(M + 1))^2 \right\}.
\end{aligned} \tag{8.64}$$

Note that the second equality from the bottom follows from equation (8.49).

9

Monitoring of Slowly Changing Ship and Engine Motions

The weather and sea conditions surrounding ships are nonstationary, and they change slowly with few sudden changes. In other words, they can be regarded as locally stationary. Although real-time methods such as the sequential estimation method are available as model estimation methods for nonstationary time series, in the case of ship motions in waves, it is more reliable to divide the time series of ship motions into batch sections and to treat them as stationary during the batch sections.

This chapter discusses a semi-online autoregressive model estimation method based on the minimum AIC method for such a **locally stationary process** and a shipboard monitoring system for ship and engine motions using this method.

9.1 Householder Transformation Method and Estimation of Time Series Model

As discussed in Chapter 8, autoregressive models can be estimated efficiently by the least squares method for parameter estimation. Here, we show a computational method that allows for semi-online least squares estimation of model parameters.

As described in the previous chapter, the least squares estimator of the coefficients, $a(m)$, of the AR model

$$y(s) = \sum_{m=1}^{M} a(m)\, y(s - m) + v(s), \tag{9.1}$$

are obtained by minimizing

$$L(a(m)) = \sum_{s=M+1}^{N} \left\{ y(s) - \sum_{m=1}^{M} a(m)\, y(s - m) \right\}^2. \tag{9.2}$$

Here, given $N - M$ data from $y(M + 1)$ to $y(N)$, denoting

$$\mathbf{y} = \begin{pmatrix} y(M+1) \\ y(M+2) \\ \vdots \\ y(N) \end{pmatrix}, \quad \mathbf{Z} = \begin{pmatrix} y(M) & y(M-1) & \cdots & y(1) \\ y(M+1) & y(M) & \cdots & y(2) \\ \vdots & \vdots & \ddots & \vdots \\ y(N-1) & y(N-2) & \cdots & y(N-M) \end{pmatrix}, \quad \mathbf{a} = \begin{pmatrix} a(1) \\ a(2) \\ \vdots \\ a(M) \end{pmatrix},$$

the AR model can be written as

$$\mathbf{y} = \mathbf{Z}\mathbf{a} + \mathbf{v}. \tag{9.3}$$

DOI: 10.1201/9781003428565-9

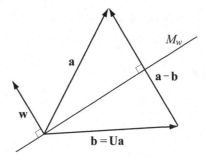

FIGURE 9.1
Reflection transformation

Then the L is given by

$$L = ||\mathbf{y} - \mathbf{Za}||^2 = ||\mathbf{v}||^2,$$

where $||\mathbf{x}||$ is the Euclidean norm of the vector $\mathbf{x}$. Now, let $\mathbf{U}$ be an arbitrary orthogonal matrix, then it holds that

$$||\mathbf{y} - \mathbf{Za}||^2 = ||\mathbf{U}(\mathbf{y} - \mathbf{Za})||^2 = ||\mathbf{Uy} - \mathbf{UZa}||^2. \tag{9.4}$$

This means that, the vector $\mathbf{a}$ which minimizes $||\mathbf{Uy} - \mathbf{UZa}||^2$ also minimizes $||\mathbf{y} - \mathbf{Za}||^2$.

So, after transforming $\mathbf{Z}$ into a computationally convenient form of the matrix $\mathbf{UZ}$ by means of the orthogonal transformation $\mathbf{U}$, by finding a vector $\mathbf{a}$ that minimizes equation (9.4), we can obtain the least squares solution of the model. Here, we consider a **Householder transformation**, which transforms an arbitrary $n \times m$ matrix $\mathbf{X}$ into an upper triangular matrix by an orthogonal matrix[91][94].

Let M_w be a plane orthogonal to an arbitrary unit vector $\mathbf{w}$ and pass through the origin. Considering M_w as a mirror, the image $\mathbf{b}$ of any vector $\mathbf{a}$ reflected in this mirror is given by $\mathbf{a} - 2\mathbf{w} \cdot (\mathbf{w}^t \mathbf{a})$. This means that it is equivalent to multiplying the vector $\mathbf{a}$ by

$$\mathbf{U} = \mathbf{I} - 2\mathbf{w}\,\mathbf{w}^t \tag{9.5}$$

from the left. This transformation is called **reflection transformation**. Let $\mathbf{a}$ and $\mathbf{b}$ be two vectors of equal length. In order for there to be a mirror transformation $\mathbf{U}$ that transfers $\mathbf{a}$ to $\mathbf{b}$, the vector $\mathbf{w}$ above should be set to

$$\mathbf{w} = \frac{\mathbf{a} - \mathbf{b}}{||\mathbf{a} - \mathbf{b}||}. \tag{9.6}$$

This is evident from Figure 9.1, where $\mathbf{w}$ should be a unit vector parallel to $\mathbf{a} - \mathbf{b}$.

Defining $\mathbf{a}_1 = (z_{11}, z_{21}, \cdots, z_{n1})^t$, $\mathbf{b}_1 = (z_{11}^{(1)}, 0, \cdots, 0)^t$, $Z_{11}^{(1)} = \mp|a_1|$ (sign is positive and negative of Z_{11}), there exists an orthogonal matrix $\mathbf{U}_1$ that transforms the first column vertical vector $\mathbf{a}_1$ of the $\mathbf{Z}$ matrix into $\mathbf{b}_1$ as mentioned earlier. As a result, all but the z_{11} elements in the first column of $\mathbf{Z}$ can be set to zero. If we repeat this mirror transformation m times for each column, we can transform the $n \times m$ matrix $\mathbf{X}$ (where $n \geq m$) into an upper triangular matrix.

Next, we show how to find the least squares estimates of the AR model using this Householder transformation. First, consider the autoregressive model of order M in equation (9.3), where

$$\mathbf{X} = [\mathbf{Z} : \mathbf{y}], \tag{9.7}$$

and repeat the Householder transformation $m + 1$ times to obtain

$$\mathbf{UX} = \mathbf{S} = \begin{pmatrix} S_{11} & \cdots & & S_{1\,m+1} \\ & \ddots & & \vdots \\ & & & S_{m+1\,m+1} \\ 0 & & & 0 \end{pmatrix}.$$

In this case, we have

$$\|\mathbf{Uy} - \mathbf{UZa}\|^2 = \left\| \begin{pmatrix} S_{1\,m+1} \\ \vdots \\ S_{m\,m} \\ S_{m+1\,m+1} \\ 0 \\ \vdots \\ 0 \end{pmatrix} - \begin{pmatrix} S_{11} & \cdots & S_{1m} \\ \vdots & \ddots & \vdots \\ 0 & \cdots & S_{mm} \\ 0 & \cdots & 0 \\ 0 & \cdots & 0 \\ \vdots & \ddots & \vdots \\ 0 & \cdots & 0 \end{pmatrix} \begin{pmatrix} a(1) \\ \vdots \\ a(m) \end{pmatrix} \right\|^2$$

$$= \left\| \begin{pmatrix} S_{1\,m+1} \\ \vdots \\ S_{m\,m+1} \end{pmatrix} - \begin{pmatrix} S_{11} & \cdots & S_{1m} \\ \vdots & \ddots & \vdots \\ 0 & \cdots & S_{mm} \end{pmatrix} \begin{pmatrix} a(1) \\ \vdots \\ a(m) \end{pmatrix} \right\|^2 + S_{m+1\,m+1}^2. \quad (9.8)$$

Since the second term on the right-hand side of the last equation is a constant unrelated to $\mathbf{a}$, it follows that to minimize $\|\mathbf{Uy} - \mathbf{UZa}\|^2$, the first term on the right-hand side should be zero. In other words, we have

$$\begin{pmatrix} S_{11} & \cdots & S_{1m} \\ \vdots & \ddots & \vdots \\ 0 & \cdots & S_{mm} \end{pmatrix} \begin{pmatrix} a(1) \\ \vdots \\ a(m) \end{pmatrix} = \begin{pmatrix} S_{1\,m+1} \\ \vdots \\ S_{m\,m+1} \end{pmatrix}. \quad (9.9)$$

That is, by solving equation (9.9), the least squares solution of $\mathbf{a}$ can be obtained, and since the coefficient matrix of this equation is upper triangular form, $a(m), \ldots, a(1)$ can be easily calculated by backward substitution. The residual of the model is given by $S_{m+1\,m+1}^2/N$. Using this method for each order, the least squares estimator of the regression coefficients $a(m), \ldots, a(1)$ and variance of the prediction error $\sigma_v^2(m)$ and AIC of the AR model can be obtained. This method provides an efficient and accurate least squares estimates for parameter estimation and model selection for AR model.

9.2 Estimation Method for the Locally Stationary Time Series Model

Suppose that N_1 data are available, and an autoregressive model of order M is identified for this data using the least squares estimates and the minimum AIC method. As a result of the Householder transformation, the matrix $\mathbf{X}$ becomes

$$\begin{pmatrix} S_{11} & \cdots & S_{1\,M+1} \\ & \ddots & \vdots \\ 0 & \cdots & S_{M+1\,M+1} \end{pmatrix}.$$

Next, suppose that $L_2 > M+1$ additional data are obtained. In this case, we add $L_2 \times (M+1)$ elements corresponding to the new data under the triangular matrix above, then we have a $(M + 1 + L_2) \times (M + 1)$ matrix as follows:

$$\begin{pmatrix} S_{11} & \cdots & S_{1M} & S_{1\,M+1} \\ 0 & \ddots & \vdots & \vdots \\ \vdots & \ddots & S_{M\,M} & S_{M\,M+1} \\ 0 & \cdots & 0 & S_{M+1\,M+1} \\ y(L_1-M+1) & \cdots & y(L_1) & y(L_1+1) \\ \vdots & \ddots & \vdots & \vdots \\ y(L_1+L_2-M) & \cdots & y(L_1+L_2-1) & y(L_1+L_2) \end{pmatrix}.$$

Considering that the Householder transformation is an orthogonal transformation, by transforming this matrix to a new triangular matrix, we can obtain the least squares estimates of the AR model parameters for merged data with $L_1 + L_2$ observations. Thus, the solution $\{a(m), m = 1, \cdots, M\}$ of the merged time series including the time series of the new batch interval is obtained[45]. Using this property, it is possible to build models from large data by partitioning the data, as well as to **batch sequential** estimation of the locally stationary time series when additional data is obtained sequentially as follows.

Suppose that L_1 time series are obtained as before, and that the estimation of the autoregressive model of this time series results in an AR_0 model, where the AIC of this model is AIC_0. Suppose further time passes and L_2 data are obtained and an AR_2 model is obtained for this time series and its AIC is AIC_2. Thus the AIC of a model in which a time series of length $L_1 + L_2$ is split into two time series of length L_1 and length L_2 (split model) is obtained by

$$AIC_D = AIC_0 + AIC_2. \tag{9.10}$$

On the other hand, let AR_1 be the model obtained from entire $L_1 + L_2$ time series (merged model) and the AIC of the pooled model be AIC_1. We then obtain the semi-online estimation method of the locally stationary AR model shown in Figure 9.2[87].

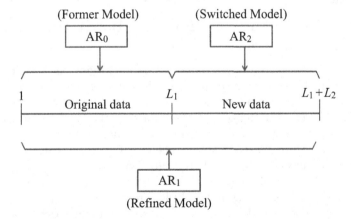

FIGURE 9.2
Semi-online estimation of locally stationary time series models using AIC

Estimation Procedure for Locally Stationary AR Model[87]

1. If $AIC_D = AIC_0 + AIC_2 < AIC_1$, then split model is adopted and new AR_2 (switched model) is adopted.

2. If $\text{AIC}_D = \text{AIC}_0 + \text{AIC}_2 \geq \text{AIC}_1$, then the merged model AR_1 (Refined Model) is adopted.

This method allows us to confirm the local stationarity of the time series and to perform online spectrum estimation of the **batch sequential data**.

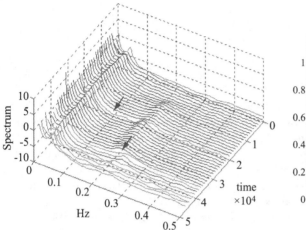

FIGURE 9.3
Spectra of batch data of rolling motion (in log-scale)

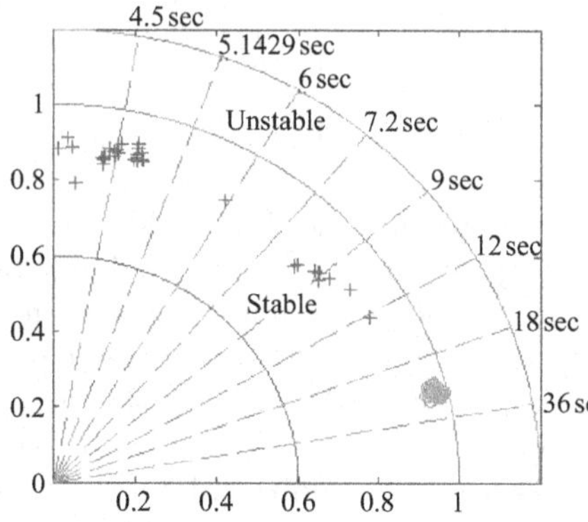

FIGURE 9.4
Distribution of major and secondary characteristic roots of the rolling batch data

9.3 Analysis of Rolling and Pitching Batch Data Using Locally Stationary Time Series Model

In the following, we present the results obtained by identifying the statistical model using the estimation method described in the previous section for 40 consecutive batch data ($N = 1200$ points, $\Delta t = 1\,\text{sec}$) of the rolling and pitching data of a certain large container ship. Figure 9.3 shows the time-varying spectrum obtained by the batch sequential estimation of rolling data, Figure 9.4 shows the distribution of the main and second spectrum peaks at each time point and Figure 9.5 shows the change in AIC for each batch during that period. The model change occurred at the 17th batch data (with 1200 seconds) starting from 19200 seconds in Figure 9.3. After 36000 seconds (31st batch), model change occurs. This can be seen from the change in AIC in Figure 9.5. In Figure 9.4, the major roots are concentrated around the 24~25 second period, where the circles are clumped together and do not move. This indicates the strength of the natural period of the rolling described in the previous section. The second peak (marked with $+$) is dispersed.

In contrast, Figures 9.6, 9.7 and 9.8 show the spectrum, the distribution of main and second characteristic roots and AIC change for batch data of pitching observed at the same time. The spectrum in Figure 9.6 has three peaks in a row, but the model switch did not occur. Figure 9.7 shows that their periods are scattered around 16, 9 and 6 seconds. It can be seen that the pitching motion is responding to waves with various periods.

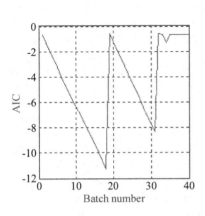

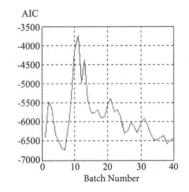

FIGURE 9.5
Variation of AIC for rolling batch data

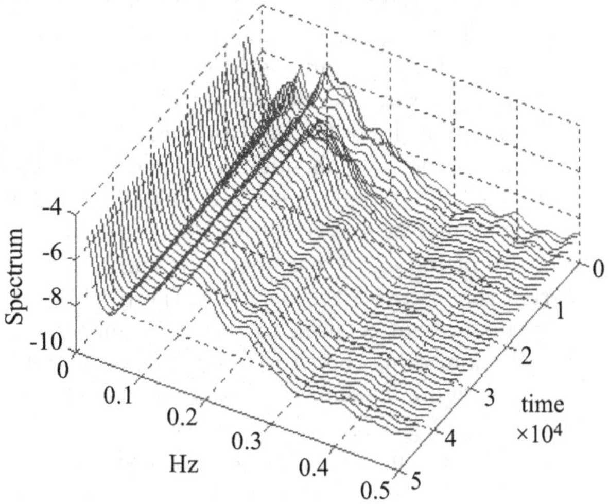

FIGURE 9.6
Spectra of batch data of pitching motion (in log-scale)

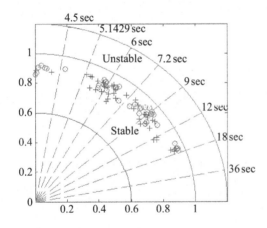

FIGURE 9.7
Distribution of major and secondary characteristic roots of the pitching batch data

FIGURE 9.8
Variation of AIC for pitching batch data

9.4 Application to Monitoring Systems

As many of the figures shown in the previous section indicate, the method of estimating locally stationary time series model described in this chapter is effective for monitoring the state of ship motions such as rolling and pitching motions, which can be regarded as stationary in a certain subinterval of the data. In particular, the period of the rolling motion provides important information to the operator on the stability of the ship's rolling.

First, the **synchronization phenomenon** between the main period of the wave disturbance and rolling must be avoided, which is important in considering the loss of stability of the rolling. If the pitching is regarded as a wave height gauge and its period is regarded as a proxy for the wave period, the synchronization phenomenon can be avoided by monitoring the periods of the rolling and pitching.

The second is a phenomenon called "**parametric rolling**," which has recently been elucidated. This phenomenon is said to occur frequently on container ships and fishing vessels in quarter seas conditions, where the hull form becomes wider toward the stern. This increase of rolling by parametric rolling occurs when such a ship navigates with a rolling period that is twice the period of the pitching period. Therefore, the method described in this chapter can be used to avoid the increase of rolling in the quarter seas condition by closely monitoring the period of rolling and pitching, and in some cases, changing course for a while. Figure 9.9 shows an example of such a monitoring system. The real-time data of a given sampling is shown on the left side, and after the batch processing, the main period (top of each figure) and the second peak period (bottom of each figure) are shown.

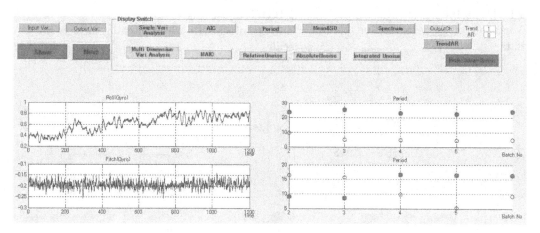

FIGURE 9.9
Example of period display on a shipboard monitor (top: rolling, bottom: pitching)

10

Analysis and Prediction of Ship and Engine Motions Using State-Space Models

Autoregressive models have a wide range of applications as an effective way to capture dynamically changing time series in the time domain. In order to further broaden the potential of the autoregressive model, we introduce the state-space model[50][45], which has been developed intensively by G. Kitagawa of the Institute of Statistical Mathematics. This method transforms the autoregressive model into a state-space model and uses a computational method called the Kalman filter to perform sequential prediction, filtering and smoothing of time series. The state-space method has a wide range of applications, as it conveys information from the past to the present of a system to the future via hidden variables called state variables. In this chapter, the Kalman filter is treated as known. However, since it has a wide range of applications, including nautical science, it will be explained in detail in Chapter 18.

10.1   What is the State-Space Method

The state-space method appropriately defines a vector of random variables called **state variable** that captures the changes in the system from moment to moment. This state variable serves as an interface to extract information from past and present observations that best describe possible future system states, as shown in Figure 10.1, and conveys the information by the following equations,

$$x(s) = F\,x(s-1) + G\,v(s). \tag{10.1}$$

This representation is called **system equation**, where s is a certain time and $s-1$ is one step before. Let $x(s-1)$ be a k-dimensional vector of random variable that best represents the state of the system of interest at time $s-1$.

The system equation expresses the value of state $x(s)$ at time s by the state variable $x(s-1)$ at time $s-1$ multiplied by a $k \times k$ matrix F called **transition matrix**. The actual value of $x(s)$ is expressed by adding $G\,v(s)$. Here, $v(s)$ is an m-dimensional vector called the system noise, whose mean vector is 0 and a variance-covariance matrix Q. The G is a $k \times m$ matrix called **driving matrix**, which is the coefficient matrix for $v(s)$. However, the observer does not observe all of this internal state, but rather observes an l-dimensional signal $y(s)$ in a form that is a transformed version of the state $x(s)$ accompanied by observation noise. To express such a situation, we use **observation equation**

$$y(s) = H\,x(s) + w(s), \tag{10.2}$$

DOI: 10.1201/9781003428565-10

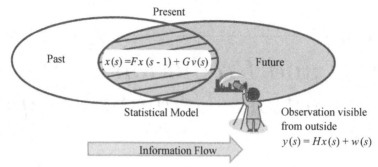

FIGURE 10.1
Concept of state-space modeling

where H is a $l \times k$ matrix, called the **observing matrix**, that connects the internal and external states. The $w(s)$ is the **observation noise**, a Gaussian white noise with zero mean vector following a variance-covariance matrix R.

Once the model is transformed into such a state-space model, the estimate of the past, present and future state variables are updated by a method called the Kalman filter to bring the estimates closer to their true values, using the observations obtained moment by moment. It should be noted here that $x(s)$ is a vector variable. This is an excellent point that broadens the range of applications of the state-space representation.

In this section, we represent the autoregressive model constructed from the observed values of the univariate time series described in Chapter 8 (hence $l = 1$) in a state-space model and reveal how to estimate $x(s)$, the information inside the system, recursively using a method called the Kalman filter.

10.2   State-Space Representation of Time Series Model

10.2.1   State-Space Representation of the Autoregressive Model

Now let $y(s)$ be a univariate time series and denote their set by $Y = \{y(1), y(2), \cdots, y(N)\}$. An autoregressive model is used to represent the changes of this time series as

$$y(s) = \sum_{m=1}^{M} a(m)\, y(s - m) + v(s), \tag{10.3}$$

and consider defining a new state variable $x(s)$ by variable transformation without changing the structure of this model and transforming it into the state-space representation (equations (10.1) and (10.2))[50][45]. However, it is known that this state-space representation of $y(s)$ is not unique. For example, when considering the transformation of an autoregressive model of order M with $v(s) \sim N(0, \sigma^2)$ into a state-space representation, the most natural method is to define the state variables as

$$x(s) = (y(s), y(s - 1), \cdots, y(s - M + 1))^t. \tag{10.4}$$

In this case, it is easy to see that

$$
F = \begin{pmatrix}
a(1) & a(2) & \cdots & a(M{-}1) & a(M) \\
1 & 0 & \cdots & 0 & 0 \\
0 & 1 & \ddots & \vdots & \vdots \\
\vdots & \vdots & \ddots & 0 & 0 \\
0 & 0 & \cdots & 1 & 0
\end{pmatrix}, \quad H = (1\ 0\ \cdots\ 0). \tag{10.5}
$$

This transformation method is called the controllable state-space model.

By the way, we can see that to compute one-step ahead prediction of the actual $y(s)$ using this transformation method, we need to compute the sum of the products of the first row of F and the vector $x(s-1)$. However, in some control problems, such as those we will consider later, we may want to compute the next control signal using the predictions and execute control as soon as possible after the observation is obtained. In this case, the following transformation method is used to store and prepare variables from past data in advance so that predicted values in the next step can be obtained with the least amount of calculation when new observed values are obtained.

In equation (10.3), let s be the current time and examine the relationship between this state and the model of the state at a future time $s + l$. Substituting $s + l$ for s, equation (10.3) becomes

$$
\begin{aligned}
y(s+l) &= \sum_{m=1}^{M} a(m)\, y(s+l-m) + v(s+l) \\
&= \sum_{i=1}^{M-l} a(l+i)\, y(s-i) + \sum_{j=0}^{l-1} a(l-j)\, y(s+j) + v(s+l). \tag{10.6}
\end{aligned}
$$

Here, the first term of the third line in equation (10.6) shows the contribution of the past data $y(s-1), y(s-2), \cdots$, etc. in the calculation of the l-step ahead prediction. Therefore, this part is denoted as

$$
x_l(s) = \sum_{i=1}^{M-l} a(l+i)\, y(s-i), \quad l = 1, \cdots, M-1.
$$

In this case, to know how the observed value $y(s-1)$ is used in the computation of the $(M-1)$-step ahead prediction, we only need to compute $x_{M-1}(s)$, so we have

$$
x_{M-1}(s) = a(M)\, y(s-1).
$$

Next, we calculate $x_{M-2}(s)$ to know the relation between $y(s-1)$ and $(M-2)$-step ahead predictor, then we obtain

$$
\begin{aligned}
x_{M-2}(s) &= \sum_{i=1}^{2} a(M-2+i)\, y(s-i) \\
&= a(M-1)\, y(s-1) + a(M)\, y(s-2) \\
&= a(M-1)\, y(s-1) + x_{M-1}(s-1).
\end{aligned}
$$

That is, we can obtain it by saving the result $x_{M-1}(s)$ in the previous stage and add $a(M-1) y(s-1)$ to it. Repeat this until $x_1(s)$, it can be shown that

$$
\begin{aligned}
x_l(s) &= \sum_{i=2}^{M-l} a(l+i) y(s-i) + a(l+1) y(s-1) \\
&= \sum_{i=1}^{M-l-1} a(l+1+i) y(s-i-1) + a(l+1) y(s-1) \\
&= x_{l+1}(s-1) + a(l+1) y(s-1), \quad l = M-2, \cdots, 1.
\end{aligned}
$$

Finally, at the moment when the observation $y(s)$ is obtained, i.e., $l = 0$, it holds that

$$
y(s) = x_0(s) = \sum_{i=1}^{M} a(i) y(s-i) + v(s) = x_1(s-1) + a(1) y(s-1) + v(s).
$$

This value represents a prediction equation for the current value based on data up to time $s-1$, and shows that an estimate of $y(s)$ can be obtained by adding $a(1) y(s-1)$ to the stored state variable $x_1(s-1)$, using only the last observed value $y(s-1)$. Thus, summarizing these relationships, we obtain

$$
\begin{cases}
x_0(s) = x_1(s-1) + a(1) x_0(s-1) + v(s) = y(s) \\
x_l(s) = x_{l+1}(s-1) + a(l+1) x_0(s-1), \quad l = 1, \cdots, M-2 \\
x_{M-1}(s) = a(M) x_0(s-1).
\end{cases} \tag{10.7}
$$

To write this relation in matrix form, if we define an M-dimensional state vector $x(s) = (x_0(s), x_1(s), \cdots, x_{M-1}(s))^t$, we obtain the state-space representation in matrix form

$$
x(s) = F x(s-1) + G v(s), \tag{10.8}
$$

where

$$
F = \begin{pmatrix}
a(1) & 1 & 0 & \cdots & 0 \\
a(2) & 0 & 1 & \cdots & 0 \\
\vdots & \vdots & \vdots & \ddots & \vdots \\
a(M-1) & 0 & 0 & \cdots & 1 \\
a(M) & 0 & 0 & \cdots & 0
\end{pmatrix}, \quad
G = \begin{pmatrix}
1 \\
0 \\
\vdots \\
0 \\
0
\end{pmatrix}. \tag{10.9}
$$

In this case, the first term of the state vector is equal to the observation $y(s)$, so the observation equation can be written as

$$
y(s) = H x(s), \tag{10.10}
$$

where

$$
H = (1 \ 0 \cdots 0). \tag{10.11}
$$

Here, note that $Q = \sigma^2$ and $R = 0$, i.e., in the state-space representation of the autoregressive model, the observation noise is zero (called a perfect information system). A state-space representation of this form is called an **observable state-space model**.

10.2.2   State Estimation by Kalman Filter

When the time series model is represented by a state-space model as described so far, there is a method for sequentially obtaining the mean and variance-covariance of the state $x(s)$[39]. Now let us denote them as

$$
\begin{cases}
x(s \mid j) \equiv E(x(s) \mid Y_j) \\
\Sigma(s \mid j) \equiv E\left(x(s) - x(s \mid j)\right)\left(x(s) - x(s \mid j)\right)^t,
\end{cases} \tag{10.12}
$$

where the notation $(a \mid b)$ indicates the event that a occurs under b. Thus $x(s \mid j)$ is the mean value of $x(s)$ at time j, given the information of $y(s)$ obtained up to time j. The Kalman filter is divided into a one-step-ahead prediction (time update) and a filter part (measurement update), which is expressed in the following subsection (see Chapter 18 for details).

10.2.3 Kalman Filter Algorithm

[**One-step-ahead prediction (time update)**]

$$\begin{cases} x(s \mid s-1) = Fx(s-1 \mid s-1) \\ \Sigma(s \mid s-1) = F\,\Sigma(s-1 \mid s-1)\,F^t + GQG^t \end{cases} \tag{10.13}$$

[**Filter (measurement update)**]

$$\begin{cases} K(s) \quad = \Sigma(s \mid s-1)\,H^t\left\{H\,\Sigma(s \mid s-1)\,H^t + R\right\}^{-1} \\ x(s \mid s) = x(s \mid s-1) + K(s)\left\{y(s) - Hx(s \mid s-1)\right\} \\ \Sigma(s \mid s) = \left\{I - K(s)\,H\right\}\Sigma(s \mid s-1) \end{cases} \tag{10.14}$$

To perform this algorithm online, start with appropriately chosen initial values $x(0 \mid 0)$, $\Sigma(0 \mid 0)$, calculate the predictions $x(1 \mid 0)$, $\Sigma(1 \mid 0)$, $K(1)$. Then, when the actual observation $y(1)$ is obtained, the filter values $x(1 \mid 1)$ and $\Sigma(1 \mid 1)$ are calculated. Based on them, we can proceed to the next stage by computing the predicted values $x(2 \mid 1)$, $\Sigma(2 \mid 1)$ and $K(2)$ and so on in succession.

10.2.4 Smoothing Algorithm

Given N time series, $Y_N = \{y(1), y(2), \cdots, y(N)\}$, the problem of **smoothing** is to estimate $x(s)$ of the state at time s where $s \leq N$. Smoothing provides more reliable estimates of past data than filter. There are three types of smoothing problems[39].

1. **fixed-point smoothing**
2. **fixed-lag smoothing**
3. **fixed-interval smoothing**

First, the fixed-point smoothing problem uses the observations $y(1), y(2), \cdots, y(k)$ to sequentially estimate the state $x(s)$ at a fixed point s. For example, it is used for the problem of estimating the initial value $x(0)$ of the state variable of the system when the observation is obtained subsequentially.

The fixed-lag smoothing is the problem of sequentially estimating the value at t using the observations up to $t + L$ for some fixed L. When $t = 0, 1, \cdots$, we always perform smoothing of signals that are delayed by L. In contrast, fixed-interval smoothing is the problem of sequentially estimating $N - 1, N - 2, \cdots, 0$ in the backward direction when N observations are obtained. Here we consider the algorithm for fixed-interval smoothing[39].

The algorithm for fixed-interval smoothing is as follows where $s = N - 1, \ldots, 1$.

$$\begin{cases} A(s) \quad = \Sigma(s \mid s)F^t\,\Sigma^{-1}(s+1 \mid s) \\ x(s \mid N) = x(s \mid s) + A(s)\,(x(s+1 \mid N) - x(s+1 \mid s)) \\ \Sigma(s \mid N) = \Sigma(s \mid s) + A(s)\,(\Sigma(s+1 \mid N) - \Sigma(s+1 \mid s))\,A^t(s). \end{cases} \tag{10.15}$$

Smoothing can estimate the past states precisely and is applied in analyzes in diverse fields, such as estimating the arrival time of P- and S-waves in seismic phenomena, estimating the trajectory and initial values of spacecraft.

10.2.5   Long-Term Prediction of a Time Series

In the prediction step, the Kalman filter gives only the one-step-ahead prediction. To make a longer-term prediction further ahead, simply repeat the prediction part of the Kalman filter. Let s be the current time and the observations up to time s, $Y_s = \{y(1), y(2), \cdots, y(s)\}$ are obtained, and consider making j-step ahead long-term prediction of $y(s+j)$ for $s+j$, $(j > 1)$.

Since no new data will be available in the long-term prediction, the innovation term in the Kalman filter, $y(s) - Hx(s)$, will not be updated by the new observed data. Therefore, this term is set to zero, and the predictive distribution at time $s+1$ obtained by the Kalman filter, i.e., the time update part, is identical to the filter distribution at time $s+1$, and we have $x(s+1 \mid s+1) = x(s+1 \mid s)$, $\Sigma(s+1 \mid s+1) = \Sigma(s+1 \mid s)$. That is, by omitting the filter step and repeating only the prediction step j times, we can obtain j-step ahead prediction and its variance-covariance matrix of state $x(s)$. Namely, we have

[**Long-term prediction of state** $x(s+j)$]

$$\begin{cases} x(s+j \mid s) = Fx(s+j-1 \mid s) \\ \Sigma(s+j \mid s) = F\,\Sigma(s+j-1 \mid s)\,F^t + GQG^t. \end{cases} \tag{10.16}$$

Now that we know the prediction of the state $x(s)$, the j-step ahead prediction of the time series $y(s)$ and its variance-covariance matrix, namely $y(s+j \mid s) \equiv E\,[y(s+j \mid Y_s)]$, $d(s+j \mid s) \equiv \mathrm{Cov}\,[y(s+j \mid Y_s)]$ are obtained as follows;

[**Long-term prediction of time series** $y(s+j)$]

$$\begin{cases} y(s+j \mid s) = E\,[Hx(s+j) + w(s+j) \mid Y_s] = Hx(s+j \mid s) \\ d(s+j \mid s) = \mathrm{Cov}\,[Hx(s+j) + w(s+j) \mid Y_s] \\ \qquad\quad = H\Sigma(s+j \mid s)\,H^t + R. \end{cases} \tag{10.17}$$

To illustrate the calculation procedure in detail, suppose that we are currently at time s and that the filtered estimate and its variance-covariance matrix of the state is given by the Kalman filter as $x(s \mid s)$ and $\Sigma(s \mid s)$. Using the result, compute the one-step ahead prediction and the variance-covariance matrix of the time series at time $s+1$ using equation (10.17). Next, the two-step ahead prediction and the variance-covariance matrix of the time series at time $s+2$ are obtained using equation (10.16) and so on.

10.3   Use of Kalman Filter in Likelihood Computation

The autoregressive model is determined if the parameters

$$\theta = \left(a(1), a(2), \cdots, a(M), \sigma_M^2\right)$$

are estimated. Such an estimation problem is called **point estimation** in statistics (see Chapter 17). To obtain the statistical model by point estimation, the maximum likelihood method, which has already been described, is used as a powerful method. In the maximum likelihood method, the likelihood function is calculated from the data obtained by sampling from the population, the parameters of the statistical model that maximize the likelihood are obtained and the parameters are considered to be their most probable values. The Kalman filter described so far facilitates the computation of the likelihood and makes modeling using the maximum likelihood method easier. In this section, we discuss application of the Kalman filter to the likelihood computation[45][60].

Suppose now that a population with a statistical structure is represented by a statistical model with the parameter θ. Suppose further that $y(1)$ is obtained as a result of independent sampling from a population with such a distribution, and denote the probability as $f(Y(1) \mid \theta)$. Next, suppose that such sampling was performed N times, resulting in the observations $Y_N = \{y(1), y(2), \cdots, y(N)\}$. The N-dimensional simultaneous density function of the time series Y_N with N observations determined by this model is denoted by $f_N(Y_N \mid \theta)$. This quantity is by definition the likelihood and is denoted by $L(\theta)$. Here assume that we divide the time series into the time series up to time $N-1$ and the values at the latest time N. Then the simultaneous probability $f_N(Y_N \mid \theta)$ is equal to the product of the conditional probability that Y_{N-1} is generated from the statistical model with the parameter θ and $g_N(y(N) \mid Y_{N-1}, \theta)$ which is the conditional distribution of $Y(N)$ under the condition that Y_{N-1} is generated from the same statistical model. In other words, we have

$$f_N(Y_N \mid \theta) = f_{N-1}(Y_{N-1} \mid \theta) \, g_N(y(N) \mid Y_{N-1}, \theta). \tag{10.18}$$

Repeatedly applying this decomposition, $L(\theta)$ can be finaly expressed as

$$L(\theta) = \prod_{s=1}^{N} g_s(y(s) \mid Y_{s-1}, \theta). \tag{10.19}$$

Here, the initial state Y_0 is assumed to be an empty set and $f_1(y(1) \mid \theta) \equiv g_1(y(1) \mid Y_0, \theta)$. In the maximum likelihood method, however, it is more convenient to use the log-likelihood defined as

$$l(\theta) = \log L(\theta) = \sum_{s=1}^{N} \log g_s(y(s) \mid Y_{s-1}, \theta). \tag{10.20}$$

Here, if Y_N is a Gaussian process, $g_s(y(s) \mid Y_{s-1}, \theta)$ is the predictive distribution of $y(s)$ given the observations Y_{s-1} up to time $s-1$, that is, one-dimentional Gaussian distribution with the mean $y(s \mid s-1)$ and the variance-covariance matrix $d(s \mid s-1)$ (see Section 17.6.2),

$$\begin{aligned}
g_s(y(s) \mid Y_{s-1}, \theta) = {} & \left(\frac{1}{\sqrt{2\pi}}\right) |d(s \mid s-1)|^{-\frac{1}{2}} \\
& \times \exp\left\{-\frac{1}{2}(y(s) - y(s \mid s-1))^t d^{-1}(s \mid s-1)(y(s) - y(s \mid s-1))\right\}.
\end{aligned}$$

Substituting this result into equation (10.20), the log-likelihood of the model is obtained by

$$\begin{aligned}
l(\theta) = {} & -\frac{1}{2}\left\{N \log 2\pi + \sum_{s=1}^{N} \log |d(s \mid s-1)| \right.\\
& \left. + \sum_{s=1}^{N}(y(s) - y(s \mid s-1))^t d^{-1}(s \mid s-1)(y(s) - y(s \mid s-1))\right\}. \tag{10.21}
\end{aligned}$$

The maximum likelihood method obtains $\hat{\theta}$ that maximizes the likelihood function (10.19) or (10.20) among the possible value of the parameter θ. It is clear that the predictions by the Kalman filter can be used to calculate $y(s \mid s-1)$ and $d(s \mid s-1)$ in this maximum likelihood method. This method was first used by Mehra to estimate the maneuverability parameters of airplanes[60]. It was further applied to the identification of ship maneuvering models by Åström, Kallström et al.[13]. Ohtsu and Kitagawa[83] discretized the vibration equation in a continuous domain such as roll motion, and then used it as a computational tool to estimate parameters by the maximum likelihood method. Since the log-likelihood optimization in the maximum likelihood method is generally nonlinear, it is necessary to use a numerical method such as nonlinear programming. This method is described in Appendix A of Chapter 11.

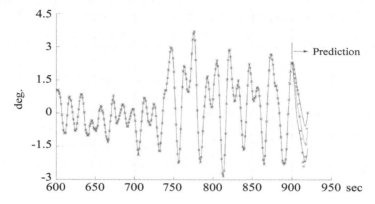

FIGURE 10.2
Prediction of the rolling motion of a container ship

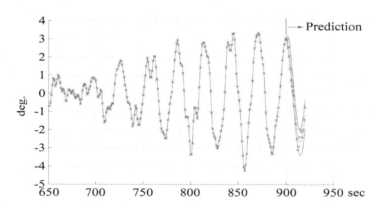

FIGURE 10.3
Prediction of the pitching motion of a container ship

10.4   Applications

10.4.1   Prediction of Stationary Time Series

Figures 10.2 and 10.3 show the results of the AR model fitting using 900 seconds of rolling and pitching data ($\Delta t = 1$ sec) of a container ship, followed by 20 seconds of prediction using the Kalman filter. The center line is the prediction (mean of the predictive distribution), the solid lines above and below around it are the $\pm\sigma$ confidence interval and the $\times$ marks indicate the actual observed values.

Both the rolling and pitching show good prediction accuracy. In particular, the rolling shows very good prediction accuracy and it is known that it maintains good accuracy even in long-term prediction. To further improve the prediction accuracy, we can use the Kalman filter based on the state-space model obtained from the multivariate autoregressive model.

10.4.2 Estimation of Missing Values

Figure 10.4 shows the application of the fixed-interval smoothing method to the estimation of missing values in the ship rolling data. The solid line in the center shows the estimated value, the standard deviation is above and below the estimated value and ∘ shows the measured values.

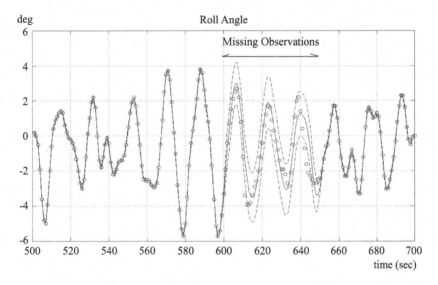

FIGURE 10.4
Interpolation of 50 seconds of missing data (large container ship)

11

Estimation of Trend and Seasonal Adjustment Models and Time-Varying Spectrum

Trend analysis is essential for system monitoring. Here, we describe a statistical model for nonstationary time series with trends and its identification method[45][46][47]. Of course, it is impossible to accurately capture time series showing complex trends only by the methods described here, but one can examine various combinations of the models described here, evaluate the goodness of the models by AIC and select a good model among various models. The estimation is performed by using a state-space representation as a computational method.

11.1 Trend Model

11.1.1 Estimation of Polynomial Trend Model

The most commonly considered model for representing the trend of a nonstationary time series is when the time series can be assumed to have a trend represented by a certain function, such as a trigonometric function or a polynomial. In this section, let us consider the case where the AIC is used as a criterion to measure the quality of the model and the polynomial is considered as a model to represent the trend.

In the continuous time domain, a polynomial time series model with time t is defined by

$$y(t) = a(0) + a(1)\, t + a(2)\, t^2 + \cdots + a(m)\, t^m.$$

This model can be considered as a regression model with the power of time as the explanatory variable.

For discrete data sampled every Δt, the discrete polynomial model is given at time $s\Delta t$ as follows

$$\begin{cases} t(s) = a(0) + a(1)\, x(s) + a(2)\, x(s)^2 + \cdots + a(m)\, x(s)^m \\ y(s) = t(s) + w(s), \end{cases} \tag{11.1}$$

where $x(s)$ is the value of time series at time $s\Delta t$. The $w(s)$ is the Gaussian white noise at time $s\Delta t$, with mean 0 and variance σ^2.

Since this is a regression model in which the objective variable $y(s)$ at time $s\Delta t$ ($s = 1, 2, \cdots, N$) is explained by the explanatory variable $x(s)^j$ ($= x(s\Delta t)^j$, $j = 0, 1, 2, \cdots, m$), the estimation of its parameter $a(j)$ is performed by applying the least squares method. Therefore, the Householder transformation method used in Chapter 9 can also be applied for estimation here. The procedure for estimating the polynomial regression model is shown below.

DOI: 10.1201/9781003428565-11

Suppose we are given observations of the objective variable $y(s)$ and explanatory variable $x(s)$ at time $s\Delta t$ ($s = 1, 2, \cdots, N$). Using those values, we first make

$$X = \begin{pmatrix} 1 & x(1) & \cdots & x(1)^m & y(1) \\ 1 & x(2) & \cdots & x(2)^m & y(2) \\ \vdots & \vdots & \ddots & \vdots & \vdots \\ 1 & x(N) & \cdots & x(N)^m & y(N) \end{pmatrix}.$$

Then, using the Householder transformation U (see equation (9.5)), we reduce X to an upper triangular $(m+2) \times (m+2)$ matrix S as follows

$$UX = \begin{pmatrix} S \\ 0 \end{pmatrix} = \begin{pmatrix} s_{11} & s_{12} & \cdots & s_{1, m+2} \\ 0 & s_{22} & \cdots & s_{2, m+2} \\ \vdots & \vdots & \ddots & \vdots \\ 0 & 0 & \cdots & s_{m+2, m+2} \\ 0 & 0 & \cdots & 0 \end{pmatrix}.$$

In this case, the estimate of the variance of the residuals of the j-th order ($j \le m$) polynomial trend model is obtained by using the upper $j + 2$ rows of the last column as follows

$$\hat{\sigma}_j^2 = \frac{1}{N} \sum_{i=j+2}^{m+2} s_{i, m+2}^2. \tag{11.2}$$

Also, since the number of parameters is $j + 2$, i.e., the $j + 1$ regression coefficients and one residual, the AIC for the j-th order polynomial regression model is given by

$$\text{AIC}(j) = N \log \hat{\sigma}_j^2 + N + 2(j + 2). \tag{11.3}$$

The regression coefficients can be obtained by solving the following equations

$$\begin{pmatrix} s_{11} & s_{12} & \cdots & s_{1, j+1} \\ 0 & s_{22} & \cdots & s_{2, j+1} \\ \vdots & \vdots & \ddots & \vdots \\ 0 & 0 & \cdots & s_{j+1, j+1} \end{pmatrix} \begin{pmatrix} a(0) \\ a(1) \\ \vdots \\ a(j) \end{pmatrix} = \begin{pmatrix} s_{1, m+2} \\ s_{2, m+2} \\ \vdots \\ s_{j+1, m+2} \end{pmatrix}. \tag{11.4}$$

Note that, since the matrix is a triangular form, this equation can be solved easily in order from the lowest coefficient, $a(j)$, to the highest one, $a(0)$.

The optimal order of the polynomial is determined by minimizing the AIC that evaluates the goodness of the model.

11.1.2   Trend Component Model

Next, consider the case where the trend line of the model representing the long-term variation in the structure of the time series is represented by a first order polynomial

$$T(t) = at + b, \tag{11.5}$$

where t is continuous time. The first and second derivatives of this model with respect to time are respectively given by

$$\begin{aligned} T'(t) &= a \\ T''(t) &= 0. \end{aligned}$$

Now let us consider this property using a difference equation model for a discrete time $s\Delta t$ sampled at time interval Δt

$$T(s) = as + b.$$

The difference between values at adjacent times is defined by $\Delta T(s) = T(s) - T(s-1) = (1 - B)\,T(s)$, where B denotes the **Box operator** and is equivalent to z^{-1} in the z transformation. Since the differential operation is a concept concerning the change of a function value in small time, in discrete time, we consider the difference operator for the function

$$T(s) = as + b.$$

In this case, the first-order difference is given by

$$\Delta T(s) = T(s) - T(s-1) = a(s+1) + b - (a(s) + b) = a.$$

Similarly, the second-order difference is obtained by

$$
\begin{aligned}
\Delta^2 T(s) &= \Delta(\Delta T(s)) = \Delta(T(s) - T(s-1)) \\
&= \Delta T(s) - \Delta(T(s-1)) \\
&= a - a = 0.
\end{aligned}
$$

The initial condition is

$$T(0) = b.$$

This indicates that if the time series is a first-order polynomial, it can be represented by the solution of the initial value problem for the second-order difference equation.

In general, in the continuous domain, the $(k\text{–}1)$-st order polynomial model is the solution of the k-th order difference equation model,

$$\Delta^k T(s) = 0,$$

where Δ^k is defined by

$$\Delta^k = (1 - B)^k = \sum_{i=1}^{k} c(i)\,(-B)^i, \quad c(i) = (-1)^{i+1}\frac{k!}{i!\,(k-i)!}. \tag{11.6}$$

Among the models that represent long-term trends within such time series, **trend component model** is considered that the k-th differences of the observed time series are on average zero, but not exactly zero, and have some variation, so that the following stochastic difference equation holds,

$$\Delta^k T(s) = v_T(s),$$

where $v_T(s)$ is a Gaussian white noise with mean 0 and variance τ_T^2. Using equation (11.6), we obtain

$$T(s) = \sum_{i=1}^{k} c(i)\,T(s-i) + v_T(s). \tag{11.7}$$

Equation (11.7) has formally the same form as the AR model, so if we put

$$
x(s) = \begin{pmatrix} T(s) \\ T(s-1) \\ \vdots \\ T(s-k+1) \end{pmatrix}, \quad
F_T = \begin{pmatrix} c(1) & c(2) & \cdots & c(k) \\ 1 & & & \\ & & \ddots & \\ 0 & & 1 & 0 \end{pmatrix}, \quad
G_T = \begin{pmatrix} 1 \\ 0 \\ \vdots \\ 0 \end{pmatrix}
$$

$$H_T = \begin{pmatrix} 1 & 0 & \cdots & 0 \end{pmatrix},$$

we obtain the state-space representation of trend component models

$$\begin{cases} x(s) = F_T\,x(s-1) + G_T\,v(s) \\ T(s) = H_T\,x(s). \end{cases} \tag{11.8}$$

11.1.3   Estimation of the Trend Model

The trend component model with $(k-1)$-th order polynomial as the trend line was a model in which a noise term was added to the k-th order difference of trend. The trend component model is considered as an internal model, to which a variable component is added to the trend part as a model of the observed values, as shown in the following model

$$y(s) = T(s) + w_T(s), \tag{11.9}$$

where $w_T(s)$ is a Gaussian white noise with mean 0 and variance σ_T^2. In other words, the trend model is defined as

$$\begin{cases} \Delta^k T(s) = v_T(s) \\ y(s) = T(s) + w_T(s), \end{cases} \tag{11.10}$$

where $v_T(s)$ is Gaussian white noise with mean 0 and variance τ_T^2 and $w_T(s)$ is Gaussian white noise with mean 0 and variance σ_T^2. This model can be expressed in terms of a state-space model as

$$\begin{cases} x(s) = F_T\, x(s-1) + G_T\, v_T(s) \\ y(s) = H_T\, x(s) + w_T(s). \end{cases} \tag{11.11}$$

The trend model has two unknown parameters τ_T^2 and σ_T^2. These parameters are obtained numerically by nonlinear programming using the maximum likelihood method with the Kalman filter as the computational tool, as described in Chapter 10. In the case of univariate time series, the Kalman filter gives the same result if $\lambda = \tau_T^2/\sigma_T^2$ is the same, so the number of parameters to search can actually be reduced by one[45]. The λ is called the trade-off parameter.

11.2   Seasonal Adjustment Model

Consider the change in daily mean temperature. The temperature fluctuates slowly in spring, summer, fall and winter due to the movement of the sun, and returns to the daily mean temperature of the same day in the previous year around spring of the following year. Similarly, the number of vessels entering a large port changes periodically, with one week as the cycle[82]. A nonstationary time series with such a cyclically changing trend is called a **seasonal time series**. That is, seasonal time series have a component in the time series that returns to its original value every period p.

11.2.1   Seasonal Component Model

The component of variation in a time series that occurs repeatedly in a periodic cycle is called the seasonal component. When annual data for monthly aggregate data are obtained over multiple years, 12 months is one cycle. In a model with such a seasonal component, the requirement for the corresponding state variable component is

$$S(s) = S(s - p), \tag{11.12}$$

where p is the length of one period, and $p = 12$ for monthly data with annual seasonality as shown above.

In such a seasonal component, let us examine the relationship requested for the values of a continuing p-time series. Now, for example, when $p = 3$, there is always a relation

$$S(s) + S(s - 1) + S(s - 2) = 0. \tag{11.13}$$

Since it must hold even if s advances to $s + 1$, then it holds that

$$S(s + 1) + S(s) + S(s - 1) = 0. \tag{11.14}$$

From equation (11.13), we have

$$S(s) + S(s - 1) = -S(s - 2).$$

Substituting this relationship into equation (11.14), we obtain

$$S(s + 1) - S(s - 2) = 0.$$

In other words, we have

$$S(s + 1) = S(s - 2).$$

Thus, in general, for equation (11.12) to hold, the following relation should hold

$$
\begin{aligned}
S(s) + S(s - 1) &+ \cdots + S(s - p + 1) \\
&= (1 + B + B^2 + \cdots + B^{p-1})\, S(s) = 0,
\end{aligned} \tag{11.15}
$$

i.e.,

$$\sum_{i=0}^{p-1} B^i\, \dot{S}(s) = 0. \tag{11.16}$$

Furthermore, if we want to assume that this condition changes slowly, just as equation (11.6) should have held in the case of the trend model, we can also assume that the estimates do not change much at intervals of the assumed period and we can assume that[45],

$$\left(\sum_{i=0}^{p-1} B^i \right)^l S(s) = 0, \tag{11.17}$$

where l is a positive integer representing smoothness, usually we put $l = 1$ or 2. The coefficients in model (11.17) are

$$\left(\sum_{i=0}^{p-1} B^i \right)^l = 1 - \sum_{i=1}^{l(p-1)} d(i)\, B^i.$$

Then, for $l = 1$, we have

$$d(i) = -1.$$

In the case of $l = 2$, we have

$$
\begin{cases}
d(i) = -i - 1, & (i \le p - 1) \\
d(i) = i + 1 - 2p, & (p \le i < 2(p - 1)).
\end{cases}
$$

The seasonal component model is a stochastic model that assumes that the above relationship holds approximately expressed as

$$\left(\sum_{i=0}^{p-1} B^i \right)^l S(s) = v_S(s), \quad v_S(s) \sim N(0, \tau_S^2). \tag{11.18}$$

Usually, the model with $l = 1$ is used

$$\sum_{i=0}^{p-1} B^i\, S(s) = v_S(s). \qquad (11.19)$$

Formally, the state-space model of this model is obtained by putting the state variable and each coefficient as

$$x(s) \;=\; \begin{pmatrix} S(s) \\ S(s-1) \\ \vdots \\ S(s-l(p-1)+1) \end{pmatrix}, \quad F_S = \begin{pmatrix} d(1) & d(2) & \cdots & d(l(p-1)) \\ 1 & 0 & \cdots & 0 \\ \vdots & \ddots & \ddots & \vdots \\ 0 & \cdots & 1 & 0 \end{pmatrix}, \quad G_S = \begin{pmatrix} 1 \\ 0 \\ \vdots \\ 0 \end{pmatrix}$$

$$H_S \;=\; \begin{pmatrix} 1 & 0 & \cdots & 0 \end{pmatrix},$$

we obtain

$$\begin{cases} x(s) = F_S\, x(s-1) + G_S\, v_S(s) \\ y(s) = H_S\, x(s) + w_S(s), \end{cases}$$

where $w_S(s)$ is $w_S(s) \sim N(0, \sigma_S^2)$.

11.2.2 Standard Seasonal Adjustment Model

Combining this seasonal component model with the trend model described earlier, we obtain **standard seasonal adjustment model**,

$$\begin{aligned} y(s) \;&=\; T(s) + S(s) + w_{TS}(s) \\ \Delta^k\, T(s) &= v_T(s) \\ \left(\sum_{i=0}^{p-1} B^i\right)^l S(s) &= v_S(s). \end{aligned}$$

In this case, define the state variable as

$$x(s) = (T(s), T(s-1), \cdots, T(s-k+1), S(s), S(s-1), \cdots, S(n-p+2))^t, \qquad (11.20)$$

and the transition matrix, the driving matrix, and the observation matrix as

$$F_{TS} = \begin{pmatrix} F_T & 0 \\ 0 & F_S \end{pmatrix}, \quad G_{TS} = \begin{pmatrix} G_T & 0 \\ 0 & G_S \end{pmatrix}, \quad H_{TS} = \begin{pmatrix} H_T & H_S \end{pmatrix},$$

we obtain the state-space model for standard seasonal adjustment model

$$\begin{cases} x(s) = F_{TS}\, x(s-1) + G_{TS}\, v_{TS}(s) \\ y(s) = H_{TS}\, x(s) + w_{TS}(s), \end{cases} \qquad (11.21)$$

where $v_{TS}(s) = (v_T(s), v_S(s))^t$.

11.2.3 Seasonal Adjustment Model with Autoregressive Component

In addition to the large trend components described above, there are also local variations such as stationary component in the time series of ship motions, especially in the time series of rolling. If we assume that such stationary time series components can be approximated by an autoregressive model, we can create a time series **decomposition model** with a wide range of applications. In the decomposition model, we assume that the state variables are divided into **trend component** $(T(s))$, **seasonal component** $(S(s))$ and **stationary autoregressive component** $(P(s))$ as

$$y(s) = T(s) + S(s) + P(s) + w_{TSP}(s), \qquad (11.22)$$

where the stationary autoregressive component is represented by the autoregressive model

$$P(s) = \sum_{m=1}^{M_P} a(m)\, P(s-m) + v_P(s),$$

with $v_P(s) \sim N(0, \tau_P^2)$. Here, define the transition matrix etc., of this autoregressive component,

$$F_P = \begin{pmatrix} a(1) & a(2) & \cdots & a(M_P) \\ 1 & 0 & \cdots & 0 \\ \vdots & \ddots & \ddots & \vdots \\ 0 & \cdots & 1 & 0 \end{pmatrix}$$

$$G_P = \begin{pmatrix} 1 \\ 0 \\ \vdots \\ 0 \end{pmatrix}, \quad H_P = \begin{pmatrix} 1 & 0 & \cdots & 0 \end{pmatrix}, \qquad (11.23)$$

the state-space representation of the seasonal adjustment model with trend, seasonal and stationary autoregressive component is obtained by

$$\begin{cases} x(s) = F_{TSP}\, x(s-1) + G_{TSP}\, v_{TSP}(s) \\ y(s) = H_{TSP}\, x(s) + w_{TSP}(s), \end{cases} \qquad (11.24)$$

where $v_{TSP}(s) = [v_T(s), v_S(s), v_P(s)]^t$ and

$$F_{TSP} = \begin{pmatrix} F_T & & \\ & F_S & \\ & & F_P \end{pmatrix}, \quad G_{TSP} = \begin{pmatrix} G_T & & \\ & G_S & \\ & & G_P \end{pmatrix}$$

$$H_{TSP} = \begin{pmatrix} H_T & H_S & H_P \end{pmatrix}, \quad Q_{TSP} = \begin{pmatrix} \tau_T^2 & & \\ & \tau_S^2 & \\ & & \tau_P^2 \end{pmatrix}. \qquad (11.25)$$

Note that, $x(s)$ is augmented at the end of the vector in equation (11.20) by M_P autoregressive components $P(s-i+1)$, $i = 1, 2, \cdots, M_P$. The estimates of the unknown parameters of these models is obtained numerically using nonlinear programming.

11.3   Example of Time Series Decomposition

11.3.1   Decomposition of a Rolling Time Series into Trend + Autoregressive Component + Noise

The bottom plot in Figure 11.1 shows a record of the rolling of a container ship. When the ship is subjected to disturbances such as waves and wind from the stern or quarter seas

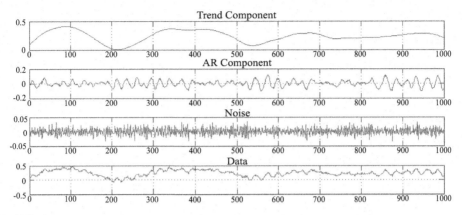

FIGURE 11.1
Decomposition of rolling data with long-period trend component

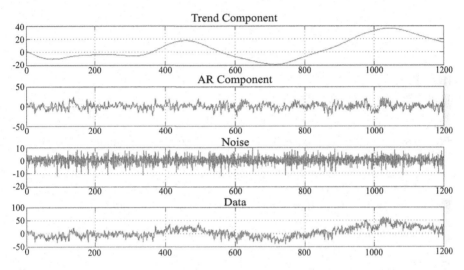

FIGURE 11.2
Decomposition of thrust data with long-period trend component

direction, the roll motion is such that the inherent short-period roll motion is added to the long-period waves. The plots in the upper three rows of Figure 11.1 show the time series additively decomposed into a long-period trend component, an autoregressive component, and a residual noise process obtained by the seasonal adjustment model and state-space representation (11.25) described in this chapter. From the top, the trend component, the autoregressive component and the noise component are shown. The order of the trend model is set to second order by model comparison by AIC. The noise component is almost white noise, indicating that the model adequately represents long-period rolling motion.

Figure 11.2 shows an example of a decomposition of the time series of the variation of the thrust generated by a main engine into trend, autoregressive and noise components.

11.3.2 Seasonal Adjustment of the Zonal Index

For mid- to long-term weather and oceanographic prediction at mid-latitudes, such as when ships navigate North American routes, an index called the **zonal index (ZI)** is used. The

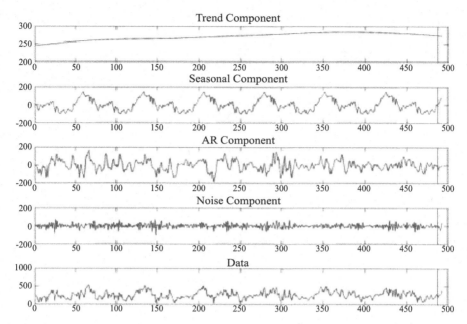

FIGURE 11.3
Decomposition of zonal index data

zonal index is the mean difference in altitude of the 500hPa isobars between 40°N and 60°N. When this index is large, westerly winds are strong, the flow is east-west and the temperature difference between north and south is large. Conversely, when it is small, the temperature difference between the equator and the North Pole becomes small, causing north-south mixing of the airflow and it is said that meandering of westerly winds will occur.

The bottom-most time series in Figure 11.3 shows 488 time series (488×5 days) of 5-day averages of daily zonal indices obtained from the data for the period from January 1, 2003 to 2009. From the top, four plots in the upper part of the figure show decomposition of the time series into the trend component (T), seasonal component (S), stationary AR component (AR) and noise component (N) by the seasonal adjustment model shown in the previous section. The trend shows an upward trend for the first five years and then a downward trend. The seasonal component is high from fall to winter and low from spring to summer. The last 488×5 days (vertical line) and thereafter show the predicted results.

11.4   Time-Varying Spectrum

11.4.1   Time-Varying Coefficient Autoregressive Model

The discussion so far has shown that the regression coefficients of the autoregressive model have been treated as not changing with time in the interval. However, when a ship navigates in large waves, for example, the hull may be located in a trough or on a ridge of the wave. In such cases, the height of the ship's center of gravity changes from moment to moment. The statistical model called **time-varying AR model** is a linear model that represents such a time-varying object.

The time-varying autoregressive model is given by

$$y(s) = \sum_{j=1}^{m} a(s,j)\, y(s-j) + w(s), \tag{11.26}$$

where $a(s,j)$ is the regression coefficient of $y(s-j)$ at time s on the value j times prior to it, called the **time varying autoregressive coefficient**. The $w(s)$ is a Gaussian white noise with mean 0 and variance σ^2. Kitagawa showed how to estimate the coefficients $a(s,j)$ and σ^2 by applying the maximum likelihood method to a signal process passing through a Kalman filter[47][44][48][49]. In this method, given N observations $Y_N = \{(y(1), y(2), \cdots, y(N)\}$, instead of using this data to estimate $a(s,j)$ by applying the least squares method, we consider the following trend model assuming $a(s,j)$ is slowly changing

$$\Delta^k a(s,j) = v(s,j), \quad j = 1, \cdots, m. \tag{11.27}$$

Here Δ is the difference operator defined by

$$\Delta a(s,j) = a(s,j) - a(s-1,j), \tag{11.28}$$

As the difference order, we consider $k = 1$ or $k = 2$. Let $v(s,j), j = 1, \ldots, m$ be m Gaussian white noises at time s, and an m-dimensional white noise vector consisting of these elements can be expressed as $v_s \equiv \big(v(s,1), v(s,2), \cdots, v(s,m)\big)^t$. The variance-covariance matrix Q of this vector is assumed to be the diagonal matrix $Q = \mathrm{diag}\,\big(\tau_{11}^2, \tau_{22}^2, \cdots, \tau_{mm}^2\big)$, with zero correlation between noises for $i \neq j$. Suppose further that all τ_{ii}^2 have equal value τ^2. In this case, let the coefficient $a(s,j)$ be the state variable $x(s,j)$ and by putting

$$\begin{cases} x(s,j) = a(s,j), & k = 1 \\ (x(s,2j-1), x(s,2j)) = (a(s,j), a(s-1,j))^t, & k = 2, \end{cases}$$

we obtain the state-space representation

$$\begin{cases} x(s,j) = F^{(1)}\, x(s-1,j) + G^{(1)}\, v(s,j) & k = 1 \\ (x(s,2j-1), x(s,2j))^t = F^{(2)}\, (x(s-1,2j-1), x(s,2j))^t + G^{(2)}\, v(s,j) & k = 2, \end{cases} \tag{11.29}$$

Here the transition matrix is given by

$$\begin{cases} F^{(1)} = G^{(1)} = 1, & k = 1 \\[4pt] F^{(2)} = \begin{pmatrix} 2 & -1 \\ 1 & 0 \end{pmatrix}, \quad G^{(2)} = \begin{pmatrix} 1 \\ 0 \end{pmatrix}, & k = 2. \end{cases}$$

In equation (11.26), if we consider the j-th term in $\sum_{j=1}^{m} a(s,j)\, y(s-j)$ to be the j component of equation (11.29), for $k = 1$, we obtain

$$a(s,j)\, y(s-j) = y(s-j)\, a(s,j) = H^{(1j)}(s-j)\, x(s,j).$$

Also, for $k = 2$, the j-th component can be written as

$$\begin{aligned} a(s,j)\, y(s-j) &= (y(s-j), 0)\, (a(s,j), a(s-1,j))^t \\ &= H^{(2j)}(s-j)\, x(s,j), \end{aligned}$$

where $H^{(1j)}(s) = y(s-j)$ and $H^{(2j)}(s) = (y(s-j), 0)$. In addition, if we put

$$H^{(1)} = 1, \quad H^{(2)} = (1, 0),$$

we obtain the representation of the observation process as

$$
\begin{aligned}
H^{(1j)}(s) &= y(s-j)\,H^{(1)} \\
H^{(2j)}(s) &= y(s-j)\,H^{(2)}.
\end{aligned}
$$

Using these expressions and the assumption $\tau_{1,1}^2 = \tau_{2,2}^2 = \cdots = \tau_{m,m}^2 = \tau^2$, we end up using the state-space representation of the time-varying autoregressive model

$$
\begin{cases}
x(s) = Fx(s-1) + G(s)\,v(s) \\
y(s) = H(s)\,x(s),
\end{cases}
\tag{11.30}
$$

where F, G and $H(s)$ are defined as

$$
F = \operatorname{diag}_{km \times km}\left(F^{(k)}, \cdots, F^{(k)}\right)
$$

$$
G = \operatorname{diag}_{km \times m}\left(G^{(k)}, \cdots, G^{(k)}\right)
$$

$$
H(s) = \left(H^{(k1)}(s), H^{(k2)}(s), \cdots, H^{(km)}(s)\right),
$$

$$
= \left(y(s-1)H^{(k)}, \cdots, y(s-m)H^{(k)}\right),
$$

and the state variable $x(s)$ is given by

$$
x(s) = \begin{cases}
(a(s,1), \cdots, a(s,m))^t, & k = 1 \\
(a(s,1), a(s-1,1), \cdots, a(s,m), a(s-1,m))^t, & k = 2.
\end{cases}
$$

The properties of the system noise and the observation noise are expressed by their variance–covariance matrices,

$$
Q = \operatorname{diag}_{m \times m}\left(\tau^2, \cdots, \tau^2\right) \quad \text{and} \quad R = \sigma^2.
$$

For example, writing down this model for $k = m = 2$, the system equations is given by

$$
\begin{pmatrix}
a(s,1) \\
a(s-1,1) \\
a(s,2) \\
a(s-1,2)
\end{pmatrix}
=
\begin{pmatrix}
2 & -1 & 0 & 0 \\
1 & 0 & 0 & 0 \\
0 & 0 & 2 & -1 \\
0 & 0 & 1 & 0
\end{pmatrix}
\begin{pmatrix}
a(s-1,1) \\
a(s-2,1) \\
a(s-1,2) \\
a(s-2,2)
\end{pmatrix}
+
\begin{pmatrix}
1 & 0 \\
0 & 0 \\
0 & 1 \\
0 & 0
\end{pmatrix}
\begin{pmatrix}
v(s,1) \\
v(s,2)
\end{pmatrix},
$$

where,

$$
\begin{pmatrix}
v(s,1) \\
v(s,2)
\end{pmatrix}
\sim N\left(
\begin{bmatrix} 0 \\ 0 \end{bmatrix},
\begin{bmatrix} \tau^2 & 0 \\ 0 & \tau^2 \end{bmatrix}
\right),
$$

and the observation model for $y(s)$ is given by

$$
y(s) = (y(s-1), 0, y(s-2), 0)
\begin{pmatrix}
a(s,1) \\
a(s-1,1) \\
a(s,2) \\
a(s-1,2)
\end{pmatrix}
+ w(s),
$$

$$
w(s) \sim w(s)N(0, \sigma^2).
$$

Thus, with the state-space representation becoming available, the problem becomes one of numerically optimizing the likelihood of the model with respect to the parameters m, k, σ^2 and τ^2. The goodness of the model is evaluated by the AIC and the smallest one is chosen as the best model. Then, after obtaining $x(s \mid s)$ and $V(s \mid s)$ by the Kalman filter, the smoothed state $x(s \mid N)$, $s = 1, 2, \cdots, N$ provides the estimate of $a(s,j)$, $s = 1, 2, \cdots, N$.

11.4.2   Time-Varying Spectrum of Pitching Time Series

As described in Chapter 8, the period of pitching motion is affected by wave disturbance. Figure 11.4 shows a time series of pitching motion of a container ship with 500 observations ($\Delta t = 1\,\mathrm{sec}$). It can be seen that the waveform changes from around 350 seconds. Figure 11.5 shows the time-varying spectrum of the time series, in which the spectrum peak with a period of 0.05 Hz to 0.1 Hz splits into two peaks.

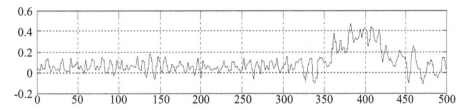

FIGURE 11.4
Time series of pitching motion of a container ship

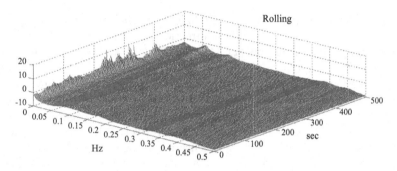

FIGURE 11.5
Time-varying spectrum of time series shown in Figure 11.4 (x-axis: period (Hz), y-axis: time (sec)).

Appendix: Numerical Optimization by Nonlinear Programming

In the identification of trend and seasonal adjustment models described in this chapter, use of numerical optimization methods is necessary when estimating the system noise variance τ^2 and the observation noise variance σ^2 by the maximum likelihood method in the framework of the Kalman filter for the state-space model described in Section 11.1. As the numerical optimization method, we can use an unconstrained nonlinear optimization method[45][29][11]. Here, we show the method called the **Davidon-Fletcher-Powell method** (or **DFP method**) without proofs.

Now consider finding the minima of a nonlinear function $f(x)$ of a k-dimensional vector $x = (x_1, \cdots, x_k)^t$ such as the likelihood function. Assume that $f(x)$ is differentiable up to the second order, and by Taylor expansion, for small d, we have an approximation

$$f(x + d) \simeq q(d) = f(x) + \nabla f(x)\, d + \frac{1}{2} d^t \nabla^2 f(x)\, d, \tag{11.31}$$

where

$$\nabla f(x) \;=\; g(x) = \left(\frac{\partial f(x)}{\partial x_1}, \frac{\partial f(x)}{\partial x_2}, \cdots, \frac{\partial f(x)}{\partial x_k} \right)^t \tag{11.32}$$

$$\nabla^2 f(x) \;=\; H(x) = \begin{pmatrix} \dfrac{\partial^2 f(x)}{\partial x_1 \partial x_1} & \cdots & \dfrac{\partial^2 f(x)}{\partial x_1 \partial x_k} \\ \vdots & \ddots & \vdots \\ \dfrac{\partial^2 f(x)}{\partial x_k \partial x_1} & \cdots & \dfrac{\partial^2 f(x)}{\partial x_k \partial x_k} \end{pmatrix}. \tag{11.33}$$

The $g(x)$ is called **gradient vector** and $H(x)$ is called **Hessian**. At the extreme value point of $f(x)$, the derivative with respect to that x must be zero, and thus it must hold $\nabla q(d) = 0$.

Thus, we have

$$\nabla q(d) = \nabla f(x) + \nabla^2 f(x)\, d = 0 \tag{11.34}$$

and from this, we obtain

$$\nabla^2 f(x)\, d = -\nabla f(x). \tag{11.35}$$

The search for the optimal point is performed along the direction d satisfying this relation. The solution of equation (11.35) shows the direction of the extrema. This method is called the **Newton algorithm**. Equation (11.34) is also known as the root-finding method for nonlinear equations.

While the Newton method has the advantage of faster quadratic convergence near the extreme value points, it also has an unexpectedly narrow range of converging initial values and is numerically unstable. In addition, it is troublesome to calculate Hessian each time. Methods that mitigate these shortcomings are collectively referred to as **quasi-Newton algorithm**. In the DFP method, to compensate for the former disadvantage, a **linear search** is performed for each search. That is, the next point to proceed to is determined by

$$x_{k+1} = x_k + \lambda d,$$

where λ is a scalar variable, defined so that $f(x_{k+1})$ is minimal. To compensate for the second drawback, we do not calculate Hessian directly, but modify it from the gradient of the previous stage, using the following DFP formula

$$H_k = H_{k-1} + \frac{(x_k - x_{k-1})(x_k - x_{k-1})^t}{(x_k - x_{k-1})^t (g_k - g_{k-1})} - \frac{H_k(g_k - g_{k-1})(g_k - g_{k-1})^t H_k}{(g_k - g_{k-1})^t H_{k-1}(g_k - g_{k-1})}, \tag{11.36}$$

where we put $H_0 = I$. If H_k is positive definite, then H_{k+1} is also guaranteed to be positive definite. However, for stability, reset to $H = I$ if either of the denominators of the last two terms in the above formula is negative.

There are several methods of linear search, but the following method is presented here.

1. Enclosing:

   The step width is doubled with $\lambda = 1, 2, 4, 8, \cdots$ and ends when $f'(\lambda) > 0$ and $f(\lambda) > f(0)$. Then, there is at least one minimum between $\lambda/2$ and λ.

2. Interval reduction:

   When there is a minimum between λ_1 and λ_2, third interpolation is performed using $f(\lambda_1)$ and $f(\lambda_2)$ to estimate the minimum point. If $f'(\lambda_3) > 0$ or $f(\lambda_3) > f(\lambda_1)$, then $\lambda_2 = \lambda_3$ because there is a minimum between λ_1 and λ_3. Otherwise, $\lambda_1 = \lambda_3$. Repeat this step until convergence.

Note that in the above method, the derivatives can be obtained by numerical differenciation[98].

12

Classification of Ship and Engine Motion Time Series

Given a group of time series (**batch data**) that can be considered stationary for a given time, but may differ from each other, consider how to classify them. If such a **Clustering** is made possible, a database of these data can be used to predict, for example, to some extent from the past database, what state the ship will be when the current state of ship and engine motions are continued. If the predicted condition is bad, it may be possible to find a way to operate the ship to avoid that condition.

12.1   Kullback-Leibler Information and Divergence

For simplicity, let us consider the case of univariate time series. Suppose we are given many univariate time series and an autoregressive model $AR_1, AR_2, \cdots$ for each of the time series. Now suppose that the autoregressive model for the data $y(1), y(2), \cdots, y(N)$ is given as

$$
\begin{aligned}
y(s) &= a(1)\, y(s-1) + a(2)\, y(s-2) + \cdots + a(M)\, y(s-M) + v(s) \\
&- \sum_{m=1}^{M} a(m)\, y(s-m) + v(s).
\end{aligned}
\tag{12.1}
$$

In fitting this model, the order M of the model is determined so as to minimize the following AIC (Akaike Information Criterion),

$$
\mathrm{AIC}(m) = N(\log 2\pi\sigma_v^2(m) + 1) + 2(m+1),
\tag{12.2}
$$

where N is the number of data and $\sigma_v^2(m)$ is the prediction error variance of the order m model.

The problem here is to classify these many models and group them together by collecting similar ones. In statistics, such a method is generally referred to as the **clustering method**. Many clustering methods have been proposed, but in any case, it is necessary to introduce a quantity that corresponds to the "distance" between the models. Euclidean distance, Mahalanobis distance, etc. are usually used as distances, but both of them are inappropriate as distances to express differences in time series groups in the frequency domain.

Then, as shown in Figure 12.1, given two statistical distributions, the "distance" between them can be measured by the **Kullback-Leibler information** (**KL information**, KL divergence)[56]. Here the KL information is a quantity that measures the closeness of an arbitrary statistical distribution $f(x)$ with respect to the true statistical distribution $g(x)$

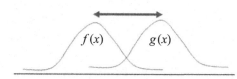

FIGURE 12.1
"Distance" between two statistical distributions

as

$$I(g; f) = E_g\left[\log \frac{g(x)}{f(x)}\right] = \int g(x) \log \frac{g(x)}{f(x)} dx \qquad (12.3)$$

$$= \int g(x) \log g(x) dx - \int g(x) \log f(x) dx, \qquad (12.4)$$

where $E_g[\cdot]$ denotes the expectation with respect to the distribution $g(x)$.

However, the KL information defined here does not satisfy the requirement of distance like the Euclidean distance. Actually, the KL information when g is considered the true distribution differs from that when f is considered the true distribution. Therefore, in order to satisfy the symmetry, if the KL information with respect to the true distribution f is given by

$$I(f; g) = E_f\left[\log \frac{f(x)}{g(x)}\right] = \int f(x) \log \frac{f(x)}{g(x)} dx,$$

the distance between the two statistical models is defined by the following **symmetrized divergence**,

$$D(f, g) = I(g; f) + I(f; g). \qquad (12.5)$$

12.2   KL Information of AR model

Suppose now that there are two univariate stationary time series and that AR_1 and AR_2 have been estimated using the minimum AIC estimation method, respectively. Also, the statistical distributions of the residuals of these two processes are assumed to follow Gaussian distributions, $N(0, \sigma^2)$ and $N(0, \tau^2)$. Namely,

$$\text{Model } f: \quad y(s) = \sum_{j=1}^{m} a(j)\, y(s-j) + v(s), \quad v(s) \sim N(0, \sigma^2) \qquad (12.6)$$

$$\text{Model } g: \quad y(s) = \sum_{j=1}^{l} b(j)\, y(s-j) + w(s), \quad w(s) \sim N(0, \tau^2). \qquad (12.7)$$

In this case, assuming that g is the true distribution, the KL information, which measures the difference between two statistical distributions is calculated as

$$I(g; f) = E_g\left[\log \frac{g(y)}{f(y)}\right].$$

The probability density function of the AR model and its logarithm are given by

$$f(y(s)) = \frac{1}{\sqrt{2\pi\sigma^2}} \exp\left\{\frac{1}{2\sigma^2}\left(y(s) - \sum_{j=1}^{m} a(j)\,y(s-j)\right)^2\right\},$$

$$\log f(y(s)) = -\frac{1}{2}\log 2\pi\sigma^2 - \frac{1}{2\sigma^2}\left(y(s) - \sum_{j=1}^{m} a(j)\,y(s-j)\right)^2.$$

Thus assuming g to be the true model, the expected value of $\log f(y(s))$ is given by

$$
\begin{aligned}
E_g[\log f(y(s))] &= -\frac{1}{2}\log 2\pi\sigma^2 - \frac{1}{2\sigma^2} E_g\left[\left(y(s) - \sum_{j=1}^{m} a(j)\,y(s-j)\right)^2\right] \\
&= -\frac{1}{2}\log 2\pi\sigma^2 - \frac{1}{2\sigma^2}\left(C^g(0) - 2\sum_{j=1}^{m} a(j)\,C^g(j) + \sum_{i=1}^{m}\sum_{j=1}^{m} a(i)\,a(j)\,C^g(|i-j|)\right) \\
E_g[\log g(y(s))] &= -\frac{1}{2}\log 2\pi\tau^2 - \frac{1}{2\tau^2} E_g\left[\left(y(s) - \sum_{j=1}^{l} b(j)\,y(s-j)\right)^2\right] \\
&= -\frac{1}{2}\log 2\pi\tau^2 - \frac{1}{2}.
\end{aligned}
$$

This results in the following KL information when g is a true distribution

$$
\begin{aligned}
I(g;f) &= E_g\left[\log \frac{g(y(s))}{f(y(s))}\right] \\
&= -\frac{1}{2}\log \frac{\sigma^2}{\tau^2} + \frac{1}{2\sigma^2}\left(C^g(0) - 2\sum_{j=1}^{m} a(j)\,C^g(j) + \sum_{i=1}^{m}\sum_{j=1}^{m} a(i)\,a(j)\,C^g(|i-j|)\right) - \frac{1}{2}.
\end{aligned}
$$

(12.8)

Similarly, the KL information when f is the true distribution is given by

$$
\begin{aligned}
I(f;g) &= E_f\left[\log \frac{f(y(s))}{g(y(s))}\right] \\
&= -\frac{1}{2}\log \frac{\tau^2}{\sigma^2} + \frac{1}{2\tau^2}\left(C^f(0) - 2\sum_{j=1}^{l} b(j)\,C^f(j) + \sum_{i=1}^{l}\sum_{j=1}^{l} b(i)\,b(j)\,C^f(|i-j|)\right) - \frac{1}{2}.
\end{aligned}
$$

(12.9)

Thus, the symmetrized KL information (the symmetrized divergence of f and g) is given by[25][50]

$$
\begin{aligned}
D(f,g) &= I(g;f) + I(f;g) \\
&= \frac{1}{2\sigma^2}\left(C^g(0) - 2\sum_{j=1}^{m} a(j)\,C^g(j) + \sum_{i=1}^{m}\sum_{j=1}^{m} a(i)\,a(j)\,C^g(|i-j|)\right) \\
&\quad + \frac{1}{2\tau^2}\left(C^f(0) - 2\sum_{j=1}^{l} b(j)\,C^f(j) + \sum_{i=1}^{l}\sum_{j=1}^{l} b(i)\,b(j)\,C^f(|i-j|)\right) - 1.
\end{aligned}
$$

(12.10)

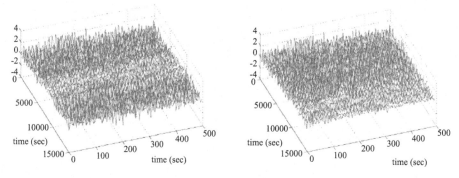

FIGURE 12.2
Sorting by KL information based on a certain batch time series

12.3   Clustering by KL Information

The left panel in Figure 12.2 shows arbitrarily arranged batch time series, each of which is a rolling time series of a long RoRo (roll-on/roll-off) ship with 600 points observed at 1-second intervals. In contrast, the right panel shows the batch time series with the first batch time series on the left as the reference, rearranged in order of closeness to the reference time series in terms of the amount of KL information. These figures do not tell us how the frequency changes, but they have been rearranged, at least with respect to variance.

Next, let us consider a method for clustering these batch time series. The clustering method used here is to calculate the amount of KL information between each element and group elements that are close to each other, i.e., those that form a **cluster**. Such clustering methods can be divided into **hierarchical methods** and **non-hierarchical methods**. The hierarchical method means that when clusters are formed, they are always subdivided as a set. That is, when the number of clusters is 2, the population set is divided into two subsets, and when these are divided into three, three new clusters are not formed, but one of the two subsets is further divided into two, forming three subsets as a whole. Therefore, the process can be represented by a **dendrogram** as shown in Figure 12.4 (see below). On the other hand, nonhierarchical clustering methods are classification methods without such restrictions, and the k-means method is a typical method. Here, we adopt the hierarchical method.

Suppose we have n objects (either individuals or variables), $O_1, O_2, \cdots, O_n$, and we have a numerical value d_{ij} representing the degree of similarity between the objects O_i and O_j. Here, we adopt the symmetrized divergence computed from the KL information so that $d_{ij} = d_{ji}$.

The basic procedure for hierarchical clustering is as follows

1. Initially, each of the n individuals is considered to form one cluster. The number of clusters $K = n$.

2. Among K clusters, find the pair with the greatest similarity (i.e., with the smallest KL information) and combine them into one cluster. Replace the number of clusters K with $K - 1$. If $K > 1$, go to step 3., otherwise exit.

3. Update the similarity matrix by computing the KL information between the newly obtained cluster and the other clusters. Return to step 2. with that updated similarity matrix.

(a) Minimum Distance Method

(b) Maximum Distance Method

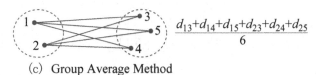

(c) Group Average Method

FIGURE 12.3
Various clustering methods

There are various methods of clustering such as the minimum distance method, the maximum distance method and the group averaging method (Figure 12.3).

We will illustrate this procedure using the minimum distance method, based on an example in the reference[36]. Suppose now that there are five individuals whose distances are given as in the following table. Let us call this the distance table.

$$D = \begin{array}{c|ccccc} & 1 & 2 & 3 & 4 & 5 \\ \hline 1 & 0 & & & & \\ 2 & 9 & 0 & & & \\ 3 & 3 & 7 & 0 & & \\ 4 & 6 & 5 & 9 & 0 & \\ 5 & 11 & 10 & 2 & 8 & 0 \end{array}$$

The minimum distance in this distance table is

$$\min_{i,k}(d_{i,k}) = d_{35} = 2,$$

as shown in bold in the table. Thus, individuals 3 and 5 are first combined and the generated group is denoted by (35). Next, we evaluate the distance between this (35) and the other individuals 1, 2 and 4, as follows

$$
\begin{aligned}
d_{(35)1} &= \min\{d_{31}, d_{51}\} = \min\{3, 11\} = 3 \\
d_{(35)2} &= \min\{d_{32}, d_{52}\} = \min\{7, 10\} = 7 \\
d_{(35)4} &= \min\{d_{34}, d_{54}\} = \min\{9, 8\} = 8.
\end{aligned}
$$

Based on this, the following new distance table is obtained.

$$\begin{array}{c|cccc} & (35) & 1 & 2 & 4 \\ \hline (35) & 0 & & & \\ 1 & 3 & 0 & & \\ 2 & 7 & 9 & 0 & \\ 4 & 8 & 6 & 5 & 0 \end{array}$$

As shown in bold in the table, the minimum distance in this table is the distance between (35) and 1, $d_{(35)1} = 3$. Thus, (35) and 1 are combined to form a new group (135). Evaluating the distance between this (135) and the other individuals, we obtain

$$d_{(135)2} = \min\{d_{(35)2}, d_{12}\} = \min\{7, 9\} = 7$$
$$d_{(135)4} = \min\{d_{(35)4}, d_{14}\} = \min\{8, 6\} = 6.$$

This results in the following distance table;

	(135)	2	4
(135)	0		
2	7	0	
4	6	**5**	0

The minimum value is $D_{24} = 5$. Thus, 2 and 4 are combined. At this point, two groups (135) and (24) are formed. Since

$$d_{(135)(24)} = \min\{d_{(135)2}, d_{(135)4}\} = \min\{7, 6\} = 6,$$

the following distance table is obtained.

	(135)	(24)
(135)	0	
(24)	6	0

(12.11)

As a result, individuals are integrated into a single group (12345). The diagrammatic representation of this procedure is called a **dendrogram** (Figure 12.4).

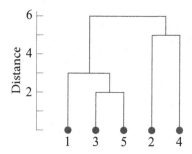

FIGURE 12.4
Dendrogram for the example

12.4 Example of Hierarchical Clustering of Rolling and Pitching Data of a Certain RoRo Ship

Here we show the results of clustering 82 batch files obtained by dividing the rolling and pitching time series of a certain coastal RoRo (roll-on/roll-off) ship for two consecutive days into sections of appropriate length, using the method described above. The basic sampling period of each data file is 1 second, and the number of data points is 2000.

Figure 12.5 shows the logarithm of spectrum of the clustering results of the batch data of rolling. Here, out of the 10 groups of the results, the groups containing two or more time series are shown. It can be seen that there are cases where the peaks are concentrated at

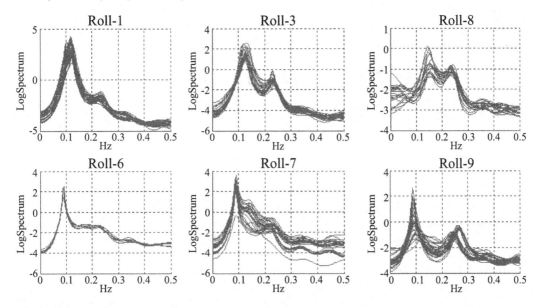

FIGURE 12.5
Clustering of rolling data

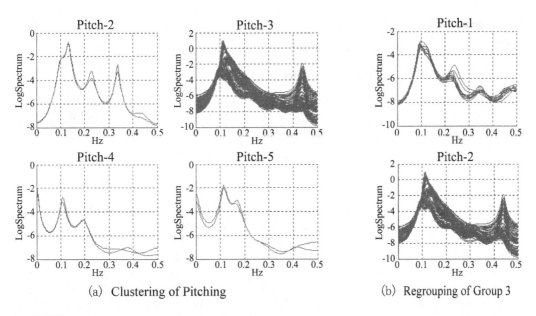

(a) Clustering of Pitching   (b) Regrouping of Group 3

FIGURE 12.6
Clustering of pitching data

the frequencies that are considered to be the natural period of the rolling, and there are also cases where other peaks exist at higher frequencies.

Figure 12.6(a) shows the clustering results for pitching time series. They appear to be concentrated in group 3 on the upper right. Therefore, only group 3 was re-clustered in Figure 12.6(b). It can be seen that the spectrum of each group is clustered similarly for both rolling and pitching motions.

13

On-Shore Simulation of Ship's Motions at Sea

In order to meet the demands of marine environmental conservation and resource conservation with a small number of seafarers, collaborative work at sea and on shore is essential. For that purpose, it would be convenient if the state of ship and engine motions occurring at sea could be seen as a simulated waveform in reality. In this case, for example, it is unrealistic to transmit time series every second. In this chapter, instead of actual time series, we consider the effective method of using ship-to-shore communication, which has developed rapidly in recent years, to send very few parameters of a statistical model representing the latest motion of a ship to shore and simulate its waveforms.

13.1   Simulation of Time Series

13.1.1   State-Space Methods and Simulations

We have learned in Chapters 10 and 11 that all linear time series models, including the stationary time series model, can be represented in the discrete-time domain in a unified, though not unique, state-space representation as follows

$$x(s) = F(s) x(s-1) + G(s) v(s) \tag{13.1}$$
$$y(s) = H(s) x(s) + w(s), \tag{13.2}$$

where $x(s)$ is a k-dimensional **state variable** obtained by an appropriate transformation of the time series. Also, $y(s)$ is the l-dimensional observed time series. F is the $k \times k$ **transition matrix** and $G(s)$ is the $k \times m$ **driving matrix**. The $v(s)$ and $w(s)$ are m-dimensional and l-dimensional Gaussian white noises, with mean 0 and variance-covariance matrices Q and R, respectively. They are respectively called system noise and observation noise. In this section, we consider how to reproduce the actual waveform given a model represented in this way. This method can also be used in Chapter 15, which deals with statistical control, as a pre-evaluation method in control system design[7].

By the way, equation (13.1), which represents the dynamics of the system, will eventually converge to a static state unless there is an external input, namely $v(s)$. On the other hand, if we input noise that has the same distribution as the original time series, the statistical structure of the state variable $x(s)$ of the system will be the same as the original time series, and we expect the two to have similar waveforms. The statistical properties of the Gaussian noise $v(s)$ are determined by giving the mean and the variance, Q. Thus, the first step in simulating a real-time series is to generate, by some means, a k-dimensional Gaussian white noise $v(1), v(2), \cdots, v(N)$ that has a mean vector 0 and a variance-covariance matrix Q. Thereafter, $x(1), x(2), \cdots, x(N)$ can be generated in turn by assuming an appropriate

DOI: 10.1201/9781003428565-13

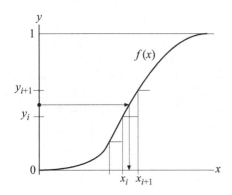

FIGURE 13.1
Generation of random numbers with arbitrary distribution

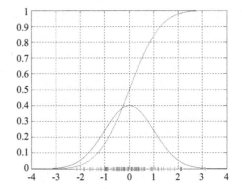

FIGURE 13.2
Generation of pseudo-random numbers with normal distribution

initial value $x(0)$ and iterative computation using the equation (13.1). Note here that for real applications, a correction for the mean value is necessary. Then, by using the generated $x(s)$ and an l-dimensional Gaussian white noise with variance R in equation (13.2), realizations $y(1), y(2), \cdots, y(N)$ of the time series are generated in turn.

13.1.2 Generating Pseudo-Random Numbers with Arbitrary Distribution

The general method for generating pseudo-random numbers with arbitrary distribution $f(x)$, such as Gaussian distribution, is as follows (Figure 13.1).

1. Generate a pseudo-random number y uniformly distributed on [0 1).

2. Using the inverse function $f(x)^{-1}$, obtain x that satisfies $y = f(x)$ for the generated y.

Therefore, it is necessary to first generate in the computer a pseudo-random number with a uniform distribution in the interval [0 m). A well-known simple method is the **linear congruential method**[97]. This method generates the random number sequentially from $i = 0$,

$$X(i+1) = aX(i) + c, \quad (\text{mod } m),$$

where (mod m) means the remainder of $aX(i)$ divided by m. Then the pseudo-random number is obtained by

$$U(i) = X(i)/m.$$

If the number of digits of the computer is k, we can take about $m = 2^k$. For a 32-bit machine, $m = 2^{31} - 1$, $a = 7^5$, $c = 0$, $X(0) = 12345$, etc. Note that, the period length of the random number is at most m. Note also that recently, algorithms with guaranteed very long periods, such as the Mersenne twister, are being used in computer simulations[45].

By finding x using the inverse function from the random numbers in the [0 1) interval thus generated, we can generate pseudo-random numbers that follow the corresponding arbitrary distribution. Figure 13.2 shows the case of a normal distribution. Since the normal distribution has a flat tail and steep curve around the mean, there is a high possibility that the inverse function of the uniformly distributed pseudo-random number will be concentrated in the center. Thus, the distribution of x thus generated takes the form of a normal distribution.

13.1.3   Generation of White Noise

Generating the pseudo-random numbers given in the previous section and considering them as a time series yields one-dimensional white noise. To transform it into a Gaussian white noise $v(s)$ with variance-covariance matrix Q, we first generate k-dimensional Gaussian random number and then multiply the square root matrix L of Q. That is, at time s, we generate k Gaussian white noise with variance mean 0 and variance 1 using random numbers in the computer and set $u(s) = (u^{(1)}(s), u^{(2)}(s), \cdots, u^{(k)}(s))^t$. Next, decompose Q as

$$Q = LL^t. \tag{13.3}$$

Using the obtained L, we can generate k-dimensional Gaussian random number by

$$v(s) = Lu(s).$$

To obtain the square root of a matrix (equation (13.3)), the **Cholesky decomposition** is often used (e.g., Ref. [94]). Figure 13.2 shows a pseudo-random number generated by the above method that follows a normal distribution with mean 0 and variance 1 with the following density function

$$g(x) = \frac{1}{\sqrt{2\pi\sigma^2}} \exp\left\{-\frac{(x-\mu)^2}{2\sigma^2}\right\}.$$

13.2   Generation of Gaussian Random Numbers

A method of generating Gaussian random numbers efficiently is the **Box-Muller method** [45], which is given as follows;

1. Generate two pseudo-random numbers U_1 and U_2.
2. Set $V_1 = 2U_1 - 1$ and $V_2 = 2U_2 - 1$.
3. Compute $S^2 = V_1^2 + V_2^2$.
4. If $S^2 \geq 1$, reject and return to step 1.
5. Obtain two independent Gaussian random numbers by

$$X_1 = V_1\sqrt{-\frac{2\log S}{S}}, \quad X_2 = V_2\sqrt{-\frac{2\log S}{S}}.$$

13.3   Example of Simulation

The upper panel of Figure 13.3 shows a real data of rolling of a container ship ($\Delta t = 1\,\text{sec}$, $N = 400$) and the lower panel shows a result of simulation using the method described in this chapter, using autoregressive coefficients and prediction error variance of the AR model of order 6 identified using the real data. Figure 13.4 shows the simulation of the yawing of the same ship obtained by using the real record of yawing and an AR model of order 8. From these two examples, it can be seen that by using only a small number of parameters, it is possible to reproduce a time series of length 400 with waveforms similar to the real data.

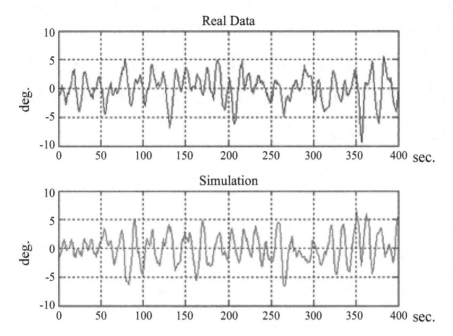

FIGURE 13.3
Real data and the simulation of rolling

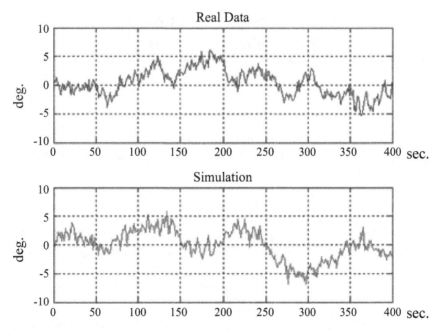

FIGURE 13.4
Real data and the simulation of yawing

14

Multivariate Autoregressive Model and Statistical Analysis of Ship and Engine Motions

In this chapter, we extend the univariate autoregressive model to a multivariate autoregressive model. Then it is shown how to obtain unbiased estimates of the impulse response function and the frequency response function. Once the multivariate autoregressive model is obtained, dynamic relationships between time series of ship and engine, the subject of this book, can be obtained and new knowledge can be obtained. Especially important is a new concept called noise contribution ratio, which shows the influence of one variable on another.

14.1 Multivariate Autoregressive Model

14.1.1 Definition of Multivariate Autoregressive Model

Let $y(s)$ be a k-variate time series $y(s) = (y_1(s), y_2(s), \cdots, y_k(s))^t$. The **multivariate autoregressive model** is an extension of the univariate autoregressive model and defined by

$$y(s) = \sum_{m=1}^{M} A(m) \, y(s - m) + v(s), \tag{14.1}$$

where $v(s)$ is the k-variate white noise given by $v(s) = (v_1(s), v_2(s), \cdots, v_k(s))^t$. We henceforth refer to this model as the **MAR model**. Written in terms of elements, MAR model can be expressed as

$$y_i(s) = \sum_{m=1}^{M} \sum_{j=1}^{k} A_{ij}(m) \, y_j(s - m) + v_i(s), \tag{14.2}$$

for $i = 1, \ldots, k$. $A_{ij}(m)$ is the (i, j) element of $A(m)$ and is the regression coefficient expressing the degree of influence from the jth variable $y_j(s-m)$ on the ith variable $y_i(s)$, i.e.,

$$A(m) = \begin{pmatrix} A_{11}(m) & A_{12}(m) & \cdots & A_{1k}(m) \\ A_{21}(m) & A_{22}(m) & \cdots & A_{2k}(m) \\ \vdots & \vdots & \ddots & \vdots \\ A_{k1}(m) & A_{k2}(m) & \cdots & A_{kk}(m) \end{pmatrix}.$$

DOI: 10.1201/9781003428565-14

$v(s)$ is a k-variate white noise process. Each element of multivariate white noise has mean 0, no correlation between components at different times and the following properties hold

$$\begin{cases} E[v(s)\,v(s)^t] = \begin{pmatrix} \sigma_{11} & \sigma_{12} & \cdots & \sigma_{1k} \\ \sigma_{21} & \sigma_{22} & \cdots & \sigma_{2k} \\ \vdots & \vdots & \ddots & \vdots \\ \sigma_{k1} & \sigma_{k2} & \cdots & \sigma_{kk} \end{pmatrix} \\ E[v(s)\,v(m)^t] = 0, \quad s \neq m, \end{cases} \tag{14.3}$$

where $\sigma_{ij} = \sigma_{ji}$. If we define a discrete δ function as

$$\delta_{l,0} = \begin{cases} 1, & l = 0 \\ 0, & l \neq 0, \end{cases}$$

equation (14.3) can be expressed as

$$E[v_i(s)\,v_j(s+l)] = \delta_{l,0}\,\sigma_{ij}. \tag{14.4}$$

Also the covariance between any $v_i(s)$ and the past observation $y_j(s-m)$, $(m = 1, 2, \cdots)$ is 0. In other words, we suppose that

$$E[v(s)\,y(r)^t] = 0, \quad s > r. \tag{14.5}$$

For example, if $M = 2$ and $k = 2$, we have

$$\begin{aligned} y_1(s) &= A_{11}(1)\,y_1(s-1) + A_{11}(2)\,y_1(s-2) + A_{12}(1)\,y_2(s-1) + A_{12}(2)\,y_2(s-2) + v_1(s) \\ y_2(s) &= A_{21}(1)\,y_1(s-1) + A_{21}(2)\,y_1(s-2) + A_{22}(1)\,y_2(s-1) + A_{22}(2)\,y_2(s-2) + v_2(s). \end{aligned}$$

If we denote the term in the equation (14.4), i.e., the covariance function between $v_i(s)$ and $v_j(s)$, by $R_{v_i v_j}(l)$ and $A_{ij}(0) = -1$ $(i = j)$, $A_{ij}(0) = 0$ $(i \neq j)$, then by substituting equation (14.2) to both v_i and v_j, we obtain

$$R_{v_i v_j}(l) = \sum_{m=0}^{M} \sum_{n=0}^{M} \sum_{r=1}^{k} \sum_{s=1}^{k} A_{ir}(m)\,A_{js}(n)\,R_{rs}(l - m + n), \tag{14.6}$$

where $R_{rs}(l)$ is the cross-covariance function between variables $y_r(s)$ and $y_s(s)$.

Equation (14.1) can be written using the z transformation as

$$\Psi(z^{-1})\,y(s) = v(s), \tag{14.7}$$

where

$$\Psi(z^{-1}) = I - A(1)\,z^{-1} - A(2)\,z^{-2} - \cdots - A(M)\,z^{-M}. \tag{14.8}$$

In this case, the characteristic equation is

$$\left| \Psi(z^{-1}) \right| = 0. \tag{14.9}$$

The stationarity condition for equation (14.1) is that all of the z^{-1} roots (characteristic roots) of this equation are **outside the unit circle**. Alternatively, it is the same if we consider that the reciprocals of the roots are **inside the unit circle**.

14.1.2   AIC and FPE for the Multivariate Autoregressive Model

The following FPE or AIC is used to determine the order M of the multivariate autoregressive model (14.1).

The FPE of the k-variate autoregressive model with order m is given by

$$\text{MFPE}(m) = \left(1 + \frac{mk+1}{N}\right)^k \left(1 - \frac{mk+1}{N}\right)^{-k} |d_m|, \qquad (14.10)$$

where d_m is the variance-covariance matrix of the prediction error (equation (14.3))

$$d_m = E[v(s)\, v(s)^t],$$

and $|d_m|$ is the determinant of the matrix d_m.

In this case, the number of parameters is $mk^2 + k(k+1)/2$ including variance-covariance matrix, therefore the AIC is given by

$$\text{AIC}(m) = Nk \log 2\pi + N \log |d_m| + Nk + 2mk^2 + k(k+1). \qquad (14.11)$$

In order to compare MAR models, the following definitions, omitting the common constant terms, can be used

$$\text{AIC}^*(m) = N \log |d_m| + 2mk^2. \qquad (14.12)$$

14.2   Parameter Estimation for Multivariate Autoregressive Models

14.2.1   Yule-Walker Equation

The cross-covariance function $R(m)$ of the k-variate autoregressive model with order M for $y(s)$ is obtained from the product of the model of $y(s)$ and $y(s - l)$ by

$$
\begin{aligned}
R(l) &= E\left[\left(\sum_{m=1}^{M} A(m)\, y(s - m) + v(s)\right) y(s - l)^t\right] \\
&= \sum_{m=1}^{M} A(m)\, E\left[y(s - m)\, y(s - l)^t\right] + E\left[v(s)\, y(s - l)^t\right].
\end{aligned}
$$

The expected value of the first term in the last equation is the cross-covariance matrix with lag $l - m$. The second term is the expected value of the product of $v(s)$ and $y(s - l)^t$. By assumption, this term is zero except for $l = 0$, for which it can be seen from equations (14.4) and (14.5) that

$$E\left[v(s)\, y(s)^t\right] = E\left[v(s)\, \{A(m)\, y(s - m) + v(s)\}^t\right] = E\left[v(s)\, v(s)^t\right].$$

Thus, if we put

$$d_M = E\left[v(s)\, v(s)^t\right],$$

then we obtain

$$R(0) = \sum_{j=1}^{M} A(j)\, R(-j) + d_M, \qquad (14.13)$$

$$R(l) = \sum_{j=1}^{M} A(j)\, R(l - j), \quad l = 1, 2, \cdots, M. \qquad (14.14)$$

This is the **Yule-Walker equation for the multivariate autoregressive model**. These equations also show that if $A(j)$ is given in some way, the covariance function can be obtained by solving these equations.

Next, let us consider how to compute the covariance functions when data are available. If k-variate observations of length N, $\{y(1), y(2), \cdots, y(N)\}$ are obtained, the cross-covariance function is calculated by subtracting the mean value of each variable from $y(s)$, and computing

$$C_{ij}(l) = \frac{1}{N} \sum_{s=1}^{N-l} (y_i(s) - m_{y_i})(y_j(s+l) - m_{y_j}), \quad l = 0, 1, 2, \cdots,$$

where m_{y_i} represents the mean value of variable $y_i(s)$. The cross-covariance matrix with these (i, j) elements can be defined as

$$C(l) = \begin{pmatrix} C_{11}(l) & \cdots & C_{1k}(l) \\ \vdots & \ddots & \vdots \\ C_{k1}(l) & \cdots & C_{kk}(l) \end{pmatrix}. \tag{14.15}$$

Substituting the cross-covariance matrix thus computed into equation (14.14), we obtain

$$\begin{aligned} C(1) &= A(1)\,C(0) + A(2)\,C(-1) + \cdots + A(M)\,C(1-M) \\ C(2) &= A(1)\,C(-1) + A(2)\,C(0) + \cdots + A(M)\,C(2-M) \\ &\vdots \\ C(M) &= A(1)\,C(1-M) + A(2)\,C(2-M) + \cdots + A(M)\,C(0), \end{aligned}$$

i.e., we obtain

$$C(l) = \sum_{m=1}^{M} A(m)\,C(l-m), \quad l = 1, 2, \cdots, M. \tag{14.16}$$

This is a linear simultaneous equation for $A(1), A(2), \cdots, A(M)$ and can be solved. Substituting the obtained $\hat{A}(m)$ into the following equation

$$C(0) = \sum_{m=1}^{M} \hat{A}(m)\,C(-m) + d_M, \tag{14.17}$$

which is obtained from equation (14.13), we obtain an estimate of the variance-covariance matrix d_M of the prediction error.

14.2.2 Levinson-Type Sequential Estimation Methods for Multivariate Autoregressive Models

When applying the Levinson-Durbin recursive estimation method for a univariate autoregressive model to multivariate autoregressive model, it is necessary to consider both forward and backward multivariate autoregressive models, since the symmetry of the Toeplitz matrix is not available. Here, the backward multivariate autoregressive model is defined by

$$y(s) = \sum_{m=1}^{M} B_m(m)\,y(s+m) + v(s).$$

The specific Levinson-type estimation method for a multivariate autoregressive model proceeds as follows:

Levinson's procedure for estimating multivariate autoregressive models

1. Set initial value at $m = 0$ and compute AIC by equation (14.12);

$$d_0 = e_0 = C(0)$$
$$\text{AIC}(0) = N \log |d_0|.$$

2. For $m = 1, 2, \cdots, M$, compute the parameters of model by the following procedure where M is the maximum order;

$$E(m) = C(m) - \sum_{i=1}^{m-1} A_{m-1}(i)\, C(m - i)$$
$$A_m(m) = E(m)\, e_{m-1}^{-1}$$
$$B_m(m) = E^t(m)\, d_{m-1}^{-1}$$
$$A_m(i) = A_{m-1}(i) - A_m(m)\, B_{m-1}(m - i), \quad i = 1, 2, \cdots, m - 1$$
$$B_m(i) = B_{m-1}(i) - B_m(m)\, A_{m-1}(m - i), \quad i = 1, 2, \cdots, m - 1$$
$$d_m = C(0) - \sum_{i=1}^{m} A_m(i)\, C^t(i)$$
$$e_m = C(0) - \sum_{i=1}^{m} B_m(i)\, C(i)$$
$$\text{AIC}(m) = N \log |d_m| + 2mk^2.$$

3. Perform the above calculation for $m = 1, 2, \cdots, M$, and the order for which $\text{AIC}(m)$ is the smallest is selected as the optimal order of the model.

14.2.3   Cross-Spectrum Estimation

Let $A(f)$, $Y(f)$ and $V(f)$ be the discrete Fourier transforms of the coefficient matrix $A(m)$, signal process $y(s)$ and noise process $v(s)$, respectively. Then, from the discrete Fourier transformation of equation (14.1), we obtain

$$Y(f) = (I - A(f))^{-1}\, V(f). \tag{14.18}$$

From this, using the basic relation for the cross-spectrum $P(f)$ described in Chapter 6, $P(f)$ is obtained by

$$P(f) = (I - A(f))^{-1}\, d_M \left(\overline{(I - A(f))}^t \right)^{-1}, \tag{14.19}$$

where d_M is the variance-covariance matrix of $v(s)$ and $\overline{X}$ denotes the complex conjugate of X. The diagonal terms of this matrix give the power spectrums of the time series, and the (i, j) element of $P(f)$ gives the cross-spectrum of the ith time series and the jth time series. Here, the (i, j) element $A_{ij}(f)$ of $A(f)$ is calculated by

$$A_{ij}(f) = \sum_{m=1}^{M} A_{ij}(m)\, e^{-i2\pi fm}. \tag{14.20}$$

14.3 System Analysis Method with Multivariate Autoregressive Model

14.3.1 Systems with Feedback Loops

Many systems, such as the motion of a human being looking at an object and grasping it, price fluctuations in response to economic policies, and the control system of a large plant, have feedback loops as shown in (b) of Figure 14.1. In ship motion, for example, feedback loops can be seen in many kinetic systems, such as the control system of the yaw angle by the rudder, the control system of the revolutions by the governor and the coupled motion that occurs between ship motions. Such a system should be in a steady state as long as there are no driving noise sources in the elements that make up the system.

When such feedback loops exist in a system, biases that are not physically feasible often occur, such as significant values appearing in the negative region of the impulse response function obtained from the analysis using the cross-spectrum in the frequency domain treated in Chapter 6[1]. The reason for this is that the system treated in Chapter 6 is an open-loop system as shown in Figure 14.1 (a), and moreover, the noise sources entering the input and output are assumed to be uncorrelated with the present and past values, whereas in a feedback system, this assumption does not hold because the noise source $n_0(s)$ that enters the output enters the input side through the block h_{01} shown in Figure 14.1, causing the input to fluctuate.

In this section, we show that by using functions derived from the multivariate autoregressive model in the time domain instead of the frequency domain for such a feedback system, it is possible to analyze the system without bias and to separate the effects from the noise sources in the system from those caused by the dynamic characteristics of the system itself.

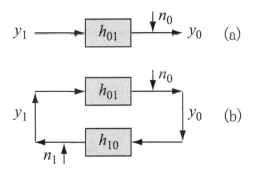

FIGURE 14.1
Systems without feedback (a) and with feedback (b). (1-input 1-output system)

14.3.2 Impulse Response Model of the System

(1) Bivariate case

First, for simplicity, let us assume that the two variables $y_0(s)$ and $y_1(s)$ are connected by a feedback loop as shown in Figure 14.1(b). These variables have an input-output relationship with each other and can be expressed using the impulse response function as

$$y_0(s) = \sum_{m=1}^{M} h_{01}(m)\, y_1(s-m) + n_0(s) \tag{14.21}$$

$$y_1(s) = \sum_{m=1}^{M} h_{10}(m)\, y_0(s-m) + n_1(s). \tag{14.22}$$

The **impulse response model** is different from the autoregressive model, for example, in equation (14.21), which does not consider the contribution from past values of $y_0(s)$. For simplicity, it is also assumed that there is no influence on $y_0(s)$ from the current value of $y_1(s)$ or in the opposite direction. The noise sources $n_0(s)$ and $n_1(s)$ are assumed to be uncorrelated with the inputs $y_0(s)$ and $y_1(s)$, respectively, but not necessarily white noise.

Suppose now that we have M initial values $\{y_0(s), y_1(s),\ s = -M+1, \cdots, 0\}$ and N observations $\{y_0(s), y_1(s),\ s = 1, 2, \cdots, N\}$. In order to capture the relationship between these two time series statistically, consider estimation of $h_{01}(m)$ and $h_{10}(m)$ using the least squares method on these observations. According to the least squares method, if we want to estimate, for example, $h_{10}(m)$, then it is obtained by finding $h_{10}(m)$ that minimizes

$$S^2 = \sum_{m=1}^{N} \left\{ y_1(s) - \sum_{m=1}^{M} h_{10}(m)\, y_0(n-m) \right\}^2.$$

Here, using the same idea as in equation (14.16), we first calculate the sum of products of y_1 and y_0 with respect to the lag l, and then we obtain

$$\sum_{s=1}^{N} y_1(s-l)\, y_0(s) = \sum_{s=1}^{N} y_1(s-l) \left(\sum_{m=1}^{M} h_{01}(m)\, y_1(s-m) + n_0(s) \right)$$

$$= \sum_{m=1}^{M} h_{01}(m) \sum_{s=1}^{N} y_1(s-l)\, y_1(s-m) + \sum_{s=1}^{N} y_1(s-l)\, n_0(s).$$

The expectation of the left-hand side of this equation converges to the cross-covariance function $C_{10}(l)$ of lag l between $y_1(s)$ and $y_0(s)$, whereas the right-hand side becomes

$$\sum_{m=1}^{M} h_{01}(m)\, C_{11}(l-m) + d, \tag{14.23}$$

where

$$d = \frac{1}{N} \sum_{s=1}^{N} y_1(s-l)\, n_0(s). \tag{14.24}$$

It should be noted here that unless $n_0(s)$ is white noise, $d = 0$ cannot be expected due to the influence from the past of n_0 on $y_1(s-l)$ that enters through the feedback loop. Therefore, simply solving the normal equation that minimizes S^2 to obtain an estimate of $h_{01}(m)$ will result in a biased estimate.

This difficulty can be resolved by linking models (14.21) and (14.22) to a two-variate autoregressive model. To do so, we first whiten $n_0(s)$ and $n_1(s)$ by using their own past values with sufficiently large L as follows;

$$n_0(s) = \sum_{l=1}^{L} c(l)\, n_0(s-l) + v_0(s) \tag{14.25}$$

$$n_1(s) = \sum_{l=1}^{L} d(l)\, n_1(s-l) + v_1(s), \tag{14.26}$$

where $v_0(s)$ and $v_1(s)$ are assumed to be white noise and uncorrelated with each other and with their own past values. The $y_0(s)$ and $y_1(s)$ are represented by the infinite sum of the present and past values of $n_0(s)$ and $n_1(s)$ from models (14.21) and (14.22). Furthermore, in general, the colored noise $n_0(s)$ and $n_1(s)$ are given by the linear infinite sum of the present and past values of $v_0(s)$ and $v_1(s)$ from equations (14.25) and (14.26). Therefore, there is no correlation between the past values of $y_0(s)$ and $y_1(s)$, $y_0(s-1)$, $y_1(s-1)$, $y_0(s-2)$, $y_1(s-2) \cdots$ and $v_0(s)$ transmitted through the feedback loop of $0 \to 1$ and $1 \to 0$. Writing this whitening operation using equation (14.25), we obtain

$$n_0(s) - \sum_{l=1}^{L} c(l) \, n_0(s-l) - v_0(s)$$

$$= y_0(s) - \sum_{m=1}^{M} h_{01}(m) \, y_1(s-m)$$

$$- \sum_{l=1}^{L} c(l) \left(y_0(s-l) - \sum_{m=1}^{M} h_{01}(m) \, y_1(s-l-m) \right) - v_0(s) = 0.$$

Therefore, we have

$$y_0(s) - \sum_{l=1}^{L} c(l) \, y_0(s-l) = \sum_{m=1}^{M} h_{01}(m) \left(y_1(s-m) - \sum_{l=1}^{L} c(l) \, y_1(s-l-m) \right) + v_0(s).$$

Rearranging this yields, we obtain a two-variate autoregressive model

$$y_0(s) = \sum_{l=1}^{L} c(l) \, y_0(s-l) + \sum_{m=1}^{M+L} A_m \, y_1(s-m) + v_0(s). \tag{14.27}$$

Here, assuming $h_{01}(m) = 0$ $(m > M)$ and $c(l) = 0$ $(l > L)$, A_m is defined by

$$A_1 = h_{01}(1)$$

$$A_m = \begin{cases} = h_{01}(m) - \sum_{l=1}^{m-1} c(l) \, h_{01}(m-l), & m = 2, 3, \cdots, M+L \\ = 0, & \text{if } l > L. \end{cases} \tag{14.28}$$

Since $v_0(s)$ in equation (14.27) is uncorrelated with $y_0(s-l)$ and $y_1(s-m)$, the A_m and $c(l)$ estimates thus obtained are unbiased. Therefore, using these unbiased estimates of $c(l)$ and A_m and solving relation (14.28) in reverse, we obtain the impulse response function successively as

$$h_{01}(1) = A_1,$$

$$h_{01}(m) = A_m + \sum_{l=1}^{m-1} c(l) \, h_{01}(m-l), \quad m = 2, 3, \cdots, M. \tag{14.29}$$

Namely, instead of directly obtaining the impulse response function using the least squares method for the impulse response model, we can obtain an unbiased estimate of the impulse response function by obtaining the regression coefficients from the two-variate autoregressive model and converting them to the impulse response function.

Since the coefficient $c(l)$ obtained in this way is the autoregressive coefficient of the model (14.25) representing the noise source $n_0(s)$ entering $y_0(s)$, we can obtain the **spectrum of**

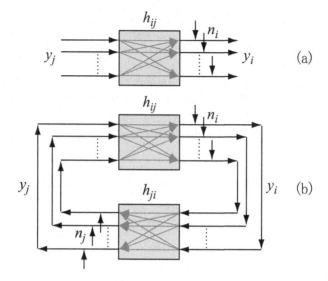

FIGURE 14.2
Systems without feedback (a) and with feedback (b). (Multiple-input, multiple-output systems)

the noise source $n_0(s)$ by using a similar formula for obtaining the spectrum from the autoregressive model.

(2) Multivariate Case

Next, let us now extend the above results to the multivariate case. Consider a k-variate time series $(y_1(s), y_2(s), \cdots, y_k(s))^t$, and suppose that these variables can be represented by the following impulse response representation,

$$y_i(s) = \sum_{m=1}^{M} \sum_{m=1}^{k} h_{ij}(m)\, y_j(s-m) + n_i(s), \tag{14.30}$$

where $h_{ij}(m)$ is the impulse response function from variable x_j to variable x_i (Figure 14.2). The impulse response model is a model that assumes that y_i is influenced by other variables y_j $(j \neq i)$ different from y_i from the past and from a unique noise source $n_i(s)$. Therefore, $h_{ii}(s) = 0$. As mentioned in (1), if we simply apply the least squares method to this model to obtain the coefficient $h_{ij}(m)$, it will be a statistically biased estimate. Therefore, by whitening of $n_i(s)$, corresponding to equation (14.27), we obtain

$$y_i(s) = \sum_{l=1}^{L} c_i(l)\, y_i(s-l) + \sum_{j=1(j \neq l)}^{k} \sum_{m=1}^{M+L} A_{ijm}\, y_j(s-m) + v_i(s). \tag{14.31}$$

Also, we can obtain

$$n_i(s) = \sum_{l=1}^{L} c_i(l)\, n_i(s-l) + v_i(s), \tag{14.32}$$

where $v_i(s)$ is the white noise. Equation (14.31) shows that this model is a special case of the ordinary autoregressive model

$$y_i(s) = \sum_{m=1}^{M} \sum_{j=1}^{k} A_{ij}(m)\, y_j(s-m) + v_i(s). \tag{14.33}$$

However, M corresponds to $M + L$ in equation (14.31). An important assumption that differs from the ordinary autoregressive model is that the off-diagonal terms in the variance–covariance matrix of the noise terms are zero, i.e., it holds that

$$
\begin{pmatrix}
\sigma_{11} & \sigma_{12} & \cdots & \sigma_{1k} \\
\sigma_{21} & \sigma_{22} & \cdots & \sigma_{2k} \\
\vdots & \vdots & \ddots & \vdots \\
\sigma_{k1} & \sigma_{k2} & \cdots & \sigma_{kk}
\end{pmatrix}
\rightarrow
\begin{pmatrix}
\sigma_{11} & 0 & \cdots & 0 \\
0 & \sigma_{22} & \ddots & \vdots \\
\vdots & \ddots & \ddots & 0 \\
0 & \cdots & 0 & \sigma_{kk}
\end{pmatrix}.
\tag{14.34}
$$

The correspondence between equations (14.31) and (14.33) is as follows. First, it is obvious that

$$
c_i(l) = A_{ii}(l).
$$

Similarly, for $i \neq j$, we have

$$
A_{ijm} = A_{ij}(m), \quad m = 1, 2, \cdots, M,
$$

where, when l and m exceed M, they are considered to be 0.

From these relations, the impulse response function $h_{ij}(m)$ is obtained by

$$
\begin{cases}
h_{ij}(1) = A_{ij}(1) \\
h_{ij}(m) = A_{ij}(m) + \displaystyle\sum_{l=1}^{m-1} A_{ii}(l)\, h_{ij}(m-l), \quad m = 2, 3, \cdots.
\end{cases}
\tag{14.35}
$$

The Fourier transform of $h_{ij}(m)$ obtained here also gives the open loop frequency response function from variable i to variable j.

14.3.3 Frequency Response Function and Noise Contribution

As obtained in the previous subsection, systems with feedback loops, which have been difficult so far, can be analyzed if equation (14.34) holds[7][3]. In this subsection, we show how to use this result to analyze the system in the frequency domain.

The Fourier transform of the impulse response function estimated using equation (14.35) yields

$$
H_{ij}(f) = \sum_{l=1}^{M} h_{ij}(l)\, e^{-i2\pi f l},
\tag{14.36}
$$

where $H_{ij}(f)$ is referred to as the **open loop frequency response function** from variable $y_i(s)$ to variable $y_j(s)$ in the system analysis.

Next, in equation (14.18), we put $B(f) = (I - A(f))^{-1}$. Namely,

$$
B(f) =
\begin{pmatrix}
b_{11}(f) & b_{12}(f) & \cdots & b_{1k}(f) \\
b_{21}(f) & b_{22}(f) & \cdots & b_{2k}(f) \\
\vdots & \vdots & \ddots & \vdots \\
b_{k1}(f) & b_{k2}(f) & \cdots & b_{kk}(f)
\end{pmatrix}
=
\begin{pmatrix}
1 - A_{11}(f) & -A_{12}(f) & \cdots & -A_{1k}(f) \\
-A_{21}(f) & 1 - A_{22}(f) & \cdots & -A_{2k}(f) \\
\vdots & \vdots & \ddots & \vdots \\
-A_{k1}(f) & -A_{k2}(f) & \cdots & 1 - A_{kk}(f)
\end{pmatrix}^{-1}.
$$

Then, the (i, j) element $b_{ij}(f)$ of $B(f)$ is the **closed loop frequency response function** of the system from variable $y_j(s)$ to variable $y_i(s)$.

Next, obtain the spectrum of the time series $y_i(s)$ according to the definition. If the Fourier transform of $y_i(s)$ is $Y_i(f)$, then the periodogram is given by

$$\frac{Y_i(f)\,\overline{Y_i(f)}}{N} = \frac{1}{N} \sum_{j_1=1}^{k} b_{ij_1}(f)\, V_{j_1}(f) \sum_{j_2=1}^{k} \overline{b_{ij_2}(f)\, V_{j_2}(f)}$$

$$= \frac{1}{N} \sum_{j=1}^{k} |b_{ij}(f)|^2\, V_j(f)\, \overline{V_j(f)} + \frac{1}{N} \sum_{j_1=1}^{k} \sum_{j_2 \neq j_1} b_{ij_1}(f)\, \overline{b_{ij_2}}\, V_{j_1}(f)\, \overline{V_{j_2}(f)}.$$

Now suppose that the noise between different elements is uncorrelated, i.e., $E\left[v_{j_1}(s)\, v_{j_2}(s)\right] = 0$, then, from definition, the spectrum of the ith variable is obtained by

$$P_{ii}(f) = \sum_{j=1}^{k} |b_{ij}(f)|^2\, \sigma_{jj}^2. \tag{14.37}$$

This indicates that the power spectrum of the variation of the ith time series at frequency f can be decomposed into k noise sources, whose magnitude is represented by $|b_{ij}(f)|^2\, \sigma_{jj}^2$. We will henceforth refer to this quantity as **contribution to spectrum**. If we define a quantity by

$$\gamma_{ij}(f) = \frac{|b_{ij}(f)|^2\, \sigma_{jj}^2}{P_{ii}(f)}, \tag{14.38}$$

this indicates the fraction of the variation in frequency f of $y_i(s)$ that is due to $v_j(s)$, the **relative noise contribution**. Also

$$R_{ij}(f) = \sum_{h=1}^{j} \gamma_{ih}(f), \quad j = 1, 2, \cdots, k \tag{14.39}$$

indicates the **cumulative noise contribution**.

14.4   Ship and Engine Motion Time Series Analysis by Multivariate Autoregressive Model

The ship's motion has six degrees of freedom, and if we consider that the rudder is also a motion, the ship is moving with seven degrees of freedom. During navigation, these motions are not independent of each other, but are related to each other. The characteristic relationships among these 7-degrees of freedom ship motions are analyzed in the way described in this chapter. In addition, these ship motions have a significant effect on the propeller torque and thrust of the main engine via the propeller. The relationship between these two has been considered as a phenomenon, but has never been clarified by actual operating data. Therefore, this complex relationship between ship motions and the main engine is analyzed using various functions obtained from a multivariate autoregressive model[107][70][73][80].

14.4.1   Response of Ship Motions in Irregular Waves in Experimental Tank Tests

Figure 14.3 shows an example of the frequency response function from wave height to pitching motion obtained in a model test of an offshore structure in irregular waves generated in an experimental tank test. The ∘ mark shows the experimental results obtained in regular

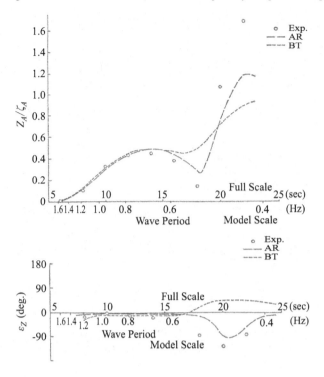

FIGURE 14.3
Comparison of frequency response functions from wave height to pitching motion of a model ship in irregular waves generated in an experimental tank

waves, the dotted line (BT) shows the response function based on the data in irregular waves obtained by the method described in Chapter 6, and the dashed line (AR) shows the results by the method described in this chapter. The upper figure shows the amplitude characteristics and the lower figure shows the phase characteristics. Both amplitude and phase characteristics obtained by the method in this chapter are closer to the experimental results in regular waves than those obtained by the method in Chapter 6. Note that the x axis in these figures show the period, not the frequency, so that the left side of the x-axis indicates higher frequencies.

Figure 14.4 shows an example of frequency response functions calculated by the method described in this chapter using wave height, pitch and roll data obtained in irregular waves generated in an experimental tank[80]. The ○ symbols are the experimental results in regular waves, the thin lines are the theoretical values obtained by the strip method and the thick lines are the results obtained by the MAR model in this chapter for the data in irregular waves. It can be seen that the method by the MAR model is in good agreement with the experimental values and shows almost satisfactory results.

14.4.2 Effects Analysis of Yawing from Other Motions Based on Actual Ship Data

Next, let us examine the effects of other ship motion components on the yawing from time series recorded in full-scale experiments using the cumulative noise contribution[107]. Using the same data as shown in Figures 1.2 and 1.3 in Chapter 1, Figures 14.5 and 14.6 show the

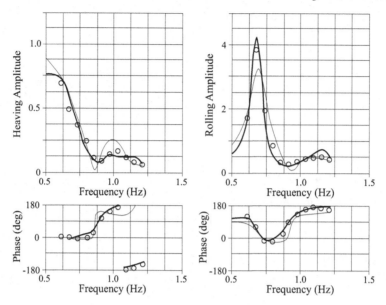

FIGURE 14.4
Frequency response functions of pitch (left) and roll (right) from the wave height of a model
ship in irregular waves generated in an experimental tank.

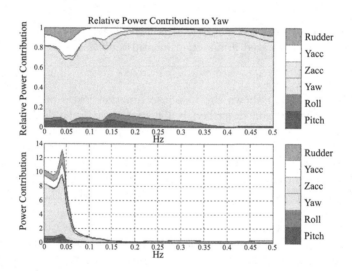

FIGURE 14.5
Cumulative noise contribution (top) and noise contribution (bottom) from pitch, roll, yaw,
Zacc, Yacc and rudder angle to yaw (in case of autopilot steering)

cumulative noise contribution and noise contribution of the effects of pitch, roll, yaw motion
itself, Zacc (vertical acceleration), Yacc (lateral acceleration) and rudder angle on yaw for
data obtained while the medium-sized container ship was being steered automatically and
manually. The upper panel shows the cumulative noise contribution (equation (14.39)), and
the lower panel shows the contribution of each variable to the yaw spectrum when converted
to a spectrum (equation (14.38)). As can be seen from Figure 14.5, autopilot steering has
the highest contribution from the yaw itself to the yaw with a period of about 22 seconds,

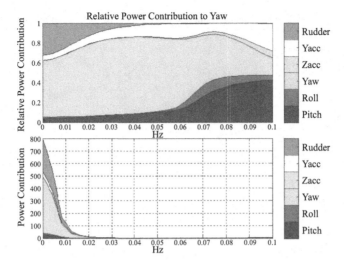

FIGURE 14.6
Cumulative noise contribution (top) and noise contribution (bottom) from pitch, roll, yaw, Zacc, Yacc and rudder angle to yaw (in case of manual steering)

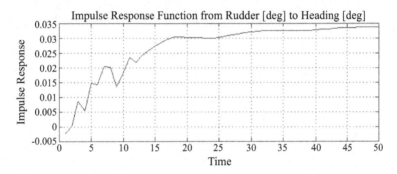

FIGURE 14.7
Impulse response functions of the steering system of a large container ship to rudder to heading (in case of manual steering)

while the rudder, Yacc and pitch contribute equally in this frequency band. In contrast, the contribution from the roll is small. On the other hand, in the case of manual steering, general tendency is the same, but the frequency is generally shifted to the longer period side than in the autopilot steering case. Although not shown in the figure, from the contribution to the rudder, it is found to feedback particularly long-period yaw, in the case of manual steering[70].

Figure 14.7 shows the impulse response function of heading to unit rudder angle for manual steering. Figure 14.8 shows the frequency response function from rudder to heading for the same large container ship. The upper panel shows the amplitude characteristics and the lower panel shows the phase characteristics.

Figure 14.9 shows the impulse response function of manual steering to heading, showing how the rudder responded to a unit change in heading. In contrast, Figure 14.10 shows the rudder's impulse response function to heading during autopilot steering[70]. It is interesting to note the distinctive differences in the way the ship is steered in the manual and automatic steering cases.

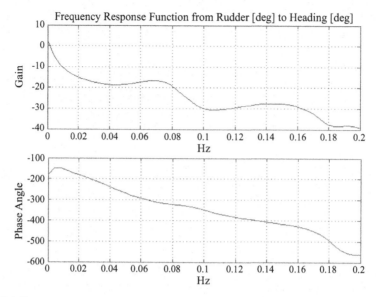

FIGURE 14.8
Frequency response function of the steering system (input: rudder angle, output: heading) of a large container ship (upper plot: amplitude characteristics in dB, lower plot: phase characteristics in deg.)

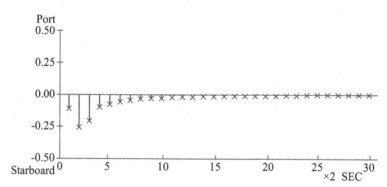

FIGURE 14.9
Impulse response function of manual steering to heading

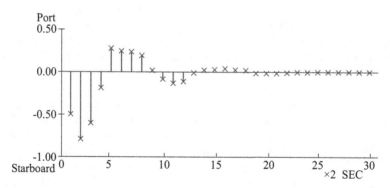

FIGURE 14.10
Impulse response function of autopilot steering to heading

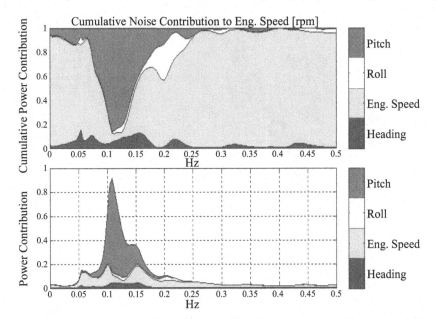

FIGURE 14.11
Noise contribution from pitch, roll, engine speed and heading to propeller revolution in headwave (top: cumulative noise contribution rate, bottom: noise contribution to spectrum)

14.4.3   Analysis of the Ship and Engine Systems of a Real Ship

Figure 14.11 shows the noise contribution of roll, pitch, propeller revolution and heading to the propeller revolution of the main engine for a large container ship sailing in head seas. The upper figure shows the cumulative noise contribution, and the bottom figure shows the cumulative noise contribution from each variable to the spectrum of propeller revolutions. It is clear from the figures that the contribution from pitch motion to the propeller revolution is high around 0.1 Hz (10 sec). This is thought to be due to the vertical movement of the propeller at the stern caused by the pitch motion, which is the cause of the propeller revolution fluctuation. Figure 14.12 shows a scatter plot of the relationship between the standard deviation of the pitch oscillation and the standard deviation of the propeller revolution from a number of batch files of the same container ship with a length of 1200 sec (1 sec sampling). It can be seen that there is a linear relationship between pitch oscillation and propeller revolution.

In the case of a ship, the propeller is equipped at the stern. Therefore, the propeller at the stern is particularly affected by the ship's pitch or vertical motion, which results in thrust fluctuations. Figure 14.13 shows an example of the frequency response function of a container ship to variations in the main engine propeller revolution in response to the ship's pitch motion. From this figure, it can be seen that the propeller revolution responds to the motion at a particular frequency of the pitch motion.

14.4.4   Steering Effects on the Engine

The ship's rudder is located behind the propeller. Figure 14.14 shows the cumulative noise contribution and contributions to spectrum from heading, rudder angle, roll, pitch and propeller revolutions themselves to the propeller revolution spectrum when a container ship

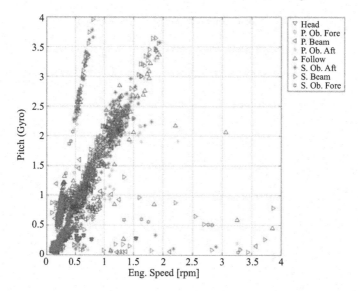

FIGURE 14.12
Scatter plot of standard deviations of pitch and propeller revolution

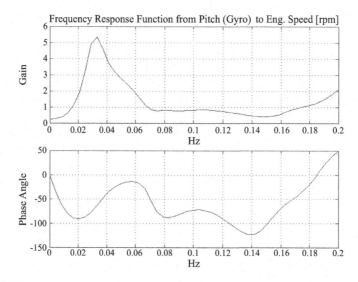

FIGURE 14.13
Frequency response function of a container ship with pitch as input and propeller revolution as output

is being steered at long periods due to a poorly adjusted autopilot. It can be seen that the long period steering angle variation around 0.002Hz causes large fluctuations of propeller revolution. When the steering area is large, additional resistance is created at the stern of the ship. This result should appear as a thrust variation. Figure 14.15 examines the relationship between rudder angle variation and thrust variation in the main period from a number of data sets for the same ship, and clearly shows that the rudder angle variation is related to the thrust variation. It can be inferred that this results in physically an increase in fuel consumption.

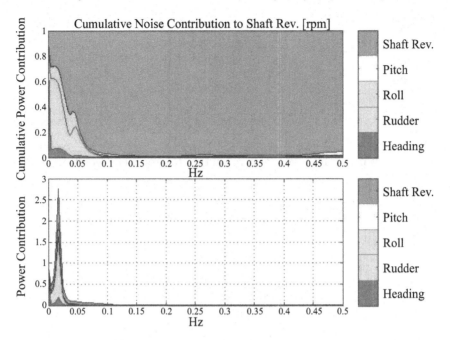

FIGURE 14.14
Cumulative noise contribution function (top) and contribution (bottom) to propeller revolution variation. (Contributions from heading, rudder angle, roll, pitch and propeller revolution variation itself, from bottom to top)

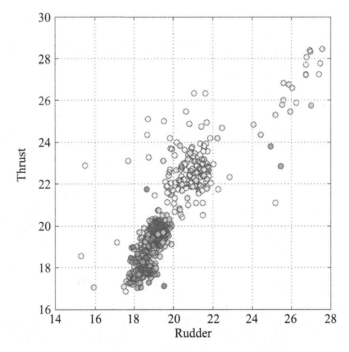

FIGURE 14.15
Relationship between rudder angle and thrust in major period

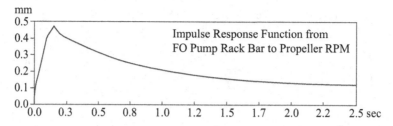

FIGURE 14.16

Impulse response function from governor revolution to propeller revolution variation

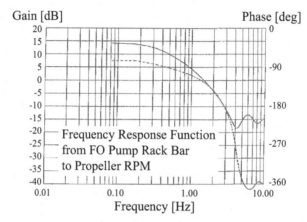

FIGURE 14.17

Frequency response function from governor to propeller revolution variation

14.4.5   Governor-Propeller Revolution Control System

Governor-propeller revolution control, which controls the fuel fed to the piston cylinder to maintain a constant propeller revolution in the main engine, is another typical feedback control system. The governor, together with the autopilot, is a control system with a long history. Recently, however, there has been a shift from centrifugal mechanical governors to electronic governors. Figure 14.16 shows the impulse response function from the governor revolution to the propeller revolution variation obtained from a time series of 10,000 observations measured at 0.2-second intervals. Figure 14.17 is the frequency response function at that time. In these figures, the FO Pump Rack Bar indicates the position of the bar that adjusts the valve that injects fuel in the governor (speed regulator). This FO Pump Rack Bar position determines the amount of valve opening and closing.

14.4.6   Bending Stress on the Hull and Load Variation

Figure 14.18 shows the effect from the hull bending moment on the load indicator by the noise contribution from the time series record of hull stress and the main engine **load indicator** measured on the T.S. Shioji Maru III of the Tokyo University of Mercantile Marine[31]. This training ship is equipped with a strain gage on the bottom plate amidship, which enables the measurement of bending stress on the hull. The left side of the figure shows the effect of bending stress on the load guideline when the ship, sailing at 12 knots,

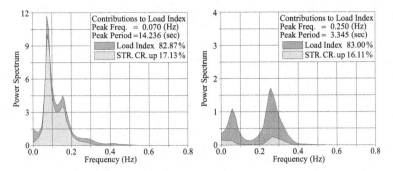

FIGURE 14.18

Contribution of bending stress on the hull to main engine load fluctuations (left: quarter seas, right: beam seas)

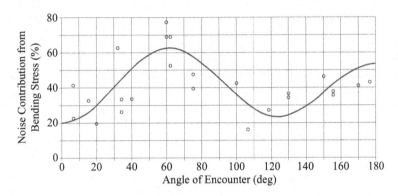

FIGURE 14.19

Effects of main engine load fluctuations due to bending stresses on the hull on the direction of encountered waves

is subjected to quarter seas. The right side of the figure shows the beam seas condition. In both cases, the thin shaded area in the lower part of the figure shows the effect of stress. In the quarter seas, the effect of the large bending stress on the load of the main engine can be seen. Figure 14.19 shows the integrated values over all frequencies of the stress affected region in Figure 14.18, depending on the direction of the wave. Each point is computed from data obtained at the same ship when an on-board wave gauge indicated the same sea state of approximately 1 m. The curve in the figure shows a combination of cosine functions obtained by the least squares method. From both figures, it can be seen that the peak frequency shifts to the low-frequency side as the wave moves from the head seas to the following seas, and the contribution is highest in the quarter seas.

14.5 Monitoring of Kinematics of Ship and Engine Motions

The MAR model treated in this chapter can also be extended, with some modifications, to the online locally stationary MAR model estimation method similar to the univariate AR model introduced in Chapter 9[45]. The bottom row buttons of the top two menu buttons

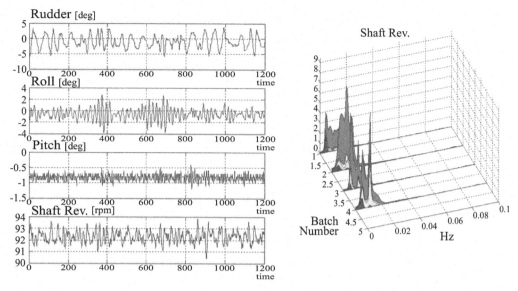

FIGURE 14.20
Contribution to the propeller shaft revolution spectrum from the rudder angle, roll, pitch and propeller revolution of a container ship

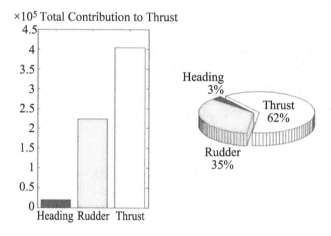

FIGURE 14.21
Percentage of integrated noise contribution of rudder angle, heading and thrust to the thrust of a container ship

in Figure 9.9 are for displaying the noise contribution and other information after the batch section is completed. The left plot of Figure 14.20 shows time series of rudder angle, roll, pitch and propeller revolution from the top while a container ship is sailing in beam seas, and the right figure shows the contributions of rudder angle, roll, pitch and propeller revolution from the bottom to the propeller shaft revolution variation spectrum at that time. In this case, the rudder angle and the roll contribution are slowly increasing because of the beam seas.

Figure 14.21 shows the noise contribution to the thrust spectrum of a container ship integrated over all frequencies and expressed as a percentage. In this case, the highest contribution to thrust is the rudder angle (the uppermost region on the right plot in Figure 14.18), except for the thrust itself. Excessive steering angle changes the surrounding flow field, which in turn causes thrust fluctuations through the propeller, which creates added resistance after all as described in Subsection 14.4.4. As this example shows, monitoring the screen allows the operator to know the causes of the fluctuations of the ship and engine motions.

15

Optimal Control of Ships by Statistical Modeling

In Chapter 14, we analyzed the time series of ship and engine motions and showed that a multivariate autoregressive model is effective in analyzing the causal relationships among these motions and in making model-based predictions. If prediction is possible, control is possible. In this chapter, we consider the application of the multi-variate autoregressive model to the design of ship control systems, as typified by the autopilot system, and the construction of actual control systems using a method called optimal control theory[75, 84].

15.1 Development of Autopilot and its Problems

In general, the control problems of a system can be broadly classified into

1. **Regulator problem** that seeks a control method to match the system output to a certain target value given in advance.

2. **Tracking problem** that seeks a control method that precisely follows the target value when the target value changes.

In the case of ship control systems, the **autopilot system**, which keeps the yawing on the desired course, and the **governor system**, which keeps the revolution of the main engine propeller constant, are typical control systems for the former regulator problem. The latter tracking system is called the **ship's tracking system**, which moves the ship along a given route (wake).

For the control problems of these systems, the following control methods have been used so far. That is, the movement of a controlled variable (such as yawing) is fed back, and for a difference from the desired value (deviation) of the **controlled variable**, an operational variable (**manipulated variable**) such as a rudder is adjusted to correct this deviation, that corrects the deviation, a signal consisting of the sum of a component proportional to the deviation, a component proportional to the time integration of the deviation and a component proportional to the time derivative of the deviation is generated, and the control method commands the actuator that actually generates the operating volume. This type of control method is called **classical control theory**. It has been commonly referred to as **PID control**, an acronym for proportional, integral and derivative. Considering the autopilot shown in Figure 15.1, such a control law is given by

$$\delta = K_P\, e + K_I \int edt + K_D\, \frac{de}{dt},$$

where e is the difference between the controlled variable, namely, the current bow direction φ and the desired bow direction φ_0, i.e., the course deviation $e = \varphi_0 - \varphi$. Also, K_P, K_I,

DOI: 10.1201/9781003428565-15

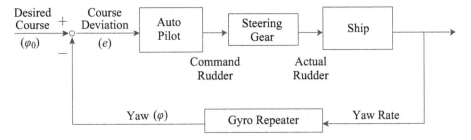

FIGURE 15.1
Ship maneuvering control system with autopilot

and K_D are called **proportional gain integral gain** and **differential gain**, respectively, which are constants to be adjusted appropriately by the designer. Minorsky wrote a paper on this method in 1914[61], which gave the mathematical basis for the method subsequently called classical control theory. Sperry Corporation and Anschütz Corporation developed a method based on this principle called **automatic steering system** (**autopilot**) based on this principle in analog form. This automated the work of the helmsman, who was always steering the ship during navigation. Since then, until 1980, autopilots have basically used the PID control method.

Today, however, the soaring fuel cost, the demand to reduce fossil fuel consumption, and international organizations' obligations to take environmental issues into consideration have led to a demand for improved ship operation methods. The development of control methods that use as little fuel as possible has been demanded for marine control equipment, such as autopilots. However, conventional classical control theory, which cannot incorporate control effects into control methods in concrete terms, cannot meet these demands. Under these circumstances, a new control theory called optimal control theory has emerged in the field of control theory, which constructs a mathematical model of the control target and optimizes the future system predicted by this model under a criterion function to evaluate the control effect. This theory is now called **modern control theory** and is discussed in distinction to the traditional classical control theory[14].

Considering ship autopilot, the crucial difference between modern control theory and classical control theory is that modern control theory uses some mathematical model to represent the motion of the yawing when steering the ship. The role of this mathematical model is to predict the future state of the control system. While classical control theory designs a control law based on the current value of the controlled variable, the design of control systems using modern control theory has the potential to achieve the following by constructing such a mathematical model;

1. It is possible to realize the intended control using predicted future values.

2. It is possible to design a control system with clearly defined control goals.

In the 1980s, when computers became much faster, autopilots using modern control theory appeared one after another[79][76][77][9][37]. Here, we introduce an autopilot design method for optimal control-type control systems that uses a multivariate autoregressive model as a mathematical model of ship steering systems developed by the authors and validate the results.

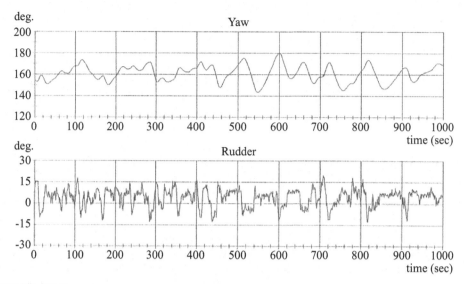

FIGURE 15.2
Response of yawing to random steering

15.2 Representation by Control Type Autoregressive Model and State-Space Representation

15.2.1 Control Type Autoregressive Model

Let us consider the rudder movement (hereafter referred to as **rudder angle** or rudder) of a ship navigating by steering (**course keeping control**) and the deviation from the desired course caused by rudder angle or disturbance (refer to this as **course deviation** or **yaw**). Such a system is called a **maneuvering control system**. Figure 15.2 is an example of such a system, in which the ship is navigating by random steering with almost no control.

As mentioned at the beginning of this chapter, modern control theory constructs a mathematical model to represent the motion of such variables. The modeling requires information from the system under study, and since we are employing a statistical model, we collect time series data of yaw in response to steering that is as random as possible without danger, as shown in Figure 15.2. Now consider these observed signals as the time series $y(1), y(2), \cdots, y(N)$ of yawing in response to N time series $u(1), u(2), \cdots, u(N)$ of rudder angle signals sampled at interval Δt. Then, for these two time series, we consider applying an autoregressive type model[7]

$$y(s) = \sum_{m=1}^{M} a(m)\, y(s-m) + \sum_{m=1}^{M} b(m)\, u(s-m) + v(s). \tag{15.1}$$

Equation (15.1) is called a one-input, one-output control model because it assumes that the one step ahead value of the yawing at time s can be expressed as the sum of the predicted value obtained as the sum of the yawing and rudder angle signal from M time ahead plus a random error $v(s)$. To extend this model to the case of multi-input multi-output control system, we can consider an r-dimensional column vector time series $y(1), y(2), \cdots, y(N)$ in response to the l-dimensional input vector time series $u(1), u(2), \cdots, u(N)$.

In the case of ship control, signals such as yaw and roll can be considered as $y(s)$, and $u(s)$ can be rudder angle and engine output. The last term on the right side, $v(s)$, is an r-dimensional vector white noise that satisfies the following conditions

$$E[v(s)] = 0, \quad E[v(s)\, v^t(s)] = R_1, \tag{15.2}$$

where R_1 is the $r \times r$ variance-covariance matrix of $v(s)$. Let $v(s)$ be uncorrelated with $y(s)$ and $u(s)$. The $r \times r$ matrix $a(m)$ is the weight coefficient from the variable m steps before, $y(s-m)$, to the present value $y(s)$, and the $r \times l$ matrix $b(m)$ is the weight coefficients from the variable m steps before, $u(s-m)$, to the present $y(s)$. Model (15.1) will henceforth be referred to as **control type autoregressive model**.

15.2.2   State-Space Representation

By the way, since the above model (15.1) includes $\sum_{m=1}^{M} a(m)\, y(s-m)$ in the model, the model is considered to be a special form of the multivariate autoregressive model described in Chapter 14,

$$Y(s) = \sum_{m=1}^{M} A(m)\, Y(s-m) + V(s). \tag{15.3}$$

Note that, here the state variable $y(s)$ defined in Chapter 14 is described as $Y(s)$.

In modern control theory, all models are converted into a state-space representation model as described in Chapter 10, and a cost function to evaluate the goodness of the control system is determined using the state-space model to design an optimal control system. In converting to this state-space representation, various methods are shown in Chapter 10, and the controllable representation method and the observable representation method are shown as representative methods of converting to the state-space model. In the following, we will use a multivariate extension of the **observable representation method**, i.e., as a transition matrix, we use

$$F = \begin{pmatrix} A(1) & I & 0 & \cdots & 0 \\ A(2) & 0 & I & \cdots & 0 \\ \vdots & \vdots & \vdots & \ddots & \vdots \\ A(M-1) & 0 & 0 & \cdots & I \\ A(M) & 0 & 0 & \cdots & 0 \end{pmatrix}. \tag{15.4}$$

The reason for this is that the observable representation method can predict the next state of the ship control system in one step when the observation signal from the ship is given. If the state can be predicted in this way, the next control command can be issued immediately using that value. After that, it can be used in a calculation to prepare for the next control cycle so that the state variable can be predicted in a single step. Let us consider again its transformation method described in Chapter 10. First, putting $s+p$ in place of s in equation (15.3) yields

$$
\begin{aligned}
Y(s+p) &= \sum_{m=1}^{M} A(m)\, Y(s+p-m) + V(s+p) \\
&= \sum_{i=1}^{M-p} A(p+i)\, Y(s-i) + \sum_{j=0}^{p-1} A(p-j)\, Y(s+j) + V(s+p). \tag{15.5}
\end{aligned}
$$

Here, the first term of the second equation in equation (15.5) is the part that shows how the past data $Y(s-1), Y(s-2), \cdots$, etc., are used in the calculation of p-step ahead prediction of the $Y(s + p)$ at present time. We put this part as

$$Z_p(s) = \sum_{i=1}^{M-p} A(p + i) Y(s - i), \quad p = 0, 1, \cdots, M - 1.$$

where, for $p = M - 2, \ldots, 1$, $Z_p(s)$ satisfies

$$
\begin{aligned}
Z_p(s) &= \sum_{i=2}^{M-p} A(p + i) Y(s - i) + A(p + 1) Y(s - 1) \\
&= \sum_{i=1}^{M-p-1} A(p + 1 + i) Y(s - i - 1) + A(p + 1) Y(s - 1) \\
&= Z_{p+1}(s - 1) + A(p + 1) Y(s - 1).
\end{aligned}
$$

Also, for $p = 0$, we have

$$Z_0(s) = Y(s) = \sum_{i=1}^{M} A(i) Y(s - i) + V(s) = Z_1(s - 1) + A(1) Y(s - 1) + V(s).$$

The last equation shows a prediction equation for the current value based on data up to time $s - 1$. Namely, it is obtained by adding $A(1) Y(s - 1)$, which is the newly obtained observation $Y(s - 1)$ multiplied by $A(1)$ to the stored state variable $Z_1(s - 1)$, to obtain the $Y(s)$ at time s to obtain an estimate of $\hat{Y}(s)$. These relations can be summarized as

$$
\begin{cases}
Z_0(s) = Z_1(s - 1) + A(1) Z_0(s - 1) + V(s) = Y(s) \\
Z_p(s) = Z_{p+1}(s - 1) + A(p + 1) Z_0(s - 1), \quad p = 1, \cdots, M - 2 \\
Z_{M-1}(s) = A(M) Z_0(s - 1).
\end{cases}
\tag{15.6}
$$

By introducing a new state variable consisting of $Z_0(s), Z_1(s), \cdots, Z_{M-1}(s)$,

$$
Z(s) = \begin{pmatrix} Z_0(s) \\ Z_1(s) \\ \vdots \\ Z_{M-1}(s) \end{pmatrix},
\tag{15.7}
$$

we obtain the system equation in equation (15.3),

$$Z(s) = F Z(s - 1) + W_0(s),
\tag{15.8}$$

where F is given by equation (15.4) and

$$
W_0(s) = \begin{pmatrix} V(s) \\ 0 \\ \vdots \\ 0 \end{pmatrix}.
$$

Since $W_0(s)$ involves only the first element, $Z_0(s)$, the topmost part is $V(s)$ and everything below it is 0.

By the way, considering model (15.3) from the control standpoint, $k \times 1$ ($k = r + l$) vector $Y(s)$ consists of r-dimensional controlled variable $y(s)$ and l-dimensional control variable

$u(s)$. Here, $u(s)$ will move in the future according to the control law, so it is meaningless to predict it. Therefore, we define the state $Y(s)$ by putting $y(s)$ above and $u(s)$ below as follows

$$Y(s) = \begin{pmatrix} y(s) \\ u(s) \end{pmatrix}. \tag{15.9}$$

Correspondingly the coefficient matrix $A(m)$, becomes as follows, where $a(m)$ and $b(m)$ are arranged upward

$$A(m) = \begin{pmatrix} a(m) & b(m) \\ * & * \end{pmatrix}. \tag{15.10}$$

Here, $*$ indicates a disused part of $u(s)$. Similarly, the upper r elements of the vector $V(s)$, $v(s)$ are used and the lower l elements are not used. Furthermore, concerning the k-dimensional vector $Z_m(s)$ of the mth element of the corresponding $Z(s)$, since only the upper r vector is relevant to the prediction, so by putting that part as $Z_m(s)$, we can express as

$$Z_m(s) = \begin{pmatrix} z_m(s) \\ * \end{pmatrix}, \quad m = 0, 1, 2, \cdots, M-1.$$

It can be seen from the first equation in equation (15.6) that

$$Z_0(s) = y(s).$$

Thus, in $Z(s)$ in (15.7), the part that is now useless is packed on top, and we redefine the vector as

$$Z(s) = \begin{pmatrix} z_0(s) \\ z_1(s) \\ \vdots \\ z_{M-1}(s) \end{pmatrix}. \tag{15.11}$$

With this rewriting, the general model of system equations in the state-space model for control is obtained as follows

$$Z(s) = \Phi Z(s-1) + \Gamma U(s-1) + W(s), \tag{15.12}$$

where

$$\Phi = \begin{pmatrix} a(1) & I & 0 & \cdots & 0 \\ a(2) & 0 & I & \cdots & 0 \\ \vdots & \vdots & \vdots & \ddots & \vdots \\ a(M{-}1) & 0 & 0 & \cdots & I \\ a(M) & 0 & 0 & \cdots & 0 \end{pmatrix}, \quad \Gamma = \begin{pmatrix} b(1) \\ b(2) \\ \vdots \\ b(M{-}1) \\ b(M) \end{pmatrix}$$

$$U(s) = u(s), \quad W(s) = \begin{pmatrix} v(s) \\ 0 \\ \vdots \\ 0 \end{pmatrix}.$$

In the above example, $a(m)$ and $b(m)$ are $r \times r$ and $r \times l$ the regression coefficient matrices obtained from the control type autoregressive model, respectively, and $v(s)$ is the r-dimensional residual vector also obtained from the same model.

15.2.3   Observation Equation

Next, we consider the observation equation. The observation equation can be clasified into the following two cases, depending on whether or not there is no noise input and the observation is accurate.

1. **Complete state information**: The system is stochastic, but the observations are complete,

2. **Incomplete state information**: Both the system and the observations contain noises from independent noise sources and only incomplete information is available.

We consider the case of an unknown Gaussian noise input to the system, i.e., consider the case where the complete information of the following linear discrete stochastic system is available

$$y(s) = HZ(s), \tag{15.13}$$

where H is given by

$$H = \begin{bmatrix} I & 0 & \cdots & 0 & 0 \end{bmatrix}. \tag{15.14}$$

I and 0 are the $r \times r$ unit matrix and zero matrix, respectively.

The information criterion AIC for model (15.1) is defined by

$$\mathrm{AIC}(M) = N \log |\Sigma_{r,M}| + 2r(r+l)M + r(r+1), \tag{15.15}$$

where N is the number of data and $\Sigma_{r,M}$ is the variance-covariance matrix of the residuals $v(s)$ that still remain after fitting an autoregressive model (15.1) of order M to the data using least squares method. The second term on the right-hand side is the number of parameters in this control-type model, and it should be noted that the number of regression coefficients for each order is not $(r+l) \times (r+l)$ as mentioned earlier.

15.2.4   Cost Function to Measure the Goodness of the Control Performance

Let us define a cost function that measures the goodness of the control system that we set as the control goal. As such **cost function**, a quadratic form of evaluation function of state variables and manipulating variables is often used. Here, the following two cost functions are considered.

1. Quadratic cost function to evaluate the variance of the state variable and the variance of the control variable

$$J_N = E \left[\sum_{s=1}^{N} Z^t(s) \, QZ(s) + U^t(s-1) \, RU(s-1) \right]. \tag{15.16}$$

2. In addition to the term 1, a cost function that adds the difference from the previous step of the manipulating variable as an evaluation item in order to suppress changes in the movement of the manipulating variable and change the control variable smoothly considering mechanical losses of the actuator,

$$J_N = E \left[\sum_{s=1}^{N} Z^t(s) \, QZ(s) + U^t(s-1) \, RU(s-1) \right.$$
$$\left. + (U(s-1) - U(s-2))^t \, T(U(s-1) - U(s-2)) \right]. \tag{15.17}$$

Here, N represents the interval to be evaluated by such a cost function, and the larger N is taken, the more control methods with control results evaluated over a long future time are obtained. Also, Q and T are nonnegative definite weighting matrices and R is a positive definite matrix for each evaluation item, and the larger this weight is, the smaller the movement of the corresponding variable becomes. As for $Z^t(s) QZ(s)$ appearing in these two functions, if we want to focus only on a particular variable in $Z(s)$, for example the term $z_0(s) = y(s) = HZ(s)$ which indicates the movement of the variable at the most recent time, we adopt a non-negative matrix Q_1 of $r \times r$ instead of Q, as follows

$$Q = H^t Q_1 H.$$

In the design of the control system, only the elements of Q_1 can be adjusted. In this case, Q is given by

$$Q = \begin{pmatrix} Q_1 & 0 \\ 0 & 0 \end{pmatrix}.$$

15.3 Optimal Control Law

In order to solve the problem described in the previous section, there are many methods, such as the variational method, the method based on Pontryagin's maximum principle, and the method using dynamic programming. In this book, we will show a solution based on the **Richard Bellman's dynamic programming**[16]. The details are given in Chapter 19, and only the results are shown here as Control Laws[76][77].

15.3.1 Control Law 1 (When Not Penalizing the Amount of Change in the Manipulating Variable)

It is assumed that the control system transitions according to the following **linear stochastic control model**

$$\begin{cases} Z(s+1) = \Phi Z(s) + \Gamma U(s) + W(s) \\ E[Z(0)] = m_Z, \end{cases}$$

where $Z(s)$ is an r-dimensional state variable vector representing the state of the system, $U(s)$ is an l-dimensional manipulating variable vector, Φ is an $r \times r$ transition matrix and Γ is an $r \times l$ driving matrix. The structure of the noise system is

$$E[W(s)] = 0, \quad E[W(s) W^t(s)] = R_1,$$

and $W(s)$ is assumed to be uncorrelated with $Z(s)$ and $U(s)$.

Assume that the observation process is undisturbed by noise and expressed as

$$y(s) = HZ(s).$$

In this case, the optimal control law that minimizes the cost function up to N periods ahead

$$J_N = E\left[\sum_{s=1}^{N} \left\{ Z^t(s) QZ(s) + U^t(s-1) RU(s-1) \right\} \right], \tag{15.18}$$

where Q and R are symmetric weight function matrices given by

$$U(N-k) = -L(k) Z(N-k). \tag{15.19}$$

Here the optimal control gain $L(k)$ is given by

$$L(k) = \left[R + \Gamma^t S(k-1)\,\Gamma\right]^{-1} \Gamma^t S(k-1)\,\Phi, \qquad (15.20)$$

where $S(k)$ starts from the initial value $S(0) = Q$, and computes recursively in order in the forward direction by

$$\begin{cases} S(k) = Q + \Phi^t M(k)\,\Phi \\ M(k) = S(k-1) - S(k-1)\,\Gamma(\Gamma^t S(k-1)\,\Gamma + R)^{-1}\,\Gamma^t S(k-1). \end{cases} \qquad (15.21)$$

For $k = N$, we have

$$L(N) = (\Gamma^t S(N-1)\,\Gamma + R)^{-1}\,\Gamma^t S(N-1)\,\Phi, \qquad (15.22)$$

which is an optimal control law that considers the state of the system up to N points ahead at $S = 1$.

Note: When the system is stationary, if the evaluation interval of the cost function is increased as $N \to \infty$, then we have $L(k) \to L$, i.e. the gain converges to a constant called **stationary gain**. In this book, except in Section 15.8, we use the fixed gain L where the optimal control input is obtained by

$$U(s) = -LZ(s). \qquad (15.23)$$

15.3.2 Control Law 2 (When the Amount of Change of the Manipulating Variable is Penalized)

If the cost function is set as equation (15.17), the optimal control law is given as follows: Consider the following **linear stochastic control model**

$$Z(s) = \Phi Z(s-1) + \Gamma U(s-1) + W(s), \qquad (15.24)$$

where $Z(s)$ is an r-dimensional vector of state variable representing the system state, $U(s)$ is an l-dimensional vector of manipulating variable, Φ is an $r \times r$ transition matrix and Γ is an $r \times l$ driving matrix. The structure of the noise system $W(s)$ is

$$E[W(s)] = 0, \quad E[W(s)\,W^t(s)] = R_1.$$

Also, $W(s)$ is assumed to be uncorrelated with $Z(s)$ and $U(s)$ and the observation system is assumed to be a complete information system.

The cost function is assumed to be

$$\begin{aligned} J_N &= E\left[\sum_{s=1}^{N} Z^t(s)\,QZ(s) + U^t(s-1)\,RU(s-1) \right. \\ &\qquad \left. + \left(U^t(s-1) - U^t(s-2)\right) T\left(U(s-1) - U(s-2)\right) \right], \qquad (15.25) \end{aligned}$$

where T is the weight of quadratic penalty term for the rate of change of the manipulating variable $U(s-1) - U(s-2)$.

In this case, the optimal control law for the step $N - k$ is given by

$$\begin{aligned} U(N-k) &= -\left\{ \Gamma^t S_Q(k-1)\,\Gamma + S_P^t(k-1)\,\Gamma + \Gamma^t S_P(k-1) + S_R(k-1) + T \right\}^{-1} \\ &\qquad \times \left\{ \left(S_P^t(k-1)\,\Phi + \Gamma^t S_Q(k-1)\,\Phi\right) Z(N-k) - TU(N-k-1) \right\} \\ &= -L_z(k)\,Z(N-k) + L_u(k)\,U(N-k-1), \qquad (15.26) \end{aligned}$$

where $L_z(k)$ and $L_u(k)$ are respectively given by

$$
\begin{aligned}
L_z(k) &= \left\{ \Gamma^t S_Q(k-1)\,\Gamma + S_P^t(k-1)\,\Gamma + \Gamma^t S_P(k-1) + S_R(k-1) + T \right\}^{-1} \\
&\quad \times \left\{ S_P^t(k-1) + \Gamma^t S_Q(k-1) \right\} \Phi
\end{aligned}
\tag{15.27}
$$

$$
L_u(k) = \left\{ \Gamma^t S_Q(k-1)\,\Gamma + S_P^t(k-1)\,\Gamma + \Gamma^t S_P(k-1) + S_R(k-1) + T \right\}^{-1} T.
\tag{15.28}
$$

To obtain $S_Q(k)$, $S_P(k)$ and $S_R(k)$, set the initial values as $P = 0$, $S_Q(0) = Q$, $S_P(0) = 0$ and $S_R(0) = R$, and then perform the following recursive computation;

$$
\begin{cases}
S_Q(k) = Q + \Phi^t M_1(k)\,\Phi \\
S_P(k) = P + M_2(k) \\
S_R(k) = R + M_3(k),
\end{cases}
\tag{15.29}
$$

where

$$
\begin{cases}
M_1(k) = S_Q(k-1) - (S_Q^t(k-1)\,\Gamma + S_P(k-1))\,A^{-1}(k)(\Gamma^t S_Q(k-1) + S_P^t(k-1)) \\
M_2(k) = \Phi^t(S_Q^t(k-1)\Gamma + S_P(k-1))\,A^{-1}(k)\,T \\
M_3(k) = T - T^t A^{-1}(k)\,T \\
A(k) = \left\{ \Gamma^t S_Q(k-1)\,\Gamma + S_P^t(k-1)\,\Gamma + \Gamma^t S_P(k-1) + S_R(k-1) + T \right\}.
\end{cases}
\tag{15.30}
$$

Note

1. If T is removed from the result, it agrees with control law 1.

2. As mentioned in the previous note, for a stationary system, $L_z(k)$ and $L_u(k)$ converge to constant values L_z and L_u as $N \to \infty$.

3. The need to add $L_u(k)\,U(N - k - 1)$ to the control law in equation (15.26) is a natural consequence of the addition of the penalty to the amount of the change of the manipulating variable.

15.4 Design of Autoregressive Type Optimal Automatic Steering System

15.4.1 Configuration of the Optimal Steering System

This section describes a method for designing an automatic steering system, i.e., a system that directs the ship's heading to a desired course, using the method described so far. The autopilot that implements this method will henceforth be referred to as an **autoregressive type autopilot system (AR autopilot)**. Figure 15.3 shows a block diagram of the steering system to be designed here.

The design of this optimal steering system proceeds in two steps as follows. Note here that since the ocean environment usually changes gradually, it is assumed to be locally stationary. Therefore, the optimal gain $L(k)$ obtained from the optimal control law can be regarded as a constant value L. As a result, in designing the autopilot, it is not necessary to calculate the gains while building the model, at each time. Rather we can proceed as follows; in the first stage, data is collected offline and the model and optimal gain L can be calculated. In the second stage, online control is realized using the gains and model

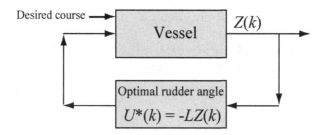

FIGURE 15.3
Block diagram of the autopilot to be designed

parameters created in the first stage. In the following, we will present the design method in detail in these two stages.

[Phase 1] Preliminary design

1. **Selection of inputs and outputs**

   The objective of ship control is to optimally steer the rudder to direct the heading to the desired course. Therefore, we consider a one-input, one-output system. In this case, the variable under control is the difference between the desired course and the current heading, and is denoted by $\varphi(s)$. On the other hand, the control input to this control system is rudder motion. The control variable is **command rudder** $\delta^*(s)$. The command rudder is issued from the optimum steering law. This command rudder angle is transmitted to the **steering gear** at the stern to generate the actual rudder angle $\delta(s)$. This actual rudder angle is called **actual rudder**.

2. **Collecting data for identifying the autoregressive model**

   The one-input, one-output controlled autoregressive model (15.1) is obtained from input-output data. The input/output data used for **fitting** the steering system should be the response signal of the yaw angle when a rudder angle signal which is as random as possible is input to the system, as shown in Figure 15.2. This is because the fitted model will be better if it is made from the response signal of the yaw to an input with as many frequencies as possible. Also, if the ship is already equipped with an automatic steering system, it is important to generate a signal by adding a random signal to the ordered steering signal to be used as the command rudder angle.

3. **Determination of the sampling period and the number of data**

   Since autoregressive type autopilot is digital system, it is important to determine an appropriate sampling period. For example, when designing a PID control system on a digital computer, the sampling period tends to be short. However, if the sampling period is too short, the model will incorporate too much motion that is considered noise in the high-frequency band and will not properly capture important low-frequency yaw motion. On the other hand, too long a sampling period can lead to aliasing. In the case of modeling of a ship, a sampling period of 1 second or longer is recommended in terms of the spectrum of ship motions. The number of data should be as large as possible because it is necessary to include many frequencies. In the case of a ship, a model should be created from data with at least 200 data.

4. **Building of the autoregressive model**

The control type autoregressive model in this case is given by

$$\varphi(s) = \sum_{m=1}^{M} a(m)\,\varphi(s-m) + \sum_{m=1}^{M} b(m)\,\delta(s-m) + v(s),$$

Namely, $r = l = 1$ in equation (15.1). In equation (15.9), k is given by $k = r + l = 2$. The size of the required regression coefficient matrix A is $k \times r = 2 \times 1$ (see equation (15.10)).

From equation (15.15), the AIC is given by

$$\text{AIC}(M) = N \log |\Sigma_{1,M}| + 2 \times 2 \times M + 2. \tag{15.31}$$

[Phase 2] On-line control on board

1. **Calculation of optimal stationary control gains**

Using the built model, set the weight coefficients Q and R (in this case scalars) appropriately, perform a calculation using the dynamic programming method with an evaluation interval I such that the value of $L(k)$ sufficiently converges to a constant L and evaluate each variable by its variance. To estimate the variance of the controlled system, we apply the white noise simulation described in Chapter 13. Q is a scalar in this case, and increasing this weight coefficient suppresses the yaw angle. On the other hand, R is a weight for the rudder angle, and increasing this weight coefficient will limit the rudder motion. In practice, the same results are obtained for the same ratio of Q/R, so for example, simulation results can be compared by setting Q to 1 and changing the value of R in various ways.

2. **Calculation of the steering law and issuance of the command steering angle**

As described in Section 15.2.2, the advantage of the observation-type state-space representation is that when a new signal, i.e., the latest $\varphi(s)$ data, is available, the next command rudder angle can be issued with very few calculations. That is, suppose we are now at time s, by that time, the first equation of $Z(s)$ is prepared as

$$Z(s) = \begin{pmatrix} * \\ z_1(s) \\ z_2(s) \\ \vdots \\ z_{M-1}(s) \end{pmatrix} \Longrightarrow \begin{pmatrix} z_0(s)(=\varphi(s)) \\ z_1(s) \\ z_2(s) \\ \vdots \\ z_{M-1}(s) \end{pmatrix}.$$

The newly obtained $\varphi(s)$ is equal to $z_0(s)$, so substituting this value at $*$ completes $Z(s)$. Then by calculating the product of the current state variable $Z(s)$ thus obtained and the fixed gain, the new command rudder can be calculated as

$$\delta(s) = -LZ(s). \tag{15.32}$$

3. **Preparation for the next step**

In the free time before the next step, we will prepare for the computation in the next step as

$$
\begin{pmatrix} \overset{*}{z_0(s+1)} \\ z_1(s+1) \\ \vdots \\ z_{M-1}(s) \end{pmatrix} = \begin{pmatrix} \overset{*}{a(1)\, z_0(s) + z_1(s) + b(1)\, \delta(s)} \\ a(2)\, z_0(s) + z_2(s) + b(2)\, \delta(s) \\ \vdots \\ a(M)\, z_0(s) + z_M(s) + b(M)\, \delta(s) \end{pmatrix}.
$$

4. **Repeat steps 1 to 3 with control period** Δt.

15.4.2 Pre-Evaluation by Simulation

Next, the control effect is examined by the simulation described in Chapter 13. In this section, variables other than the yaw angle are considered as candidates of output variables in order to investigate the possible effects of not only the yaw angle but also other related ship motions. Here, we show how to systematically search for appropriate Q and R. Suppose now that a control type autoregressive model has been obtained by selecting appropriate output variables and using response time series to input signals as random as possible. Assuming that a state-space model has been built, then the results of the simulation can be evaluated based on

$$
\begin{cases} Z(s) = \Phi Z(s-1) + \Gamma U(s-1) + W(s) \\ U(s) = -LZ(s), \end{cases} \tag{15.33}
$$

where $W(s)$ is the white noise using random numbers generated in the computer, and its variance is assumed to match the variance of the residual term in the fitted model.

Using the following procedure, the values of the weight coefficient matrices Q and R can be selected almost automatically according to the design policy.

1. Initialize the matrices Q and R and a scalar integer K. For example, take as Q a diagonal matrix consisting of the inverses of the diagonal elements of the matrix $\Sigma_{r,M}$ and R a diagonal matrix consisting of the inverses of the permissible limits of the control inputs.

2. Multiply the diagonal components $Q(i,i)$ of the matrix Q by $1 + K^{-1}$ one after another, compare the resulting state variables and determine the most effective component j.

3. Multiply $Q(j,j)$ by $1 + K^{-1}$ and increase K appropriately (for example, $K = K + 1$).

This procedure is repeated to obtain satisfactory Q and R while checking the control effect on the variance of the state variables.

The following is an example of variable selection for a control system simulated using this method. The test data is the time series data of a medium-sized container ship shown in Chapter 1, which was sailing under manual steering as shown in Figure 1.2. The data consists of pitch, roll, yaw, Yacc (lateral acceleration), Zacc (vertical acceleration) and rudder angle. Figure 15.4 shows an example of a simulation using the method described in this section. The left half of the plot shows the simulation of the manual steering data, and the right half plot shows the simulation of yaw and actual rudder angle when we set $Q = (2, 7, 35, 1)$, $T = 3$ and $R = 5$. The elements of Q are, from left to right, weights to pitch, roll, yaw and lateral acceleration. Figure 15.5 compares their spectra. From this comparative study, we find that[79][76][77].

1. The model fitted to the real data adequately represents the real data.

2. Rudder is effective for the control of yaw and roll, but not so effective for pitch and lateral acceleration.

3. The weight T to penalize the amount of changes of steering is effective.

It suggests that roll is also a good candidate for multivariate model. In particular, it is noteworthy that in addition to yaw, roll may also be reduced by a proper steering.

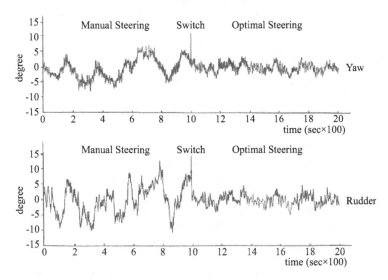

FIGURE 15.4
Simulation of optimal control method. Left: simulation of the original data, Right: simulation of an AR-type optimal steering system

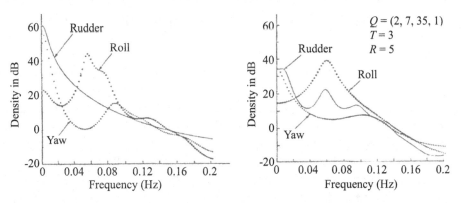

FIGURE 15.5
Digital simulation of AR autopilot designed based on manual steering data of a container ship. (Left plot: spectrum of manual steering, right plot: spectrum of AR autopilot steering)

15.4.3 Preparation for a Full-Scale Ship Experiment with an Autoregressive Model

Here, we present an example of a full-scale ship experiment in which an autoregressive type autopilot system was actually designed using the method described in this chapter. In the

TABLE 15.1

Principal dimensions of the experimental ship

	Shioji Maru II	Shioji Maru III
Length	41.73m	49.93
Width	8.00m	10.00m
Gross Tonnage	331.37G.T.	425.5G.T.
Square coefficient	0.555	0.543

following experiments, two training ships, T.S. Shioji Maru II and T.S. Shioji Maru III, affiliated with the Tokyo University of Mercantile Marine, were used as test subjects. Table 15.1 shows the principal dimensions of these two ships. Both ships were equipped with conventional autopilots designed according to classical control theory. For the experiment, the autopilot equipped on the ship was used, and a switch was installed before entering the controller shown in Figure 15.1 to establish an experimental system that can switch between experimental mode and conventional mode.

The yaw angle signal was taken bypassing the gyrocompass signal used by the ship's existing autopilot as the ship's heading angle. The rudder angle signal was taken from the ship's actual rudder angle meter. After constructing such an observation system, the T.S. Shioji Maru was navigated, and manual steering signals as randomly as possible were generated. In this section, we first describe the experiments conducted on the T.S. Shioji Maru II. In the experiment, the sampling period was set to 1 second and 600 synchronized data were collected. Based on the data, a control-type autoregressive model was built according to the procedure described in Section 15.4.1, and several optimal control gains were created offline.

15.4.4   Full-Scale Ship Experiment

Figure 15.6 shows the analog records of full-scale ship experiment of the AR autopilot designed in this way[79][76][77].

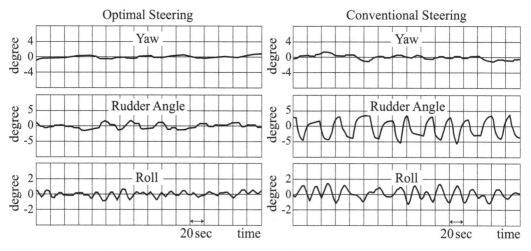

FIGURE 15.6

Experimental results by T.S. Shioji Maru

The right plots of the figure show the time series recordings by the conventional autopilot, and the left plots show the analog recordings of the AR autopilot steering. In the experiment, the ship was first navigated using the AR autopilot, and immediately afterward, a test was conducted using the conventional autopilot on the same sea conditions and heading. Therefore, the weather and sea conditions were almost the same for both autopilots. The figures show that the AR autopilot maintains the desired course with a smaller rudder angle and less yaw angle than the conventional autopilot. The bottom plot of each column shows the roll angle time series taken at the same time. It should be noted that the rudder motion causes not only the yawing but also the rolling, as shown in the results of the conventional autopilot.

15.4.5   Test for the Effects of Control Parameters and Environmental Changes

Next, we analyze the performance of the AR autopilot, designed under various conditions [72][73].

Figure 15.7 shows the results of the standard deviation around the mean of the rudder angle and yaw time series when $T = 0$ and the Q/R ratio (the ratio of the weight Q for yaw to the weight R for rudder angle in the cost function) is increased. It can be seen that the rudder angle increases as the Q/R ratio is increased, but the deviation of the yaw angle remains almost the same above $Q/R = 60$. Figure 15.8 shows the results of full-scale ship experiments with various gains based on these results in the form of standard deviations around the mean value of each variable.

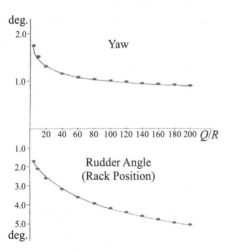

FIGURE 15.7
Simulation results with different Q/R ratios

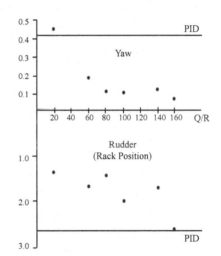

FIGURE 15.8
Results of full-scale ship experiments using various Q/R ratio gains

Figure 15.9 shows an example of a full-scale ship experiment of the AR autopilot with $Q/R = 160$. Figure 15.10 shows the results of a full-scale ship experiment with a conventional autopilot conducted shortly thereafter. From these figures, it can be seen that

1. The deviation of the yaw angle in the case of the AR-type autopilot is less than that of the yaw deviation angle by the conventional system, and the system has excellent course-keeping performance.

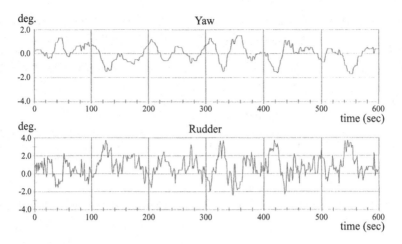

FIGURE 15.9
Course-keeping experiment with AR autopilot with Q/R=160

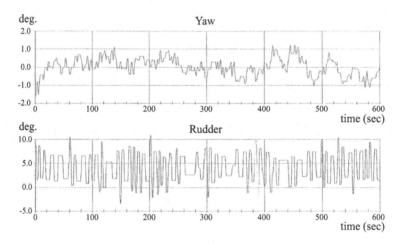

FIGURE 15.10
Course-keeping experiment with PID autopilot

2. The required rudder angle at that time is, despite the slightly higher weighting of the yaw deviation angle, $Q/R = 160$, the ship shows excellent course-keeping ability with a smaller rudder angle than the conventional autopilot.

3. The results of the full-scale ship experiment are almost consistent with the simulation results.

Figures 15.11 and 15.12 show the spectra of yaw and rudder angle for various control gains. As Q/R increases, short-period rudder motion increases, resulting in a decrease in the long-period peak, but a slight increase in the peak around 0.15Hz or 0.20Hz.

15.4.6 Effects from Disturbances

Disturbances at sea are not constant and change from time to time. Further, the direction of the wave and wind is also changed by changing the course. Since the AR autopilot is a fixed gain system, it is necessary to cope with all the changes in disturbance and ship dynamics

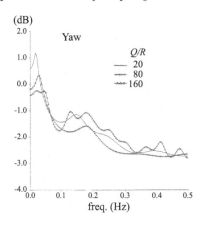

FIGURE 15.11
Yaw spectra for various Q/R ratios (full-scale ship experiment)

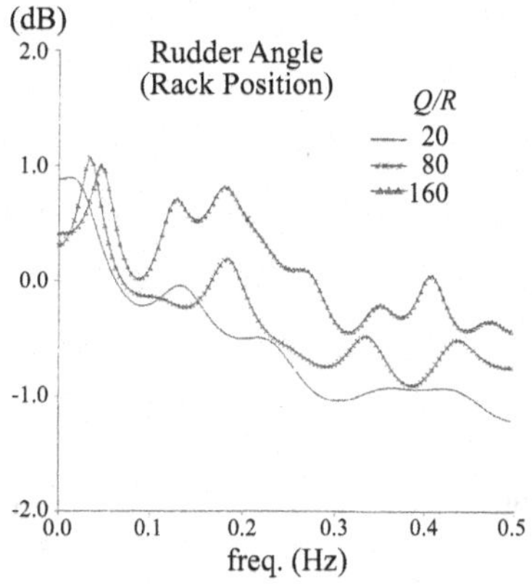

FIGURE 15.12
Rudder spectra for various Q/R ratios (full-scale ship experiment)

at sea with a single gain. Figure 15.13 shows an example of how the control characteristics change depending on the direction of the disturbance, evaluated by changing the ship's heading for a short period of time. The best results were obtained in the case of head seas, and the worst results were obtained in the case of follow seas. This is because in follow seas, especially in quarter seas, the ship motion induced by the waves increases and the fluid inflow rate to the rudder decreases, resulting in poor rudder effectiveness. However, the degree of deterioration in course-keeping ability shown in this figure is acceptable in practical use.

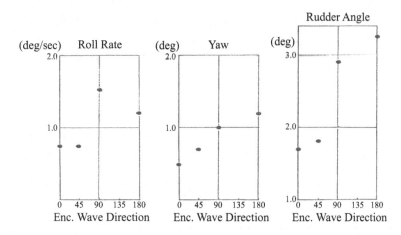

FIGURE 15.13
Variation of AR autopilot with wave directions (standard deviation)

15.4.7 Course Change Ability

The purpose of the automatic steering system is to keep the heading constant while the ship is sailing on a fixed course. If the ship changes course, it is required to move to the new course quickly and without overshooting. Figure 15.14 shows an example of a 20-degree turn to the right at around 500 seconds and return to the original course after 600 seconds. It can be seen that the course changes to the new course extremely quickly without overshooting.

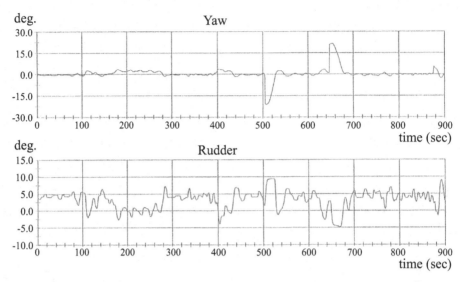

FIGURE 15.14
Course change performance test of AR autopilot

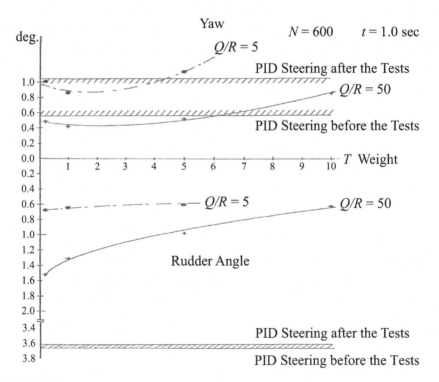

FIGURE 15.15
Effect of weight coefficient T

15.4.8 Effect of T

Too fast fluctuation of the **steering gear** that follows the command signal from the computer leads to the destruction of the steering gear. Figure 15.15 shows, to investigate the effect of T in equation (15.25), we fixed $Q/R = 50$ and $Q/R = 5$ and examined the control performance while varying T. It can be seen that the system has the same or better course keeping performance with less than 1/2 of the rudder angle of the conventional automatic steering system.

Many other full-scaled ship tests were conducted with even larger ships, and the increase in resistance due to the rudder is significant, and in this sense, the introduction of T is expected to suppress the amount of rudder angle.

These results show that the AR autopilot can be designed with a control period of about 1 second, and that by using the results of appropriate random steering experiments and appropriate gain selection, it is possible to design an automatic steering system with extremely high course-keeping performance and excellent steering change performance.

15.4.9 Features of the AR Autopilot from the Viewpoint of Impulse Response

Let us now look at the characteristics of the rudder response to yawing by means of the AR autopilot compared to the PID type, using the impulse response function obtained from the multi-variate autoregressive model described in Chapter 14.

Figure 15.16 shows the impulse response function of the rudder angle to the yawing of the AR autopilot. On the other hand, Figure 15.17 shows the impulse response function of the rudder angle by the PID autopilot. Comparing both figures, the following characteristics can be seen.

1. In the case of the AR autopilot, the initial rudder angle response is large when yaw is detected, and the rudder takes a small turnaround in a short time.

2. In the case of PID autopilot, the initial response is small and a bit slow. The rudder is then kept large for a long time.

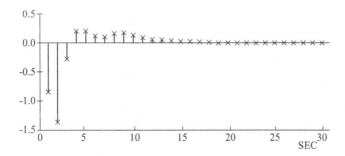

FIGURE 15.16
Impulse response of rudder angle to yawing by means of AR autopilot

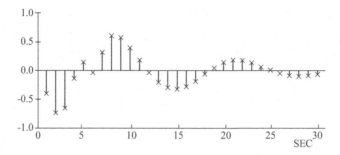

FIGURE 15.17
Impulse response of rudder angle to yawing by means of PID autopilot

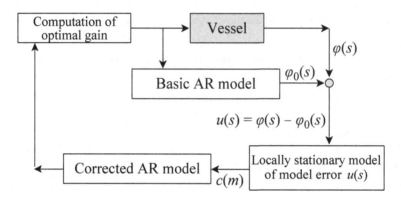

FIGURE 15.18
Configuration of NADCON

15.5   Design of Noise Adaptive Autopilot

In previous experiments, the model representing the ship's steering system was considered fixed. The autoregressive type autopilot system designed in the previous section has the ability to adapt to various types of disturbances. However, in order to further improve the course-keeping performance by taking into account the nonstationarity of disturbances and changes in the dynamic characteristics of the ship itself, it is desirable to have a control system with **adaptivity**, that is, the ability to configure a model that can adapt to such temporal changes in the system.

In Chapter 9, we presented a semi-automatic model-building method for such time series that is nonstationary in long intervals but can be regarded as locally stationary. In this section, we will show how to design a semi-automatic control system that follows this idea, which is called a "**noise adaptive control system**" (**NADCON**)[78][89].

Figure 15.18 shows a block diagram of NADCON, in which two autoregressive models are used in a multi-layered fashion. The first basic autoregressive model is a control-type autoregressive model with intentionally lowered order

$$\varphi(s) = \sum_{m=1}^{M} a(m)\,\varphi(s-m) + \sum_{m=1}^{L} b(m)\,\delta(s-m) + u(s), \qquad (15.34)$$

where $u(s)$ denotes the effect of the disturbance. However, this effect of the disturbance is not necessarily white because the autoregressive model is restricted to a low order. The reason for keeping the autoregressive model low order is that the control type autoregressive model is a statistical model, which should include noise components as much as possible in addition to the regularly moving parts. Therefore, since

$$u(s) = \varphi(s) - \left(\sum_{m=1}^{M} a(m)\,\varphi(s-m) + \sum_{m=1}^{L} b(m)\,\delta(s-m) \right)$$

is generally colored noise, we obtain an autoregressive expression of $u(s)$ as

$$u(s) = \sum_{m=1}^{K} c(m)\,u(s-m) + v(s). \tag{15.35}$$

Substituting equation (15.35) into equation (15.34), we obtain

$$\varphi(s) = \sum_{m=1}^{M+K} A(m)\,\varphi(s-m) + \sum_{m=1}^{L+K} B(m)\,\delta(s-m) + v(s), \tag{15.36}$$

where

$$A(m) = \sum_{j=0}^{m} c(j)\,a(m-j), \quad B(m) = -\sum_{j=0}^{m-1} c(j)\,b(m-j). \tag{15.37}$$

Assume here that $a(0) = c(0) = -1$, $a(i) = 0$ for $i > M$ and $c(i) = 0$ for $i > L$. Now assume that the noise process $u(s)$ is furthermore locally stationary but changes its stochastic structure in the long term. Such a locally stationary process can be fitted online with an optimal model for each batch interval using the procedure already described in Chapter 9.

First, immediately after the start of control, optimal steering is performed using a low-order autoregressive model estimated from the response data to random steering that was obtained in advance as the steering motion. Then, as shown in Figure 9.2, when a data set consisting of N data is obtained, the residuals $u(1), \cdots, u(N)$ are computed for this data set. The K_0-order autoregressive model AR_0 is fitted to this noise process, and a modified autoregressive model is obtained by equation (15.13). Suppose that N new datasets $u(N+1), \cdots, u(2N)$ are obtained. In this case, we can use AIC to determine whether the two datasets are homogeneous or heterogeneous as follows. First, to verify the homogeneity assumption, an autoregressive model AR_C is fitted to the merged data of two datasets, $u_1, \cdots, u_{N+N}$. Then the goodness of the model is evaluated by

$$AIC_C = (2N) \log \sigma_1^2 + 2(K_1 + 1), \tag{15.38}$$

where σ_1^2 is the residual variance of fitted model and K_1 is the order of the model. In contrast, the AIC for heterogeneity is $AIC_D = AIC_0 + AIC_2$, where AIC_0 is the goodness of the model AR_0 in the first part, and AIC_2 is the goodness of the autoregressive model AR_2 for the noise process in the second half. Namely AIC_D is given by

$$AIC_D = N \log \sigma_0^2 + N \log \sigma_2^2 + 2(K_0 + K_2 + 2), \tag{15.39}$$

where σ_0^2 and σ_2^2 are the variances of the residual processes of the models AR_0 and AR_2 and K_0 and K_2 are the orders of the respective models. Then these AIC_C and AIC_D are compared at the end of the batch interval and

1. If $AIC_D \leq AIC_C$, switch from model AR_0 to the new model AR_2.

2. If $\text{AIC}_D > \text{AIC}_C$, switch to the merged model AR_C.

This model fitting method is called the **batch sequential method**. Using this method, the model is switched one after another, locally adapting to changes of the state of disturbance, and immediately after the model is obtained, a new control gain is calculated, and the next interval is steered with the new gain. The calculation of the coefficients of the autoregressive model by the least squares method can be done efficiently in a small area by using the Householder transformation method described in Section 9.1. The state-space representation in this case is given by

$$\begin{cases} Z^t(s) = [\varphi(s)\ \varphi(s-1)\cdots\varphi(s-M+1)\ \delta(s-1)\cdots\delta(s-L+1)] \\ W^t(s) = [v(s)\ 0\cdots0], \end{cases} \tag{15.40}$$

where

$$\Phi = \begin{pmatrix} a(1) & a(2) & \cdots & a(M-1) & a(M) & b(2) & b(3) & \cdots & b(L-1) & b(L) \\ 1 & 0 & \cdots & 0 & 0 & 0 & 0 & 0 & \cdots & 0 \\ 0 & 1 & \ddots & 0 & 0 & 0 & 0 & 0 & \cdots & 0 \\ \vdots & \cdots & \ddots & \vdots & \vdots & \vdots & \ddots & \vdots & \cdots & \vdots \\ 0 & 0 & \cdots & 1 & 0 & 0 & 0 & 0 & \cdots & 0 \\ 0 & 0 & \cdots & 0 & 0 & 0 & 0 & 0 & \cdots & 0 \\ 0 & 0 & \cdots & 0 & 0 & 1 & 0 & 0 & \cdots & 0 \\ 0 & 0 & \cdots & 0 & 0 & 0 & 1 & 0 & \cdots & 0 \\ \vdots & \vdots & \ddots & \vdots & \vdots & \vdots & \vdots & \ddots & \vdots & \vdots \\ 0 & 0 & \cdots & 0 & 0 & 0 & 0 & \cdots & 1 & 0 \end{pmatrix}, \tag{15.41}$$

where

$$\Gamma = \begin{bmatrix} b(1) & 0 & \cdots & 0 & 1 & 0 & \cdots & 0 \end{bmatrix}.$$

Then the state-space representation is given by

$$\begin{cases} Z(s) = \Phi Z(s-1) + \Gamma Y(s-1) + W(s) \\ \varphi(s) = HZ(s). \end{cases} \tag{15.42}$$

15.5.1 Full-scale Ship Experiment of Noise Adaptive Autopilot

To realize the method described in the previous section, two programs need to be run simultaneously: one program sends commands to the rudder in real time. The other program runs in the background at regular intervals, creates models and control gains and sends the obtained models and gains to the former program.

This method has already been commercialized by YDK Technologies Co., Ltd. as **BNAAC (Batch Noise Adaptive Autopilot Controller)**. The left side of Figure 15.19 shows an example of a full-scale ship experiment for a large container ship obtained by BNAAC. The experiment was conducted with $N = 200\,\text{sec}$ and a basic control period of $\Delta t = 1\,\text{sec}$. The right-hand side shows the record of the company's conventional autopilot, which was switched immediately after BNAAC was designed to improve course-keeping performance at as small a steering angle as possible to realize energy-saving effects. As can be seen by comparing the left and right sides of the figure, BNAAC has less deviation of the yaw angle from the ordered course at small rudder angles.

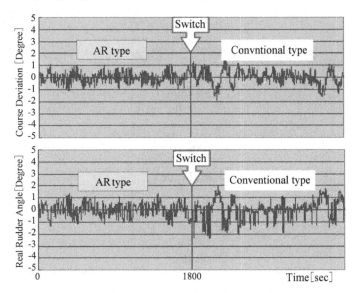

FIGURE 15.19
Full-scale ship experiment with noise-adaptive autopilot (container ship)

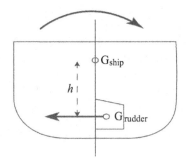

FIGURE 15.20
Causes of rolling induced by rudder

15.6 Design of a Rudder Roll Control Automatic Steering System

The role of the ship's rudder is to direct the bow to the desired direction. However, it has been pointed out that not only does the rudder cause yawing, but it also induces rolling. Figure 15.6 in Section 15.4.3 shows this phenomenon very clearly. Figure 15.20 illustrates this phenomenon. That is, when the rudder is steered, it receives hydrodynamic forces from the surrounding fluid, creating pressure on the rudder surface. As a result, a rolling moment is generated around the point of contact of this pressure and the ship's center of gravity. It is expected that this moment can be used to reduce the roll motions. Simulations had also confirmed this effect of reduction of rolling by appropriate selection of gains[79]. Noting this, Oda et al. have developed, a rudder roll control automatic steering system (**rudder roll control system: RRCS**) with rudder-induced roll reduction effect[69].

The difficulty with this type of autopilot is that since it also controls the roll motions by rudder movement, the rudder may be taken up frequently and significantly when the roll

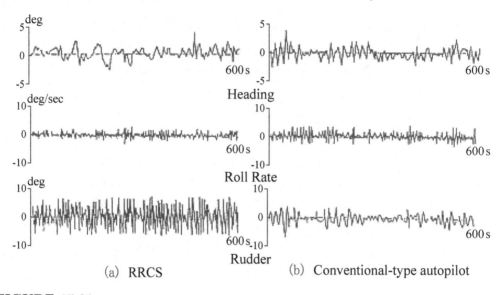

FIGURE 15.21
Example of a full-scale ship experiment of the rudder role control system

becomes severe. As a result, the ability to maintain the course, which is the original function, is reduced. The trade-off between the two should be found by selecting an appropriate gain through repeated simulations with many gains.

Figure 15.21 shows an example of the experiment conducted on the Shioji Maru III described above. Plot (a) shows the result by the RRCS control and plot (b) shows the result by the conventional autopilot of the ship. The upper panel shows the yaw, the middle panel shows the roll rate and the lower panel shows the movement of the rudder angle at that time. Since the RRCS must suppress yawing while reducing rolling, it can be seen that it is moving finely.

However, if the rudder is moved too fast and too often, the actuator will be destroyed. Therefore, we have now introduced a weight T in the cost function (15.25) to address this problem.

To date, this system has been put into practical use by Akishima Laboratory (Mitsui Zosen) Inc., and the following results have been obtained;

1. The longer the distance between the hydrodynamic force acting on the rudder, i.e., the rudder direct pressure, and the center of gravity of the hull, as shown in Figure 15.20, the greater the roll reduction effect.

2. The faster the rudder speed, the greater the roll reduction effect.

3. The effect of roll reduction increases for twin-screw vessels rather than single-screw vessels.

Figure 15.22 compares the results of rudder roll control system (RRCS) and yaw control for a medium-size ship underway at 15 knots in calm seas when rolling is forced by the rudder and then switched to RRCS or yaw control mode. The figure shows that with RRCS, the roll damping is faster and the yaw deviation is smaller. Small and medium-sized ships, such as coastal ships and fishing boats, may capsize when they become parallel to the wave head. In such cases, it is forbidden to steer in a direction that increases the roll, and a rudder roll automatic steering system is considered to be effective.

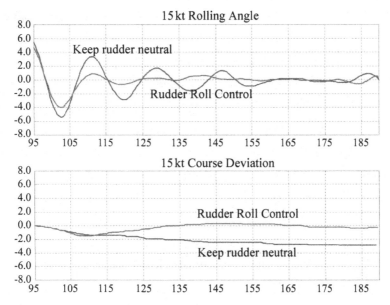

FIGURE 15.22
Roll reduction effect of rudder role control during forced rolling (mid-size ship, 15 knots)

15.7 Design of Main Engine Governor System

15.7.1 Objectives of the Main Engine Governor System for Marine Engine

Next, let us consider the case where the main engine propeller revolution, which is an important control system in marine systems along with the automatic steering system, is controlled by this method using an engine governor[32][33]. The propeller revolution of a marine engine is inherently unstable and the revolution cannot be kept constant without some control. The purpose of the **engine governor** is to control the fluctuations in propeller revolution by adjusting the amount of fuel injected into the piston. This adjustment is ultimately made by adjusting the **rack bar** connected to the fuel valve. Therefore, computer control of the propeller revolution is possible by converting the revolution fluctuation of the propeller shaft into an electrical signal, which is input to the computer, and by sending the electrical signal output from the computer to the actuator that operates this rack bar.

By the way, as described in Chapter 14 and elsewhere, the propeller revolution is affected sensitively by the ship's pitching and heaving motions. Taking this into consideration, Nakatani et al.[65] devised the system shown in Figure 15.23. The lower half of the figure shows the feedback system that gives commands to the fuel pump. The upper half of the figure shows the part of the system that tries to suppress fluctuations by feed-forward control, taking into account ship motions such as pitching and heaving that change slower than the engine revolution control cycle.

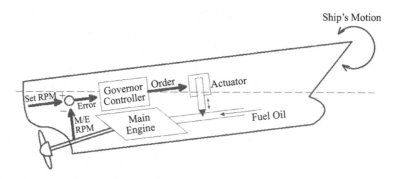

FIGURE 15.23
Configuration of governer control system

15.7.2 Design of a Main Engine Revolution Control System with a New Optimal Governor

In this section, we first consider a one-input one-output system in which the rack bar of the governor is used as the input and the propeller revolution is used as the output.

In the experiment, for safety reasons, the control by the ship's governor is first transferred to the computer governor using the PID control law, and then switched to this experiment. Next, data are obtained to build an autoregressive model in which the input is the signal to the rack bar of the governor and the output is the propeller revolution fluctuation. As mentioned in the design of the autopilot, the input should be as random as possible. However, if it were completely random, the engine would be unstable, so a random number is added to the PID control signal, which is the control law when the engine is switched over, and the RPM fluctuation is measured at that time.

Unlike the case of autopilot, sampling must be fast. The design procedure is the same as that used in Section 15.4. However, for the state variables in the autoregressive model, the objective variable of the control is the main engine revolution, and the input variable is the command signal to the rack bar of the governor.

First, 1000 points time series data of the main engine revolution, pitching and the rack bar position are collected at a period of 0.1 second. Next, this data is used to carry out simulations with various gains. With these preparations, the results of a full-scale ship experiment on the T.S. Shioji Maru III are shown below[32][33].

TABLE 15.2
Results of full-scale experiments with optimal AR governor

Item	Variance (rpm)			Variance (FO Pump rack position)		
Controller	Gain			Gain		
	A0001	A0005	A001	A0001	A0005	A001
Optimal control	7.1952	2.2512	1.2512	0.01963	0.02541	0.07563
Ship's governor		4.0120			0.171	

Table 15.2 shows the results of all experiments at this time, with the variance around the mean. The Q/R ratio indicates the ratio of the weight coefficient for the revolution to the weight coefficient for the rack bars in the cost function. In the table, for example, the experiment number A0005 indicates that the gain ratio to the revolution/rack bar motion is

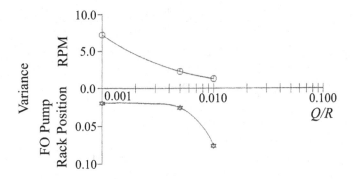

FIGURE 15.24

Comparison of variances of propeller revolution and fuel pump rack bar position in optimal AR governor experiments using various Q/R ratio gains

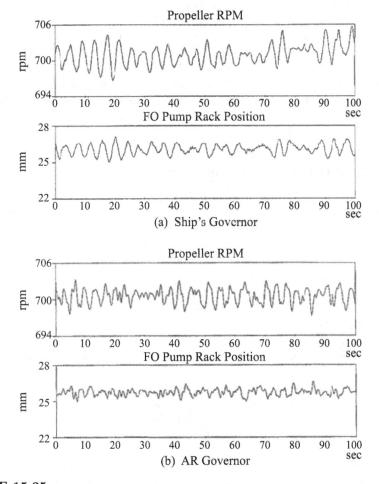

FIGURE 15.25

Comparison of the results of full-scale ship experiments using the ship's governor(a) and the AR governor (b)

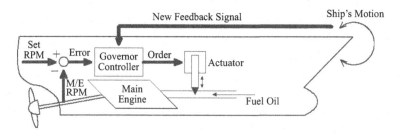

FIGURE 15.26
Conceptual diagram of a new optimal type AR governor that takes into account of ship motions

$Q/R = 0.005/1$. Figure 15.24 plots the gain ratio on the horizontal axis, with the variance of the propeller revolution on the top and the variance of the governor input on the bottom. Figure 15.25 also compares the full-scale ship experimental results of (a) control by the conventional governor of the ship and (b) the AR governor with $Q/R = 0.01/1$. In each figure, the upper panel shows the propeller revolution and the lower panel shows the rack bar position of the fuel pump. The experiment with the optimal AR governor was carried out immediately after the experiment with the ship's conventional governor. It can be seen that the gain is too strong in the case of $Q/R = 0.001/1$ in experiment number A0001, but in other cases, the fluctuation of the optimal AR governor is reduced compared to that of the conventional one.

15.7.3   Design of Optimal Governor Considering Ship Motions

As shown in Chapter 14, fluctuations in propeller revolution are affected by the heaving and pitching of the ship. Therefore, it is conceivable to design a governor that expands the optimal AR governor described in the previous section and adds the heaving and pitching of the ship to the model to improve the prediction of the RPM fluctuation. Figure 15.26 illustrates the concept of the method. Nakatani et al.[65] have carried out an experiment on the T.S. Shioji Maru III, in which a new governor was optimally controlled by adding pitching motions as a predictor variable and predicting RPM fluctuations.

Figure 15.27 shows the record at that time, where Type 1 on the left considers no ship motions and Type 2 on the right considers pitching. However, since we do not have to think about controlling the pitching by the governor, we set the weight to pitching in the cost function to 0. As can be seen from the figure, Type 2 on the right reduces the propeller revolution fluctuation.

Here, the governor does not control the pitching, but the pitching can be controlled by slightly varying the RPM. In the future, such a system may be considered for small ships.

In addition, it may be applicable to the reduction of thrust fluctuation by controlling the blade angle of a controllable pitch propeller.

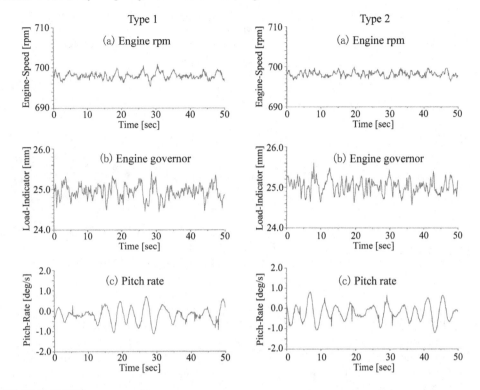

FIGURE 15.27
Effect of governor considering pitching rate (left: without consideration, right: with consideration)

15.8 Design a Tracking Control System

15.8.1 Purpose of Tracking Control

As mentioned at the beginning of this chapter, the purpose of autopilot is to maintain an ordered course. In control engineering, on the other hand, the problem of tracking the operator-directed track is called the **tracking problem**. In today's world where position on earth can be accurately measured with the help of the **GPS (global positioning system)**, computer control of ship tracking is also easily realized. Figure 15.28 illustrates the definition of tracking. In the figure, the arrow at the top indicates the desired trajectory. The ship, on the other hand, has a deviation from the course by Y_d. Here, Y_d is a perpendicular line drawn from the position of the ship to the desired trajectory, and is referred to as **crosstrack error**. The goal of tracking control is to keep Y_d as small as possible by successfully taking the rudder δ.

15.8.2 Tracking Control System

Figure 15.29 shows an overview of the tracking control system. As can be seen from the figure, the latitude X and longitude Y of the ship for each Δt measured from the DGPS are input to the navigation calculation unit. In the navigation calculation unit and ship state calculation unit, the ship position is filtered using the Kalman filter (see Chapter 18), and

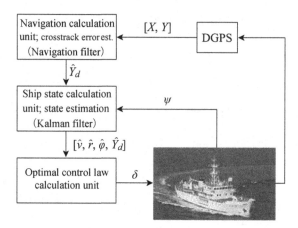

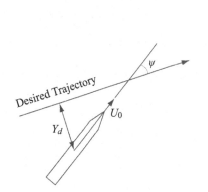

FIGURE 15.28
Definition of ship tracking

FIGURE 15.29
GPS-based ship tracking system

then the estimated values such as the yawing $\varphi(s)$, the yaw angular velocity $r(s)$ and the swaying distance $Y_d(s)$ shown in the figure are input to the optimal control law calculation unit. The optimal control law calculation unit commands the angle to the rudder at the next time on the basis of this information. Tracking control systems usually do not give the rudder angle, but rather the angle of the sight line to the next position checkpoint. However, this system is unique in that it commands the rudder directly. As a result, the amount of required rudder angle, etc. can be directly limited using a cost function[24][62].

The model for optimal control is a state-space representation of a control type autoregressive model with an additional term representing the transition of cross-track errors, as in the following equation:

$$
\begin{pmatrix} \varphi(s) \\ \varphi(s-1) \\ \vdots \\ \varphi(s-M+1) \\ \delta(s) \\ \delta(s-1) \\ \vdots \\ \delta(s-L+1) \\ Y_d(s) \end{pmatrix} = \begin{pmatrix} a(1) & a(2) & \cdots & a(M) & b(2) & b(3) & \cdots & b(L) & 0 \\ 1 & 0 & \cdots & 0 & 0 & 0 & \cdots & 0 & 0 \\ \vdots & \ddots & \ddots & \vdots & \vdots & \vdots & \cdots & \vdots & 0 \\ \vdots & \cdots & \ddots & \ddots & 0 & 0 & \cdots & \vdots & 0 \\ 0 & 0 & \cdots & 0 & 0 & 0 & \vdots & \vdots & 0 \\ 0 & 0 & 0 & 0 & 1 & 0 & 0 & 0 & 0 \\ \vdots & \vdots & \vdots & \ddots & \ddots & \ddots & \ddots & \vdots & \vdots \\ 0 & 0 & \cdots & \cdots & \cdots & \cdots & 1 & 0 & 0 \\ U_0\Delta t & 0 & \cdots & \cdots & \cdots & \cdots & \cdots & 0 & 1 \end{pmatrix} \begin{pmatrix} \varphi(s-1) \\ \varphi(s-1) \\ \vdots \\ \varphi(s-M) \\ \delta(s-1) \\ \delta(s-2) \\ \vdots \\ \delta(s-L) \\ Y_d(s-1) \end{pmatrix} + \begin{pmatrix} b(1) \\ 0 \\ \vdots \\ 0 \\ 1 \\ 0 \\ \vdots \\ 0 \\ 0 \end{pmatrix} \delta(s)
$$

(15.43)

Here, except for the bottom row, the autopilot model described in the previous section is used. However, in this case, the transition matrix Φ is a control-type transition matrix, unlike in the case of autopilot. The bottom row represents the course deviation $Y_d(s)$ at time s. U_0 is the ship's speed. It means that the ship is sailing in the direction of the desired track for time Δt at a speed of $U_0 \sin \varphi(s-1)$, where it is approximated as $\sin \varphi(s-1) \simeq \varphi(s-1)$.

As the cost function, we adopt

$$
J_T = \frac{1}{2} Y_d(N_f) \, Q_f^t \, Y_d(N_f) + \frac{1}{2} \sum_{i=1}^{N_f-1} \left[\left(Y_d(i)^t \; \delta(i)^t \right) \begin{pmatrix} Q & 0 \\ 0 & R \end{pmatrix} \begin{pmatrix} Y_d(i) \\ \delta(i) \end{pmatrix} \right],
$$

(15.44)

where the first term on the right side is the cost of the deviation at the time N_f steps ahead of the current time, the second term is the cost up to the end time, Q is the weight to the deviation and R is the weight to the steering angle.

Bryson and Ho solve the optimal control problem for the minimum of equation (15.44) by using the **calculus of variations**. The result obtained is a real-time solution of the following sequential computation in the reverse direction[19][57]:

$$\begin{cases} \delta(i) = \delta_F(i) - K(i) \\ \delta_F(i) = -\left[\Gamma^t S(i+1)\,\Gamma + R\right]^{-1} \Gamma^t g(i+1) \\ K(i) = \left[\Gamma^t S(i+1)\,\Gamma + R\right]^{-1} \Gamma^t S(i+1)\,\Phi, \end{cases} \tag{15.45}$$

where

$$\begin{cases} S(i) = \Phi^t S(i+1)[\Phi - \Gamma K(i)] + Q, \quad S(N_f) = 0 \\ g(i) = [\Phi - \Gamma K(i)]^t g(i+1) \\ g(N_f) = 0. \end{cases} \tag{15.46}$$

Here, $\delta_F(i)$ is a **feedforward control**. By the computer, the sequential formulas (15.46) are repeated N_f times in the reverse direction from N_f step ahead time, for every Δt, and then assigns the obtained result to equation (15.45) to make the command rudder angle at time i.

As a result of the rapid increase of the speed of the computer in recent years, it is possible to issue the command rudder angle while calculating the gain for every Δt by the control law shown above. This control law is called a **variable gain control**, which changes the gain every moment.

15.8.3 Full-Scale Ship Experiment of Tracking

Figure 15.30 shows the results of sailing T.S. Shioji Maru III (460 tons), a training ship belonged to Tokyo University of Mercantile Marine, on a pre-given polygonal route by an optimal tracking system with an AR model[42]. For a ship of the size of T.S. Shioji Maru, winds of 12 m/sec are a strong disturbance, but even in such winds, the ship is able to turn to the next line (leg) with accuracy and with an error of only the width of the ship. Even after the turn, the ship is able to follows the leg with an error within half the width of the ship (5 m).

By the way, when the route is curved or for an observation ship, it is sometimes necessary to track on a given curve even in such a turn. In order to perform this control with high accuracy, it is better to predict the size of Y_d ahead to some extent and move the rudder to minimize the amount. This is the same as when driving a car on a highway and turning a steering wheel to look ahead to some extent when turning a large curve. Figure 15.31 shows the result of tracking T.S. Shioji Maru on an arc of radius 500 m. Again, the ship is tracked with an accuracy of less than half the width of the ship[41].

Figure 15.32 shows the results of optimal tracking control using a maneuvering motion model called a linearized model (see Chapter 16) with variable gain control. T.S. Shioji Maru temporarily entered a tidal channel off the Boso Peninsula, Japan and experienced a large tide, but after exiting the tidal channel, the ship returned to its original course[62]. Here, the lateral velocity v and the rate of change of the course deviation d_y are taken into control method. So far, this system has been used for tracking control of a 10,000-ton class coastal cement carrier, and has a track record of stable tracking control over a long period of time.

Figure 15.33 shows the results of using the bow thruster of T.S. Shioji Maru III as an auxiliary maneuvering device and stopping it toward a fixed point while changing the ahead and astern speed with a variable-pitch propeller.

Finally, Figure 15.34 shows an example of a 10,000 ton class cement carrier sailing along a pre-designated route toward a fixed point at a reduced speed using bow thrusters, etc.[74].

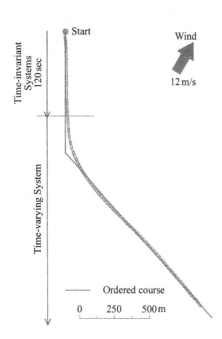

FIGURE 15.30
Tracking along a polygonal line route

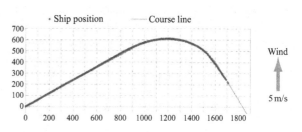

FIGURE 15.31
Tracking of a 500m radius arc route

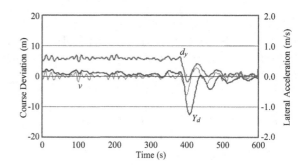

FIGURE 15.32
Retracing the course of a ship entering a tidal channel

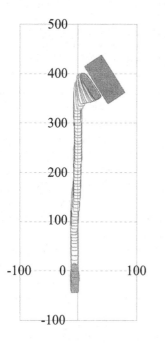

FIGURE 15.33
Berthing experiment by T.S. Shioji Maru

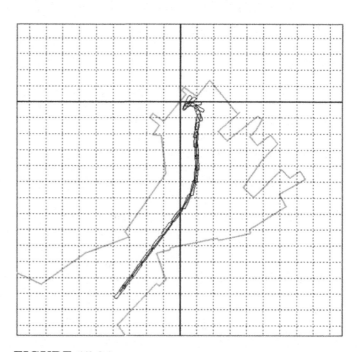

FIGURE 15.34
Berthing experiment on a cement carrier

Part III

For a Better Understanding

16

Introduction to Maneuvering, Ship Motions and Propulsion Theory

This chapter provides an overview of the minimal theories of dynamics of ship and engine motions for a broader understanding of the contents of this book. Readers wishing to gain a more in-depth understanding are referred to the references cited in this chapter.

16.1 Ship Steering Equations

16.1.1 Ship's Maneuvering Motions in Ahead Operation

In a ship that is maneuvering in ahead operation, fluid flowing in from the fore direction exerts various forces (ship's fluid dynamics) on the hull (**body hydrodynamic force**) (Figure 16.1), generating a wake current that reaches the stern. At the stern, the propeller placed in the hull wake generates propulsive force, accelerating the fluid, some of which flows aft and some of which flows to the rudder. Part of it hits the rudder and generates steering force[63].

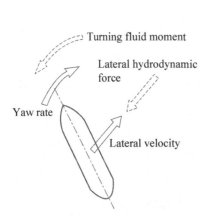

FIGURE 16.1
Hydrodynamic forces to which the hull is subjected

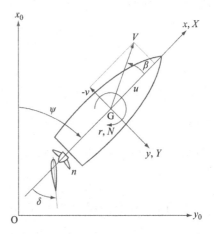

FIGURE 16.2
Maneuverability coordinates

In addition, there are forces called interference forces exerted on the hull by the propeller and rudder. The earth-fixed coordinate system describing the maneuvering motion of a vessel underway at a steady speed V is O-$x_0 y_0$ as shown in Figure 16.2, and the coordinate system viewed from a point fixed at the hull center of gravity is G-xy. At time $t = t_0$, the

DOI: 10.1201/9781003428565-16

hull center of gravity is assumed to be located at x_{0G} and y_{0G}, and the heading at that time is ψ. In this case, the equation of ship's motions seen from the O-$x_0 y_0$ coordinates is

$$\begin{cases} m\ddot{x}_{0G} = X_0 \\ m\ddot{y}_{0G} = Y_0 \\ I_{zz}\ddot{\psi} = N, \end{cases} \qquad (16.1)$$

where m is the mass (ship's displacement), I_{zz} is the moment of inertia around the z-axis taken perpendicularly downward from G to the xy-axis, X_0 and Y_0 are the hydrodynamic forces acting in the x_0 and y_0 directions in the hull, respectively, and N is the hydrodynamic force moment based on hull rotation around the z-axis taken perpendicularly downward from the center of gravity G. Transforming these equations of motions into the coordinate system of motion around the hull's center of gravity, the hydrodynamic forces X_0 and Y_0 are given by

$$\begin{cases} X = X_0 \cos\psi + Y_0 \sin\psi \\ Y = Y_0 \cos\psi - X_0 \sin\psi. \end{cases} \qquad (16.2)$$

Assuming that u and v are the velocities of V in the x and y directions in the fixed hull coordinates, the velocities $\dot{x}_{0G}$ and $\dot{y}_{0G}$ of the hull center of gravity are

$$\begin{cases} \dot{x}_{0G} = u\cos\psi - v\sin\psi \\ \dot{y}_{0G} = u\sin\psi + v\cos\psi, \end{cases} \qquad (16.3)$$

and the acceleration $\ddot{x}_{0G}$ and $\ddot{y}_{0G}$ of the hull center of gravity are given by

$$\begin{cases} \ddot{x}_{0G} = \dot{u}\cos\psi - \dot{v}\sin\psi - (u\sin\psi + v\cos\psi)\dot{\psi} \\ \ddot{y}_{0G} = \dot{u}\sin\psi + \dot{v}\cos\psi + (u\cos\psi - v\sin\psi)\dot{\psi}. \end{cases} \qquad (16.4)$$

Substituting the results of equations (16.3) and (16.4) into equation (16.1), the equation of maneuvering motions in the fixed center of gravity coordinates is obtained as

$$\begin{cases} m(\dot{u} - vr) = X \\ m(\dot{v} + ur) = Y \\ I_{zz}\dot{r} = N. \end{cases} \qquad (16.5)$$

Here, the yaw rate $\dot{\psi}$ is customarily set to

$$r = \dot{\psi}.$$

The $-mvr$ and mur appearing in the equation are the centrifugal force terms in each direction.

16.1.2   Ship's Hydrodynamics in Maneuvering Motions

The right side of the three equations in equation (16.5) are the fluid forces, or moments, from the surrounding fluid for the displacement, velocity and acceleration motions of the hull and rudder or other equipment on the hull in the respective directions[59]. So, let us express them as functions as follows

$$\begin{cases} X = X(u, v, \dot{u}, \dot{v}, r, \dot{r}, \delta, \dot{\delta}) \\ Y = Y(u, v, \dot{u}, \dot{v}, r, \dot{r}, \delta, \dot{\delta}) \\ N = N(u, v, \dot{u}, \dot{v}, r, \dot{r}, \delta, \dot{\delta}). \end{cases} \qquad (16.6)$$

Among these, it is customary to consider the lower two equations and the first equation in equation (16.5) separately, assuming that the maneuvering motions in the x direction, i.e., fore and aft direction, and the motion in the y direction, or turning direction, occur independently. Then, first consider the coupled motion of the lower two equations in equations (16.5) and (16.6). Now, assuming that a ship sailing in calm water with $u = V$, $v = r = \dot{u} = \dot{v} = \dot{r} = 0$ changes by a small variation Δu, Δv, Δr, $\Delta \dot{u}$, $\Delta \dot{v}$, $\Delta \dot{r}$, $\Delta \delta$, $\Delta \dot{\delta}$ respectively, Taylor expansion yields, for example, for Y

$$Y(V + \Delta u, \Delta v, \Delta r, \Delta \dot{u}, \Delta \dot{v}, \Delta \dot{r}, \Delta \delta, \Delta \dot{\delta})$$
$$= Y(V, 0, 0, \cdots, 0) + \frac{\partial Y}{\partial u} \Delta u + \frac{\partial Y}{\partial v} \Delta v + \frac{\partial Y}{\partial r} \Delta r$$
$$+ \frac{\partial Y}{\partial \dot{u}} \Delta \dot{u} + \frac{\partial Y}{\partial \dot{v}} \Delta \dot{v} + \frac{\partial Y}{\partial \dot{r}} \Delta \dot{r} + \frac{\partial Y}{\partial \delta} \Delta \delta + \frac{\partial Y}{\partial \dot{\delta}} \Delta \dot{\delta} + \cdots .$$

The last $\cdots$ denotes nonlinear terms of second order or higher. In maneuverability studies, nonlinear terms up to the third order are considered significant, but here we consider only linear terms (up to the first two terms). The same linear expression can be used for N. In the above equation, for example, the second term on the right-hand side $\frac{\partial Y}{\partial u} \Delta u$ in the differential coefficient representing such a hydrodynamic force is written as $Y_u u$ by convention. Here u is equal to Δu, which represents the micro velocity displacement in the fore and aft direction from the sailing speed V. The same expressions are used for other forces.

Next, let us consider $Y_{\dot{v}} \dot{v}$ among these linear hydrodynamic coefficients. As shown in the top of Figure 16.3, let us assume that the ship moves with an acceleration $\dot{v}$ in the lateral direction indicated by the arrow. At this time, the ship is subjected to a fluid force from the surrounding fluid in the opposite direction of this direction. Plotting the fluid force $Y_{\dot{v}} \dot{v}$ caused by this force $\dot{v}$ with the acceleration $\dot{v}$ on the horizontal axis, the plot below is obtained.

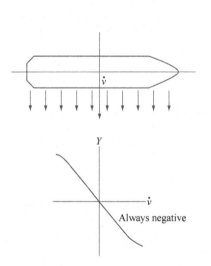

FIGURE 16.3
Lateral acceleration hydrodynamic force of the hull in the lateral direction

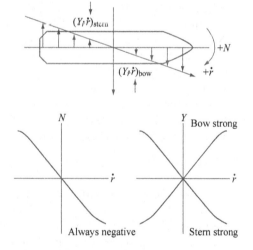

FIGURE 16.4
Yaw rate hydrodynamic force in the heading turning direction

In other words, a negative force $Y_{\dot{v}} \dot{v}$ proportional to the lateral acceleration $\dot{v}$ acts in the region where the lateral acceleration is small. However, since this term includes an

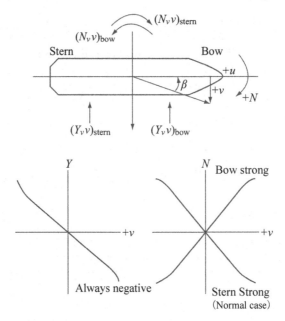

FIGURE 16.5
Hull hydrodynamic force due to lift at hull drift angle β

acceleration term, it is combined with the inertia force $\dot{v}$ on the left side of the equation of motion (16.5), and the sign becomes positive when transferred from the right side to the left side. As a result, the motion itself appears to have increased in mass during the lateral acceleration motion. Therefore, in particular, this term can be written as $m_y = -Y_{\dot{v}}$ and is called **added mass** in the lateral direction and $m + m_y$ is called **apparent mass** in the lateral direction. The lateral added mass of the hull can be seen to be about $m_y = m$.

Next, let us consider the turning moment $Y_{\dot{r}}$ and $N_{\dot{r}}\dot{r}$ for the acceleration of yaw rate. In this case, the ship moves with acceleration of yaw rate $\dot{r}$, and the contribution of the turning moment from the fore and aft of the hull will always have a negative slope, as shown by the up and down arrows in the top of Figure 16.4. According to the equations of motion, the term summarized as acceleration is transferred to the left side and has a positive sign. Therefore, the motion itself appears to have an increased moment of inertia during the acceleration motion. From this, in particular, this term can be written as

$$J_{zz} = -N_{\dot{r}}\dot{r},$$

and is called **added moment of inertia**. $I_{zz} + J_{zz}$ is called **apparent moment of inertia**, where I_{zz} is the moment of inertia around the object in the z direction. The added moment of inertia of the hull in the turning direction can be seen to be about $J_{zz} = I_{zz}$. Also, $Y_{\dot{r}}$ depends on the difference between the fore and aft hydrodynamic forces, but it can be seen to be a small quantity.

The lateral velocity v is related to the drift angle β shown in Figure 16.5, since $v = -u\sin\beta$ when the drift angle is β and the ship velocity is u. Suppose that the fluid now flows toward the hull at an angle β as shown in the top of Figure 16.5. The situation is the same as when the fluid flows into the wing at an attack angle β. That is, $Y_v v$ is equal to the combined force of lift and drag in the lateral direction acting on the hull during oblique sailing.

Moreover, the main component of this force can be seen as a lift force. Since the lift force acts at right angles to the direction of inflow, this lateral force $Y_v v$, with v on the horizontal axis is shown in the bottom left of Figure 16.5. Thus, the sign of $Y_v v$ is always negative. From the origin of such a force, we consider β as a maneuvering motion element, and this force can also be expressed as

$$Y_\beta \beta.$$

This expression is used especially for small motions. The fact that N_v is small can be seen from the relation in the bottom right of the figure.

If we indicate the linear terms that are significant in shipbuilding engineering among these linear ship force differential coefficients, the equations of motion in the y and z directions are simplifies as

$$(m + m_y)\dot{v} \;=\; Y_v v + (Y_r - mu)r + Y_\delta \delta \tag{16.7}$$

$$(I_{zz} + J_{zz})\dot{r} \;=\; N_v v + N_r r + N_\delta \delta. \tag{16.8}$$

Further, by putting

$$\begin{cases} A = (m + m_y)(I_{zz} + J_{zz}) - Y_{\dot{r}} N_{\dot{v}} \\ B = -(m + m_y)N_r - Y_v(I_{zz} + J_{zz}) - Y_{\dot{r}} N_v - (Y_r - mV)N_{\dot{v}} \\ C = Y_v N_r - (Y_r - mV)N_v \\ D = (m + m_y)N_\delta + Y_\delta N_{\dot{v}} \\ E = Y_\delta N_v - Y_v N_\delta, \end{cases} \tag{16.9}$$

since these two equations have two unknowns, eliminating the hard-to-measure variable v, for example, we obtain

$$T_1 T_2 \ddot{r} + (T_1 + T_2)\dot{r} + r = K\delta + K T_3 \dot{\delta}, \tag{16.10}$$

where

$$T_1 T_2 = \frac{A}{C}, \quad (T_1 + T_2) = \frac{B}{C}, \quad T_3 = \frac{D}{E}, \quad K = \frac{E}{C}. \tag{16.11}$$

Furthermore, equation (16.10) can be approximated by a first order system as

$$T\dot{r} + r = K\delta, \tag{16.12}$$

where

$$T = T_1 + T_2 - T_3. \tag{16.13}$$

The two equations (16.10) and (16.13) are called **Nomoto's maneuvering motion equations**, where T is the **initial turning index** or **time constant**, K is the **steady turning ability index** or **gain** and T and K together are called the **maneuverability indices**[68].

16.2 Propulsion Efficiency

16.2.1 Propulsion Efficiency in Calm Water

Propulsion efficiency means the performance of the propeller from the main engine through the shaft to the propeller, where the propeller generates thrust[71][102]. Consider the balancing equation when the ship is sailing at a steady speed u.

The **brake horse power** (**BHP**) main engine purely generated is defined as the work required to move an object weighing 75 kg per second by 1 m before the SI units were adopted, and is also called **main engine power**. In this case, the efficiency is reduced when the output is transmitted to the propeller through the shaft. This is called **delivered horse power** (**DHP**). Letting η_T denote the efficiency at this time

$$\text{DHP} = \eta_T \text{BHP},$$

where the relation between DHP and the torque Q (kg·m) generated by the propeller is

$$\text{DHP} = 2\pi n Q \ \text{(horsepower)}, \tag{16.14}$$

where n is the number of the propeller revolutions, and the unit for revolutions per second is **rps**.

On the other hand, the output generated by the propeller must eventually be balanced with the ship's resistance. The horsepower required at this point is called the **effective horse power** (**EHP**). Assuming that the ship's resistance is proportional to the square of the ship's velocity u, then it is given by

$$R = k \times u^2, \tag{16.15}$$

where k is the resistance coefficient. Therefore

$$\text{EHP} = \frac{u \cdot R}{75} \ \text{(horsepower)}. \tag{16.16}$$

Here, when propeller rotation energy is converted to forward-backward energy in this way, efficiency must be considered. In this case, the **propulsion coefficient** η is defined by

$$\eta = \frac{\text{EHP}}{\text{DHP}}.$$

This η is the product of the efficiency of many intermediate mechanical systems. From the relationship between equations (16.15) and (16.16), we can say that the *engine power required to obtain the velocity* $\boldsymbol{u}$ *is proportional to the cube of the velocity* u.

As the flow approaches the stern, the velocity is affected by the shape of the hull, viscosity and other factors. The flow velocity just before entering the propeller is different from the velocity near the bow.

As the hull moves, the fluid around the hull also moves. At this time, a flow with velocity in the forward direction is generated near the stern. This following flow is called a "wake." In the case of no propeller, three types of following wake

1. Streamline wake

2. Frictional wake (existence of boundary layer)

3. Wake due to stern wave

are considered. Of these, the streamline wake is the wake effect caused by the return of the stern streamlines toward the bow, resulting from the flow of an object placed in an ideal non-viscous fluid (this is called **potential flow**). The frictional wake is the effect of the flow caused by the friction between the water and the hull surface. The wake caused by stern waves is the effect of the forward component of the rotational motion of water particles caused by wave making at the stern. Let u_a be the velocity measured by a pitot tube or current meter at a fixed point on the hull. Also, let u be the ship's velocity. In this case, the velocity of the wake is defined by

$$\text{Wake velocity} = u - u_a.$$

Since the propeller is operating in such a wake velocity, the wake rate is defined as

$$w = \frac{u - u_a}{u} \implies \frac{u_a}{u} = 1 - w.$$

The wake differs significantly between the case without and with a propeller. The hull-only wake has a larger wake at the top of the stern, the so-called V-shaped wake (Figure 16.6).

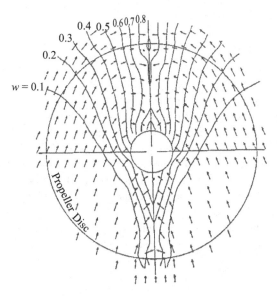

FIGURE 16.6
Distribution of propeller wake rates

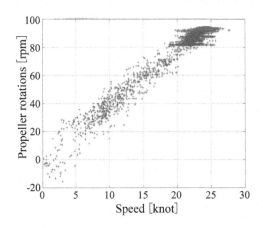

FIGURE 16.7
Relationship between propeller revolution and speed of a certain container ship

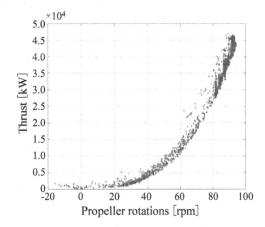

FIGURE 16.8
Relationship between propeller revolution and thrust of the same container ship

When the propeller is equipped with the hull and works, it generates induced velocity. This induced velocity works to push the fluid backward. As a result, the propeller generates **thrust**, giving the ship the power to move forward at a certain speed. The *propeller revolution in calm water is proportional to the ship's speed.* Figure 16.7 shows the relationship

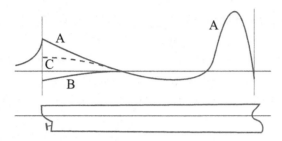

FIGURE 16.9
Thrust reduction due to the distribution of pressure in the bow-stern direction

between propeller revolution and speed measured on a certain container ship. Figure 16.8 shows the relationship between propeller revolution and thrust on the same vessel.

By the way, while the ship is going ahead, the pressure near the stern decreases just behind the propeller due to the increased backward speed. Curve A in Figure 16.9 shows the pressure distribution without a propeller, but with a propeller, the pressure decreases at the stern as shown in curve B, resulting in a pressure distribution as shown in curve C. As a result, the hull appears to have added resistance. This phenomenon is called **thrust reduction**. In this case, the increased resistance is expressed as a fraction of the thrust T. That is, when the thrust reduction coefficient is t, tT is the amount of thrust reduction. Thus only $(1-t)T$ is used to overcome the ship's towing resistance R. In other words

$$R = (1-t)T. \tag{16.17}$$

As a result, to move a ship forward at a certain speed, it must be given a thrust greater than the hull resistance created at that speed.

The two coefficients w and t define the hull resistance and propeller thrust and the inflow velocity at the propeller position. The **hull efficiency** is defined from these coefficients as

$$\eta_H = \frac{1-t}{1-w}.$$

The greater the wake and the smaller the thrust reduction, the greater the hull efficiency. Finally, it is necessary to consider the efficiency of the propeller itself, placed at the stern of the ship, i.e., the **behind propeller efficiency** defined by

$$\eta_B = \frac{Tv_a}{2\pi nQ},$$

where v_a is the flow velocity behind propeller of the ship. Let T_0 be the thrust and Q_0 be the torque when the propeller is in a uniform flow velocity v_a with no hull in front of it, then the propeller alone efficiency η_0 is given by

$$\eta_0 = \frac{T_0 v_a}{2\pi Q_0}. \tag{16.18}$$

The ratio of the propeller efficiency of the propeller independent test to the efficiency when the propeller is installed at the stern of the ship is called the relative rotative efficiency, and is defined as

$$\eta_R = \frac{\eta_B}{\eta_0}.$$

As a result, the propulsion coefficient is given as the product of these three efficiencies as

$$\eta = \eta_B \times \eta_H = \eta_R \times \eta_0 \times \eta_H. \tag{16.19}$$

16.2.2 Self-Propulsion Analysis

The propulsion coefficient of a ship is verified by a **self propulsion test** conducted in a shipyard water tank with the propeller attached to the aft of a scale model ship. The self-propulsion analysis is performed by calculating the propeller advance velocity V_a and the thrust T measured in that way as a function of the propeller advance velocity V_a, and the propeller advance rate as a function of the **advance ratio.**

$$J_P = \frac{v_a}{nD_P} \tag{16.20}$$

(where D_P is the propeller diameter) by the propeller-only torque-thrust curve, i.e., the propeller open performance curve described in equation (16.18), and the point at which it coincides with the propeller-only performance curve (the thrust matching method (**K_T thrust identity**)) (When a torque curve is used, it is called the **torque matching method** (**K_Q identity**)). In the case of the thrust matching method, consider the case where the experimentally determined thrust T matches the propeller open performance in a uniform flow at T_0. Using this T_0, calculate the thrust coefficient K_T,

$$K_T = \frac{T_0}{\rho n^2 D_P^4}. \tag{16.21}$$

Next, identify the thrust coefficient on the propeller open test curve that matches this K_T, and read the advance ratio J_P, propeller open water efficiency η_0, and at the same time torque coefficient

$$K_Q = \frac{Q}{\rho n^2 D_P^5}. \tag{16.22}$$

At this time, η_0 is

$$\eta_0 = J_P \frac{K_T}{2\pi K_Q}. \tag{16.23}$$

From the advance ratio J_P, the wake coefficient $1 - w$ can be calculated by

$$1 - w = \frac{J_P \, n \, D_P}{v_m},$$

where v_m is the advance speed of the model ship. From the hull resistance and propeller thrust obtained by the resistance test, the coefficient of thrust reduction can also be obtained by

$$1 - t = \frac{R}{T}. \tag{16.24}$$

16.2.3 Equation of Ship's Surging Motion

The dynamic equation of ship's surging motion can be obtained based on a static self-propulsion test as follows.

The surging component of the hydrodynamic force due to the surging acceleration motion is expressed as $X_{\dot{u}}$, which is negative, but to summarize it as an acceleration term, if the added mass in the surging direction is defined by

$$m_x = -X_{\dot{u}},$$

it is given by $(m + m_x)\dot{u}$ when taken together with the inertia term on the left side. The $m+m_x$ is the coefficient called the apparent mass in the surging direction, and its magnitude

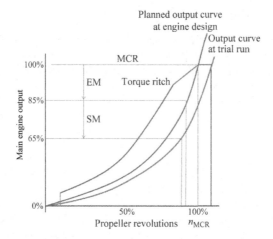

FIGURE 16.10

Engine performance curves and sea margins

is about $1.1m$, since the added mass in the surging direction is about $0.1m$ in the case of a ship.

Taking these into account, the equation of motion in the surging direction can be expressed as

$$(m + m_x)\dot{u} = (1 - t)T - ku^2 - X_\delta \delta, \qquad (16.25)$$

where m_x is the added mass in the surging direction, about $0.1m$, and X_δ is the resistance coefficient of the rudder in the forward direction.

In this case, the equation of the rotational torque of the main engine that provides the rotational force (torque) to the propeller is given by

$$2\pi(I_E + J_{pp})\dot{n} = Q_E + Q_P, \qquad (16.26)$$

where J_{pp} is the added mass related to the propeller rotation, Q_E is the torque generated by the main engine, Q_P is the rotational torque generated by the propeller given by

$$Q_P = -\rho n^2 D_P^5 K_Q(J_P),$$

and the propeller revolutions fluctuate to satisfy equation (16.26).

16.2.4   Sea Margin

As described in Section 16.2.1, the power output in calm water is proportional to the cubic of the propeller revolution. The rightmost curve in Figure 16.10 is the propeller revolution vs. power curve in calm water at shipbuilding. When planning the engine power at shipbuilding, the curve is designed to be slightly to the left of this value, which is the **margin**. However, since the navigational speed in rough seas must be guaranteed, a further margin is designed in addition to this. The engine can operate within this closed curve. If the load is higher than this, a condition called **torque rich** occurs.

At this time, the quantity called **sea margin** defined by

$$SM = \frac{T - T_0}{T_0} \qquad (16.27)$$

is important for operation, where T_0 is the required thrust in calm water. The sea margin varies depending on the number of years after shipbuilding, ship's bottom fouling, sea conditions and other factors.

16.3 Equation for Ship Motions

16.3.1 Center of Gravity, Center of Buoyancy, Metacenter

Next, the equation of ship motions with restoring force is explained using the example of rolling motion[59]. The equations for other types of pitching and heaving are the same as those for rolling motion, with the only difference being the coefficients. Rolling of a ship is caused by the fact that the gravitational force acting downward through the center of gravity G and the buoyancy force acting upward through the geometric center of the liquid (center of buoyancy) from which the hull is excluded are not on a straight line (Figure 16.11).

B′ in the figure represents the center of buoyancy when tilted. The upward buoyancy force from B′ and the downward gravity force from G exert a couple of force around G in the direction of left rotation. This direction is the rotational contingent force that tries to return the ship to the neutral axis extending upward from G. In this case, M in the figure is called the metacenter.

16.3.2 Nonresistance Free-Rolling Equation

Assuming that nonresistance proportional to the roll-rate is exerted on the ship's rolling motion, the ship free-rolling equation under the initial condition of inclination in calm water is expressed as

$$(I_{xx} + J_{xx})\frac{d^2\theta}{dt^2} + W \cdot \mathrm{GM} \cdot \theta = 0, \tag{16.28}$$

where the left-hand bracket is the apparent radius of gyration given by

$$I = I_{xx} + J_{xx} = \frac{W}{g}k_x^2 + \frac{W}{g}k_x'^2 = \frac{W}{g}(k_x^2 + k_x'^2) = \frac{W}{g}k_x''^2. \tag{16.29}$$

k_x and k_x' are the radius of gyration and the **added radius of gyration**.

Also, k_x'' is the apparent radius of gyration given by

$$k_x''^2 = k_x^2 + k_x'^2.$$

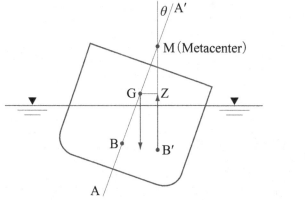

FIGURE 16.11
Metacenter concept

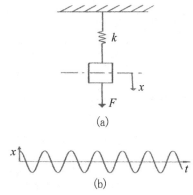

FIGURE 16.12
Spring system with spring constant k

GM is the length between the hull's center of gravity G and the metacenter M, which is the point where a line drawn vertically upward from the center of buoyancy B′ shown in the figure intersects the neutral line AA′ that passes through the center of gravity. In this case, the free-rolling equation can be written as

$$\frac{d^2\theta}{dt^2} + \frac{g \cdot GM}{k_x^2 + k_x'^2}\,\theta = 0. \tag{16.30}$$

If we put

$$\omega_n^2 = \frac{g \cdot GM}{k_x^2 + k_x'^2} = \frac{g \cdot GM}{k_x''^2}, \tag{16.31}$$

we can also write

$$\frac{d^2\theta}{dt^2} + \omega_n^2\theta = 0. \tag{16.32}$$

The dynamical system in this case is equivalent to the spring system shown in Figure 16.12. The solution is

$$\theta = \frac{H}{2}\,\sin(\omega_n t + \phi) = \Theta \sin(\omega_n t + \phi), \tag{16.33}$$

where H is the maximum total amplitude of the rolling motion, Θ is the single amplitude given by

$$\Theta = \frac{H}{2}. \tag{16.34}$$

Let ω_n be the rolling angle frequency and T be the period, it is given by

$$T = \frac{2\pi}{\omega_n} = 2\pi\sqrt{\frac{k_x''^2}{g \cdot GM}} = 2.01\frac{k_x''}{\sqrt{GM}}. \tag{16.35}$$

According to the experiments, by setting B to be the ship's breadth and

$$k_x'' = cB, \tag{16.36}$$

then the apparent radius of gyration c is approximately given as shown in Table 16.1.

Or, as an experimental formula, it is given as

$$\frac{k_x'}{B} = C_0 + 0.0227(B/d) - 0.0043(L/10) \tag{16.37}$$

$$C_0 = \begin{cases} 0.3725 & \text{normal ship type} \\ 0.3085 & \text{Inclined-ship type, flare-ship type.} \end{cases}$$

Therefore, it is approximately given as

$$T = 2.01\frac{k_x''}{\sqrt{GM}} \simeq \frac{0.8B}{\sqrt{GM}}. \tag{16.38}$$

TABLE 16.1
Values of c

Cargo	Full	$0.32 \sim 0.35$
	Light	$0.37 \sim 0.40$
Tanker	Full	$0.35 \sim 0.39$
	Light	$0.37 \sim 0.47$
Passenger		$0.38 \sim 0.43$
Fishing Boat		$0.38 \sim 0.43$
Battle Ship		$0.38 \sim 0.43$

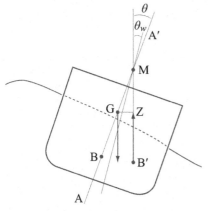

FIGURE 16.13
Froude-Krylov force assumption

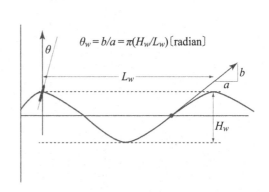

FIGURE 16.14
Concept of wave inclination

From this, GM is obtained by

$$GM = \frac{0.64B^2}{T^2}. \tag{16.39}$$

16.3.3 Nonresistance Forced Rolling

Suppose a sinusoidal wave is input as an external force to a free-rolling ship, and find the equation for this case.

The following concept of **effective wave slope** is used, based on the assumption of the Froude-Krilov force shown in Figure 16.13. Suppose that a wave attacked with a curve of the wavefront shown in Figure 16.14,

$$y = \frac{H_w}{2} \cos\left(2\pi \frac{x}{L_w}\right). \tag{16.40}$$

The angle of inclination of the wavefront to the horizontal θ_w is

$$\theta_w \simeq \tan\theta_w = \frac{dy}{dx} = \pi \frac{H_w}{L_w} \sin\left(2\pi \frac{x}{L_w}\right). $$

At this time

$$\Theta_w = \pi \frac{H_w}{L_w} \tag{16.41}$$

is called the maximum effective wave slope and is the maximum value of the effective wave slope that usually satisfies

$$\frac{1}{30} < \frac{H_w}{L_w} < \frac{1}{10}.$$

In the case of deep-sea waves, the difference between the period and the wavelength of the wave satisfies the relation

$$T_w = \sqrt{\frac{2\pi L_w}{g}} = 0.8\sqrt{L_w}. \tag{16.42}$$

If the ship is stopped at $x = 0$, the time variation of the wave slope is

$$\theta_w = \pi \frac{H_w}{L_w} \sin\left(2\pi \frac{t}{T_w}\right) = \Theta_w \sin\left(2\pi \frac{t}{T_w}\right), \tag{16.43}$$

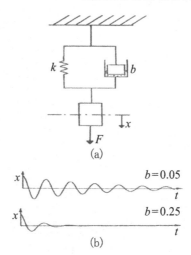

FIGURE 16.15
Resistance spring system with resistance coefficient $b = 2\gamma$

where T_w is the wave period. Assuming that the ship is tilted by θ_w due to the waves by the assumption of the Froude-Krylov force, the restoring force is given by

$$W \cdot \text{GM} \cdot (\theta - \theta_w) = W \cdot \text{GM} \cdot \left\{ \theta - \Theta_w \sin\left(2\pi \frac{t}{T_w} \right) \right\}. \tag{16.44}$$

As a result, the ship forced free-rolling equation can be written as

$$\frac{W}{g} k_x'' \frac{d^2\theta}{dt^2} + W \cdot \text{GM}\{\theta - \Theta_w \sin(\omega_e t)\} = 0,$$

where $\omega_e = 2\pi/T_w$. Alternatively, one can use

$$\frac{1}{g} k_x'' \frac{d^2\theta}{dt^2} + \text{GM}\{\theta - \Theta_w \sin(\omega_e t)\} = 0. \tag{16.45}$$

To make this equation easier to understand, if we put

$$\omega_n^2 = \frac{g \cdot \text{GM}}{k_x''},$$

we obtain

$$\frac{d^2\theta}{dt^2} + \omega_n^2 \theta = \omega_n^2 \Theta_w \sin(\omega_e t). \tag{16.46}$$

16.3.4   Resistance Free Rolling

The ship rolling equations obtained so far are for the case of nonresistive rolling, but in reality, there is resistance proportional to the roll rate. In a spring system, this is a spring with added resistance, as shown in Figure 16.15.

In the case of a ship, if γ is **roll damping coefficient**, such a term is a reaction force proportional to the velocity, so the resistance-free roll equation can be written as

$$\frac{d^2\theta}{dt^2} + 2\gamma \frac{d\theta}{dt} + \omega_n^2 \theta = 0. \tag{16.47}$$

16.3.5 Resistance Forced Rolling

Resistance forced rolling in waves is given by

$$\frac{d^2\theta}{dt^2} + 2\gamma\frac{d\theta}{dt} + \omega_n^2\,\theta = \omega_n^2\,\Theta_w \sin(\omega_e t). \tag{16.48}$$

Note: *In the book on oscillation [64], we often write* $\zeta = \frac{\gamma}{\omega_n}$. *In this case,* $2\zeta\omega_n$ *replaces* 2γ. *That is, for example, it can be written as (see Chapter 5);*

$$\frac{d^2\theta}{dt^2} + 2\zeta\omega_n\frac{d\theta}{dt} + \omega_n^2\,\theta = \omega_n^2\,\Theta_w \sin(\omega_e t). \tag{16.49}$$

The analytical solution of equation (16.46) is generally given by

$$\theta = Ae^{-\gamma t}\cos\left(\sqrt{\omega_n^2 - \gamma^2}\,t + \varepsilon\right) + \frac{\omega_n^2\,\gamma\,\Theta_w}{\sqrt{(\omega_n^2 - \omega_e^2)^2 + 4\omega_e^2\,\gamma^2}}\cos(\omega_e\,t + \varepsilon'). \tag{16.50}$$

The first term on the right-hand side is free rolling, which decays with time. The second term is the forcing term due to waves, and if we put

$$\begin{aligned}
\theta_1 &= \frac{\omega_n^2\,\gamma\,\Theta_w}{\sqrt{(\omega_n^2 - \omega_e^2)^2 + 4\omega_e^2\,\gamma^2}}\cos(\omega_e\,t + \varepsilon') \\
&= \mu\,\gamma\,\Theta_w\cos(\omega_e\,t + \varepsilon'),
\end{aligned}$$

μ indicates how many times the effective wave slope is oscillated, and this is called the **magnification factor**. Here

$$\begin{aligned}
\mu &= \frac{1}{\sqrt{\left\{1 - \left(\dfrac{\omega_e}{\omega_n}\right)^2\right\}^2 + \left(\dfrac{2\gamma}{\omega_n}\right)^2\left(\dfrac{\omega_e}{\omega_n}\right)^2}} \\
&= \frac{1}{\sqrt{(1 - \xi^2)^2 + \kappa^2\xi^2}}, \quad \xi = \frac{\omega_e}{\omega_n} = \frac{T}{T_w}, \quad \kappa = \frac{2\gamma}{\omega_n},
\end{aligned}$$

where ξ is called **tuning factor** in shipbuilding engineering and is synchronized by $\xi \simeq 1$.

16.3.6 Variation of Coefficients in the Rolling Motion Equation and Estimation of GM

(1) Rolling in Smooth Water

In the rolling motion equation with resistance, in the case of rolling motion in smooth water, the damping coefficient of the rolling motion is weak and the period may be regarded as a simple harmonic oscillation determined by the coefficient ω_n related to the restoring force. In this case, equation (16.38) holds and isochronism appears.

Therefore, if the rolling period can be measured, the GM can be estimated with good accuracy.

(2) Rolling in Waves

The larger the angle of rolling in waves, the less proportional the restoring force becomes to the angle. Also, the apparent mass (moment of inertia) changes and isochronism is lost.

(3) Speed Effect

Ships are less likely to sway when they have speed. This is because the damping coefficient increases with speed due to the lift effect.

17

Basics of Probability

This chapter summarizes the minimum knowledge of probability and statistics required to understand this book. For details, reference[22][20] are famous books. Books written in Japanese include[51][54], and engineering applications include[95].

17.1 Fundamentals of Probability Theory

17.1.1 Combinatorial Probability

Consider a **trial** involving chance, such as the roll that appears in a dice toss or the two sides of a coin that appear in a coin toss[22]. The matters that appear in these trials are called **elementary event** and are denoted by ω, and the whole set of elementary events is called **sample space** and written as Ω. The **compound event** is the collection of ω, i.e., a subset of Ω, and is denoted by A. The **probability** $P(A)$ of the event A is the degree of the occurrence of the event A on this sample space, and the probability of the sample space Ω is 1. Probability has the following properties;

1. The probability of a subset A of Ω satisfies

$$0 \leq P(A) \leq 1. \tag{17.1}$$

   In particular, it holds that
$$P(\Omega) = 1. \tag{17.2}$$

2. If A and B are subsets of Ω, we have

$$P(A \cup B) = P(A) + P(B) - P(A \cap B). \tag{17.3}$$

   In particular, if $A \cap B = 0$, then we have

$$P(A \cup B) = P(A) + P(B).$$

3. For the null set ϕ containing nothing, we have

$$P(\phi) = 0. \tag{17.4}$$

4. Probability of the complement A^c of an event A is given by

$$P(A^c) = 1 - P(A). \tag{17.5}$$

DOI: 10.1201/9781003428565-17

When a variable $X(\omega)$ is a real-valued function defined for any event ω of this sample space Ω, and a probability $P(X(\omega))$ satisfying the above axiom of probability can be assigned to any value of $X(\omega)$, $X(\omega)$ is called a **random variable**.

[**Example of a random variable**] When normal dice are thrown at random, the realized values 1, 2, 3, 4, 5 and 6 of the random variable X appear with probability 1/6. In this case, ω is 6 numbers.

For some real number x, consider the set $(\omega \mid X(\omega) \leq x)$ of events smaller than x. Then a function that satisfies

$$F(x) = P(\omega \mid X(\omega) \leq x) \tag{17.6}$$

is called the **cumulative distribution function** of the random variable X. $F(x)$ is monotone non-decreasing function and has the property that $F(\infty) = P(\Omega) = 1$. Then the derivative of $F(x)$, that intuitively corresponds to the probability in the infinitesimal dx,

$$f(x) = dF(x)/dx \tag{17.7}$$

is called the **probability density function**.

17.1.2 Expectation and Variance

Given a probability density $f(x)$ of a random variable X,

$$m_X = E[X] = \int_\Omega x\, f(x) dx \tag{17.8}$$

is called the mean value or **expectation** and is sometimes denoted as $\bar{x}$.

The second-order moment around some real value a is defined by

$$\sigma_a^2 = \int_\Omega (x - a)^2 f(x) dx. \tag{17.9}$$

When choosing the mean value m_X as a, i.e.,

$$\sigma_{m_X}^2 \equiv \mathrm{Var}(X) = E\left[(X - E[X])\right]^2 = \int_\Omega (x - m_X)^2 f(x) dx \tag{17.10}$$

is called **variance**. The square root of the variance is called the **standard deviation** (**SD**).

Variance is the smallest among the moments for any number of a. This can be verified as follows. Since

$$
\begin{aligned}
\sigma_a^2 - \sigma_{m_X}^2 &= \int_\Omega (x - a)^2 f(x) dx - \int_\Omega (x - m_X)^2 f(x) dx \\
&= \int_\Omega \left\{ 2x(m_X - a) + (a^2 - m_X^2) \right\} f(x) dx \\
&= 2(m_X - a) \int_\Omega x\, f(x) dx + (a^2 - m_X^2) \int_\Omega f(x) dx \\
&= 2(m_X - a)m_X + (a^2 - m_X^2) = (a - m_X)^2 \geq 0, \tag{17.11}
\end{aligned}
$$

σ_a^2 is minimized when we choose the mean value m_X as a.

Example 17.1 *Example of expectation. The expectation of a dice roll is*

$$1 \times \frac{1}{6} + 2 \times \frac{1}{6} + 3 \times \frac{1}{6} + 4 \times \frac{1}{6} + 5 \times \frac{1}{6} + 6 \times \frac{1}{6} = \frac{7}{2} = 3.5.$$

As for the calculation of the expectation, we have

$$E[X + c] = E[X] + c, \quad E[cX] = cE[X], \quad \text{if } X = c \text{ then } E[X] = c.$$

From the first formula, it follows that

$$E[X - E[X]] = 0.$$

17.2   Conditional Expectation and Bayes' Theorem

For example, let N be the population of a city, and let N_F be the population of female. Also, let N_A be the number of people with color-blindness. Let F be the event of being a female and A be the event of being color-blind. When one person is extracted from the people in this city, the probability that this person is a female is

$$P(F) = \frac{N_F}{N},$$

and the probability that the person is colorblind is given by

$$P(A) = \frac{N_A}{N}.$$

Next, we take one person out of the entire female population and consider the probability that this person is color-blind. This probability is given by

$$\frac{N_{FA}}{N_F},$$

where N_{FA} is the number of those who are F and A, and N_F is the female population. This probability is called the **conditional probability** of A under event F, denoted by $P(A \mid F)$. To be precise, when H is a subset of the sample space Ω and its probability is assumed to be positive, for any event (subset) A,

$$P(A \mid H) = \frac{P(A \cap H)}{P(H)} \tag{17.12}$$

is called the conditional probability given the event H. Here $P(A \cap H)$ is the probability that events A and H occur simultaneously. This conditional probability has the following properties;

1. For any event A, we have
$$0 \leq P(A \mid H) \leq 1. \tag{17.13}$$

   In particular, it holds that
$$P(\Omega \mid H) = 1.$$

2. If it holds that $A \cap B = 0$, then we have

$$P(A \cup B \mid H) = P(A \mid H) + P(B \mid H). \tag{17.14}$$

Also, from the definition of conditional probability, we have

$$P(A \cap H) = P(A \mid H) \, P(H). \tag{17.15}$$

This is called the **multiplication theorem of probability**. This relation can be extended as follows;

1. Given three events A, B and C, if $P(A \cap B) > 0$, then

$$P(A \cap B \cap C) = P(C \mid A \cap B) \, P(B \mid A) \, P(A) \tag{17.16}$$

2. Assume that $\Omega = B_1 + B_2 + \cdots + B_n$ with $B_i \cap B_j = 0$ ($i \neq j$), $P(B_i) > 0$, $i = 1, 2, \cdots, n$, that is, all events are exclusive events and an event F is given and $P(F) > 0$. In this case, in equation (17.12), if we put $A = B_i$ and $H = F$, then it holds

$$P(B_i \mid F) = \frac{P(B_i \cap F)}{P(F)}.$$

Here, for the numerator, it holds that

$$P(B_i \cap F) = P(F \mid B_i) \, P(B_i).$$

Since B_i is an exclusive event, the denominator $P(F)$ is given as the sum of the probabilities of event F occurring under the condition that each event B_i occurred;

$$P(F) = \sum_{i=1}^{n} P(F \mid B_i) \, P(B_i). \tag{17.17}$$

Therefore, we obtain

$$P(B_i \mid F) = \frac{P(B_i \cap F)}{P(F)} = \frac{P(F \mid B_i) \, P(B_i)}{\sum_{j=1}^{n} P(F \mid B_j) \, P(B_j)}. \tag{17.18}$$

This relationship is called **Bayes' theorem**[54].

Bayes' theorem (equation (17.18)) states that *"If a cause B_i is known to occur with a probability distribution $P(B_i)$ and the conditional probability of its outcome F occurring for each value of the cause B_i is given as $P(B_i \mid F)$, then after obtaining the result F, we can compute the probability of the cause B_i."* In this sense, $P(A)$ is called **prior probability** or a-priori probability and $P(B_i \mid A)$ is called **posterior probability** or a-posteriori probability.

17.3 Basic Distributions

17.3.1 Uniform Distribution

A distribution that takes any value in an interval $[a, b]$ with equal probability but 0 outside that interval;

$$f(x) = \begin{cases} \dfrac{1}{b-a} & a \leq x \leq b \\ 0 & \text{otherwise} \end{cases} \tag{17.19}$$

is called the **uniform distribution**.

17.3.2   Binomial Distribution

Independent trials that are repeated and whose outcomes are two exclusive events E_1 and E_2 are called **Bernoulli trials**. Denote the probability of two trials by p and $q = 1 - p$. For example, in the case of a coin toss, if E_1 is heads, then E_2 is tails and $p = q = 1/2$. Next, when n such trials are made, find the probability $b(k : n, p)$ that event A appears k times. Since the event E_1 can appear anywhere in n trials, this combination is the number of ways to distribute k things into n bins $\binom{n}{k}$, and their probabilities are $p^k q^{n-k}$ for each of them. Therefore, $b(k : n, p)$ is given by

$$b(k : n, p) = \binom{n}{k} p^k q^{n-k} = \frac{n!}{k!(n-k)!} p^k q^{n-k}. \tag{17.20}$$

This distribution is called the **binomial distribution**. The mean m of the binomial distribution is $np = \lambda$ and the variance σ^2 is npq.

17.3.3   Multinomial Distribution

In a trial in which one of the exclusive events $E_1, E_2, \cdots, E_r$ occurs with probabilities $p_1, p_2, \cdots, p_r$, $(p_1 + p_2 + \cdots + p_r = 1)$, then out of n trials, the probability that event $E_1, E_2, \cdots, E_r$ will occur $k_1, k_2, \cdots, k_r$, $(k_1 + k_2 + \cdots + k_r = n)$ times is given by the following **multinomial distribution**,

$$\frac{n!}{k_1! \, k_2! \cdots k_r!} p_1^{k_1} p_2^{k_2} \cdots p_r^{k_r}. \tag{17.21}$$

17.3.4   Poisson Distribution

As n increases in the binomial distribution, the computational complexity of $n!$ etc. increases and the computation of $\binom{n}{k}$ becomes more complicated. However, when n is large but p is small and therefore $\lambda = np$ is not large, the binomial distribution can be approximated by

$$b(k : n, p) \cong \frac{\lambda^k e^{-\lambda}}{k!}.$$

This distribution is called the **Poisson distribution**. The mean and variance of the Poisson distribution are given by $\lambda = np$.

17.3.5   Gaussian Distribution

If $t = k/n$ and $n \to \infty$ in the binomial distribution, the binomial distribution becomes a function that is concentrated only at the point λ and zero at other points. This property is called **law of large number**. Here, instead of k/n, we put

$$t = \frac{k - \lambda}{\sigma_k},$$

where $\sigma_k = \sqrt{npq}$, and use an approximate formula for $n!$ when n is large, called **Stirling's formula**,

$$n! \cong \sqrt{2\pi} \, n^{n+\frac{1}{2}} e^{-n},$$

the binomial distribution approaches

$$f(t) = \frac{1}{\sqrt{2\pi}} e^{-\frac{t^2}{2}}. \tag{17.22}$$

Since $\int_{-\infty}^{\infty} e^{-\frac{t^2}{2}} dt = \sqrt{2\pi}$, $f(t)$ satisfies

$$\int_{-\infty}^{\infty} f(t)dt = 1,$$

which indicates that $f(t)$ is a probability density function. The fact that the mean is 0 and the variance is 1 can be confirmed by computing

$$\int_{-\infty}^{\infty} x\, f(t)dt \quad \text{and} \quad \int_{-\infty}^{\infty} x^2 f(t)dt.$$

Returning to the binomial distribution, this result implies that

$$b(k : n, p) \implies \frac{1}{\sqrt{2\pi}\sigma_k} \exp\left\{-\frac{(k - m_k)^2}{2\sigma_k^2}\right\}.$$

This property is called the **central limit theorem** for the binomial distribution.

This distribution is the basic statistical distribution and is called the **Gaussian distribution** (or **normal distribution**), and is generally expressed as follows for a random variable X. The probability that the random variable X takes the value x is given by

$$f(x) = \frac{1}{\sqrt{2\pi}\,\sigma} e^{-\frac{(x-m)^2}{2\sigma^2}}, \tag{17.23}$$

and is denoted by $X \sim N(m, \sigma^2)$, where m is the mean value of X and σ^2 is the variance. Given m and σ^2, this distribution is determined. Given X, the distribution of Y defined by

$$Y = \frac{X - m}{\sigma} \tag{17.24}$$

becomes the standard normal distribution, namely, $Y \sim N(0, 1)$.

17.4   Variable Transformation

Given a probability distribution $p(x)$ of a one-dimensional random variable x, what happens to the distribution $q(y)$ of y, if we transfer the distribution on the y axis by the function $y = f(x)$?[51] For simplicity, let $f(x)$ be a monotone single-valued function. In this case, the probability that a point falls in the interval $(x, x + dx)$ is proportional to $p(x)$ and dx, so $p(x)dx$. When this is shifted to the y-axis by $y = f(x)$, the probability is equal to the probability of the point falling on $(y, y + dy)$, $q(y)dy$, so it holds that

$$p(x)dx = q(y)dy \quad \Rightarrow \quad q(y) = p(x)\frac{dx}{dy} = p(\varphi(y))\frac{d\varphi(y)}{dy}, \tag{17.25}$$

where $\varphi(y)$ is the inverse function of $y = f(x)$,

$$x = f^{-1}(y) = \varphi(y). \tag{17.26}$$

If $f(x)$ is a two-valued function, we can immediately find that it is the sum of two points of x that give the same probability as $q(y)$.

Assume that two variables x and y with two-dimensional distribution $p(x, y)$ can be expressed as using auxiliary variables u and v

$$\begin{cases} x = x(u, v) \\ y = y(u, v). \end{cases} \tag{17.27}$$

Then since $dx\,dy$ can be expressed by using the Jacobian as

$$dx\,dy = \left| \frac{\partial(x, y)}{\partial(u, v)} \right| du\,dv,$$

the distribution $q(u, v)$ of u and v is expressed as

$$q(u, v) = p(x(u, v), y(u, v)) \left| \frac{\partial(x, y)}{\partial(u, v)} \right|. \tag{17.28}$$

The same is true in the multivariable case.

17.5 χ^2 Distribution and Rayleigh Distribution

When x has a normal distribution $p(x) = \frac{1}{\sqrt{2\pi}} e^{-x^2/2}$, since the inverse function of $f(x) = x^2$ is two-valued, the distribution $y = x^2$ is given by

$$q(y)dy = p(x)dx + p(-x)dx = 2p(x)dx, \quad (y \geq 0).$$

Further, since $dy = 2x\,dx$, we have

$$q(y)\,2x = 2p(x).$$

Therefore, the distribution of $q(y)$ is given by

$$q(y) = \frac{1}{\sqrt{2\pi}} e^{-x^2/2} y^{-\frac{1}{2}}, \quad (y \geq 0). \tag{17.29}$$

This distribution is called the χ^2 distribution with one degree of freedom.

When k random variables $X_1, X_2, \cdots, X_k$ are independent and follow a Gaussian distribution $N(0, 1)$ with mean 0 and variance 1, the distribution of

$$\chi^2 = X_1^2 + X_2^2 + \cdots + X_k^2$$

follows a χ^2 distribution with k degrees of freedom. The density function of the χ^2 distribution is given by

$$f(\chi^2) = \left\{ 2^{k/2} \Gamma\left(\frac{k}{2}\right) \right\}^{-1} (\chi^2)^{k/2-1} \exp(-\chi^2/2), \tag{17.30}$$

where Γ is the gamma function.

The mean and variance of the χ^2 distribution are

1. Mean$[\chi^2] = k$

2. $\text{Var}[\chi^2] = 2k$.

From this we obtain $k = 2\,\text{Mean}[\chi^2]^2/\text{Var}[\chi^2]$.

As the relative magnitude of the variation of X relative to the expected value of the random variable X, **coefficient of variation** is defined by

$$\text{C.V.} = \frac{\sqrt{\text{Var}[X]}}{E[X]}. \tag{17.31}$$

In the case of χ^2 variable, it is given by

$$\text{C.V.} = \sqrt{\frac{2}{k}}. \tag{17.32}$$

In other words, *the larger the **degree of freedom** k, the smaller the variation.*

In particular, the χ^2 distribution with 2 degree of freedom is called the **Rayleigh distribution** and its density function is given by

$$f(x) = \frac{x}{\sigma^2}\,e^{-x^2/2\sigma^2} \quad (0 < x < \infty). \tag{17.33}$$

17.6 Multivariate Distributions

17.6.1 Independence, Covariance and Correlation Coefficient

Consider two random variables X and Y with probability densities $f(x)$ and $f(y)$, respectively. The probability $P(X = i, Y = j)$ that the random variable X takes the value i and the random variable Y takes the value j is called **joint probability** and its simultaneous probability distribution is written as $f(x, y)$.

As with the definition of variance

$$\begin{aligned}
\sigma_{XY} &\equiv \text{Cov}[X, Y] = E[(X - m_X)(Y - m_Y)] \\
&= \int\!\!\int_{\Omega} (x - m_X)(y - m_Y)\,f(x, y)dx\,dy \\
&= E[XY] - m_X\,m_Y
\end{aligned} \tag{17.34}$$

is called the **covariance**. If it holds that

$$f(x, y) = f(x)\,f(y), \tag{17.35}$$

X and Y are said to be **independent**. At this time, from equation (17.34), it holds that

$$\sigma_{XY} = 0. \tag{17.36}$$

For X and Y, if we have

$$E[XY] = E[X]\,E[Y], \tag{17.37}$$

we say that X and Y are uncorrelated.

If X and Y are independent, then X and Y are uncorrelated (the converse does not hold). If X and Y are uncorrelated, then $\text{Cov}[X, Y] = 0$.

Example 17.2 *Examples of independent trials.* *For n rolls of the dice we considered earlier, they are independent, since*

$$P\{X_1 = i_1,\ X_2 = i_2,\ \cdots,\ X_n = i_n\} = \frac{1}{6^n}, \quad (1 \le i_p \le 6,\ p = 1, 2, \cdots, n).$$

The **correlation coefficient** is defined by

$$\rho_{XY} = \frac{\sigma_{XY}}{\sqrt{\sigma_{XX}\,\sigma_{YY}}}. \tag{17.38}$$

The correlation coefficient has the following properties;

1. $|\rho_{XY}| \le 1$.
2. If X and Y are independent, then $\rho_{XY} = 0$.
3. $Y = aX + b \Longleftrightarrow |\rho_{XY}| = 1$.

[Proof of 1.] For a real number t, it always holds

$$T = E\left[\{t(X - m_X) + (Y - m_Y)\}^2 \right] \ge 0.$$

Since

$$
\begin{aligned}
T &= t^2 E\left[(X - m_X)^2\right] + 2tE[(X - m_X)(Y - m_Y)] + E\left[(Y - m_Y)^2\right] \\
  &= \sigma_{XX}\,t^2 + 2\sigma_{XY}\,t + \sigma_{YY}
\end{aligned}
$$

and $\sigma_{XX} \ge 0$, the discriminant becomes

$$\sigma_{XY}^2 - \sigma_{XX}\,\sigma_{YY} \le 0.$$

From this, it follows that $|\rho_{XY}| \le 1$.

[Proof of 2.]
   It is clear from equation (17.36).

[Proof of 3.]
   If $Y = aX + b$, then it holds that

$$
\begin{cases}
m_Y &= am_X + b \\
\sigma_{YY} &= a^2\,\sigma_{XX}.
\end{cases}
$$

Therefore, we have

$$
\begin{aligned}
\sigma_{XY} &= \int\!\!\int (x - m_X)(y - m_Y)\,p(x, y)dx\,dy \\
            &= \int\!\!\int a(x - m_X)^2\,p(x, y)dx\,dy = a\int_{-\infty}^{\infty}(x - m_X)^2\,q(x)dx = a\,\sigma_{XX}.
\end{aligned}
$$

This results in

$$|\rho_{XY}| = \left| \frac{\sigma_{XY}}{\sqrt{\sigma_{XX}\,\sigma_{YY}}} \right| = \left| \frac{a\sigma_{XX}}{a\sqrt{\sigma_{XX}\,\sigma_{XX}}} \right| = 1.$$

On the contrary, consider

$$T = E\left[(t(X - m_X) + (Y - m_Y))^2\right] = \sigma_{XX}\,t^2 + 2\sigma_{XY}\,t + \sigma_{YY}.$$

Since $|\rho_{XY}| = 1$, we have

$$\sigma_{XY}^2 - \sigma_{XX}\,\sigma_{YY} = 0.$$

Therefore, there exists a real number t_0 such that

$$E\left[(t_0(X - m_X) + (Y - m_Y))^2\right] = 0.$$

Such t_0 exists only when it holds that

$$t_0(X - m_X) + (Y - m_Y) = 0.$$

Thus, Y is expressed as a linear function of X.

17.6.2   Multivariate Normal Distribution

From equation (17.23), the univariate normal probability density function with mean β and variance $1/\alpha$ can be written as

$$k\,e^{-\frac{1}{2}\alpha(x-\beta)^2} = k\,e^{-\frac{1}{2}(x-\beta)\,\alpha(x-\beta)}, \tag{17.39}$$

where α is a positive number and k is adjusted so that the integral of the above equation is 1. With this in mind, we define a multivariate normal probability distribution for p random variables $X_1, X_2, \cdots, X_p$. The x and β in equation (17.39) are change to

$$x = (x_1, x_2, \cdots, x_p)^t, \quad b = (b_1, b_2, \cdots, b_p)^t.$$

Also, α is assumed to be a positive definite symmetric matrix

$$A = \begin{pmatrix} a_{11} & a_{12} & \cdots & a_{1p} \\ a_{21} & a_{22} & \cdots & a_{2p} \\ \vdots & \vdots & \ddots & \vdots \\ a_{p1} & a_{p2} & \cdots & a_{pp} \end{pmatrix}.$$

With this change, $(x - \beta)\,\alpha(x - \beta)$ can be written as

$$(x - b)^t\, A(x - b) = \sum_{i,j=1}^{p} a_{ij}(x_i - b_i)(x_j - b_j).$$

Also, the simultaneous probability distribution is given by

$$f(x_1, x_2, \cdots, x_p) = Ke^{-\frac{1}{2}(x-b)^t A(x-b)}. \tag{17.40}$$

Since A is positive definite, we have

$$\frac{1}{2}(x - b)^t A(x - b) \geq 0.$$

From this, it follows that

$$f(x_1, x_2, \cdots, x_p) \leq K.$$

Next, K is determined so that the integral of equation (17.40) is 1. For that, we evaluate

$$K^* = \int_{-\infty}^{\infty} \cdots \int_{-\infty}^{\infty} e^{-\frac{1}{2}(x-b)^t A(x-b)}\,dx_p \cdots dx_1. \tag{17.41}$$

By the way, according to linear algebra, if A is positive definite, then it is known that there exists a non-singular matrix C such that

$$C^t A C = I. \tag{17.42}$$

Using this C, we transform the variables as follows

$$x - b = Cy, \tag{17.43}$$

where $y = (y_1, y_2, \cdots, y_p)^t$. Then we have

$$(x - b)^t A(x - b) = y^t C^t A C y = y^t y \tag{17.44}$$

and the Jacobian of this transformation is given by

$$J = |C|,$$

where $|C|$ is the absolute value of the determinant of C. Thus, from equation (17.41), we obtain

$$K^* = |C| \int_{-\infty}^{\infty} \cdots \int_{-\infty}^{\infty} e^{-\frac{1}{2}y^t y} dy_p \cdots dy_1 = |C|(2\pi)^{p/2}. \tag{17.45}$$

From equation (17.42), we obtain

$$|C^t| \cdot |A| \cdot |C| = |I|. \tag{17.46}$$

However, from $|C^t| = |C|$ and $|I| = 1$, we have

$$|C| = \frac{1}{\sqrt{|A|}}. \tag{17.47}$$

It follows that K is obtained as

$$K = 1/K^* = \sqrt{|A|}\,(2\pi)^{-\frac{1}{2}p}. \tag{17.48}$$

Thus, the multivariate normal density function can be written as

$$\frac{\sqrt{A}}{(2\pi)^{\frac{1}{2}p}}\, e^{-\frac{1}{2}(x-b)^t A(x-b)}. \tag{17.49}$$

Next, we determine b and A. From equation (17.43), taking the expected value of

$$X = CY + b, \tag{17.50}$$

we have

$$E[X] = CE[Y] + b.$$

Here we evaluate $E[Y]$. Since $Y_i\, e^{-\frac{1}{2}y_i^2}$ is an odd function, we have

$$E[Y_i] = \frac{1}{\sqrt{2\pi}} \int_{-\infty}^{\infty} y_i\, e^{-\frac{1}{2}y_i^2}\, dy_i = 0.$$

Thus, $E[Y] = 0$, and hence we have

$$m_Y = E[Y] = b. \tag{17.51}$$

Then, we compute the covariance $\mathrm{Cov}(X, X^t)$:

$$\mathrm{Cov}(X, X^t) = E\left[(X - m_X)(X - m_X)^t\right].$$

From the variable transformations in equation (17.43) and equation (17.51), we obtain

$$E\left[(X - m_X)(X - m_X)^t\right] = E\left[CYY^tC^t\right] = CE\left[YY^t\right]C^t.$$

Here, consider the expected value $E[YY^t]$ of the i-th and j-th elements. When $i = j$,

$$E\left[Y_i^2\right] = \frac{1}{\sqrt{2\pi}} \int_{-\infty}^{\infty} y_i^2\, e^{-\frac{1}{2}y_i^2}\, dy_i = 1.$$

If $i \neq j$, we obtain

$$E[Y_i Y_j] = \frac{1}{\sqrt{2\pi}} \int_{-\infty}^{\infty} y_i\, e^{-\frac{1}{2}y_i^2}\, dy_i \int_{-\infty}^{\infty} y_j\, e^{-\frac{1}{2}y_j^2}\, dy_j = 0.$$

From these results, we end up with

$$E\left[YY^t\right] = I. \tag{17.52}$$

Therefore, we obtain

$$E\left[(X - m_X)(X - m_X)^t\right] = CE[I]C^t = CC^t. \tag{17.53}$$

From equation (17.46), since $A = (C^t)^{-1}(C)^{-1}$, we have

$$CC^t = A^{-1}.$$

As a result, it can be seen that the variance-covariance matrix of X is given by

$$\Sigma = E\left[(X - m_X)(X - m_X)^t\right] = A^{-1}. \tag{17.54}$$

Summarizing these results, we have the following;

1. If the probability density function of the p-dimensional random variable X is expressed as

$$\frac{\sqrt{A}}{(2\pi)^{\frac{1}{2}p}}\, e^{-\frac{1}{2}(x-b)^t A(x-b)},$$

then the expected value of X is b and the variance-covariance matrix is A^{-1}.

2. Conversely, given a vector m_X and a positive definite matrix Σ, a multivariate normal distribution with the mean vector m_X and the variance-covariance matrix Σ exists and its density function is given by

$$(2\pi)^{-\frac{p}{2}}|\Sigma|^{-1}e^{-\frac{1}{2}(x-m_X)^t\Sigma^{-1}(x-m_X)}. \tag{17.55}$$

Equation (17.55) may be written as $X \sim N(m_X, \Sigma)$.

17.6.3   Two-Dimensional Normal Distribution

In the previous subsection, if $p = 2$, then

Mean vector
$$E\begin{pmatrix} X_1 \\ X_2 \end{pmatrix} = \begin{pmatrix} m_{X_1} \\ m_{X_2} \end{pmatrix} \tag{17.56}$$

Covariance matrix
$$\Sigma = E\begin{pmatrix} (X_1 - m_{X_1})^2 & (X_1 - m_{X_1})(X_2 - m_{X_2}) \\ (X_2 - m_{X_2})(X_1 - m_{X_1}) & (X_2 - m_{X_2})^2 \end{pmatrix}$$
$$= \begin{pmatrix} \sigma_{11} & \sigma_{12} \\ \sigma_{21} & \sigma_{22} \end{pmatrix} = \begin{pmatrix} \sigma_{11} & \sqrt{\sigma_{11}\sigma_{22}}\,\rho_{12} \\ \sqrt{\sigma_{22}\sigma_{11}}\,\rho_{12} & \sigma_{22} \end{pmatrix} \tag{17.57}$$

Inverse of covariance matrix
$$\Sigma^{-1} = \frac{1}{1 - \rho_{12}^2}\begin{pmatrix} \frac{1}{\sigma_{22}} & -\frac{\rho_{12}}{\sqrt{\sigma_{11}\sigma_{22}}} \\ -\frac{\rho_{12}}{\sqrt{\sigma_{22}\sigma_{11}}} & \frac{1}{\sigma_{11}} \end{pmatrix}. \tag{17.58}$$

Therefore, the density function of the two-dimensional normal distribution is given by

$$f(x_1, x_2) = \frac{1}{2\pi\sqrt{\sigma_{11}\sigma_{22}}\sqrt{1 - \rho_{11}^2}}\exp\left\{\frac{-1}{2\sqrt{1 - \rho_{11}^2}}\left[\frac{(x_1 - m_{x_1})^2}{\sigma_{11}} - 2\rho\frac{(x_1 - m_{x_1})(x_2 - m_{x_2})}{\sqrt{\sigma_{11}\sigma_{22}}} + \frac{(x_2 - m_{x_2})^2}{\sigma_{22}}\right]\right\}, \tag{17.59}$$

where $\sigma_{ij} = E[(X_i - m_{X_i})(X_j - m_{X_j})]$ and $\rho_{ij} = \frac{\sigma_{ij}}{\sqrt{\sigma_{ii}\sigma_{jj}}}$.

17.6.4   Estimation

1. **Point Estimation and Interval Estimation**

   The purpose of statistics is to know the characteristics of a population from a sample. Suppose now that samples $X_1, X_2, \cdots, X_N$ are obtained independently one by one from a population with an unknown but constant value of the parameter θ. In this case, $X_1, X_2, \cdots, X_N$ are random variables that take the value $x_1, x_2, \cdots, x_N$. Therefore, $\hat{\theta}(X_1, X_2, \cdots, X_N)$ calculated from this sample is also a random variable and is called **statistics**. Since the $\hat{\theta}$ is a random variable, its value also varies from sample to sample. The objective of estimation is to seek the true parameter θ of the population using $\hat{\theta}(X_1, X_2, \cdots, X_N)$. The method of defining an appropriate $\hat{\theta}(x_1, x_2, \cdots, x_N)$ and using it as an estimate of the true parameter θ of the population is called **point estimation**. On the other hand, the method of choosing two statistics $\hat{\theta}_1$ and $\hat{\theta}_2$ and estimating that θ lies in the interval $\hat{\theta}_1 < \theta < \hat{\theta}_2$ is called **interval estimation**. The estimation method used in this book is point estimation.

2. **Desirable Properties of Point Estimation**

   (a) **Consistency**

   When the number of samples N increases, the statistics for which the estimate $\hat{\theta}_N$ obtained from the sample approaches the true value θ is called the consistent statistics. For the consistent statistics, for any positive small number ε, it holds that

   $$\lim_{N \to \infty} p\left[|\hat{\theta}_N - \theta| < \varepsilon\right] = 1. \tag{17.60}$$

(b) **Unbiasedness**

A statistics for which the expected value of the statistics $\hat{\theta}_N$ is equal to the true value θ is called an unbiased statistics. In other words, for an unbiased statistics, it holds that

$$E[\hat{\theta}_N] = \theta. \tag{17.61}$$

(c) **Efficiency**

Among unbiased statistics, the statistics with the smallest variance of $\hat{\theta}_N$, i.e., a statistics that satisfy

$$E\left[(\hat{\theta}_N - \theta)^2\right] = \min \tag{17.62}$$

is called an efficient estimator.

(d) **Sufficiency**

Let θ be the true parameter and $\hat{\theta}_1$ be its statistics. Also, let $\hat{\theta}_2$ be another statistics that is independent of $\hat{\theta}_1$. The simultaneous probability distribution of $\hat{\theta}_1$ and $\hat{\theta}_2$ is $f(\hat{\theta}_1, \hat{\theta}_2 \mid \theta)$ and the probability that $\hat{\theta}_1$ appears when the true value is θ is $p(\hat{\theta}_1 \mid \theta)$, and the probability that $\hat{\theta}_2$ appears under the condition that $\hat{\theta}_1$ appears when the true value is θ is $q(\hat{\theta}_2 \mid \hat{\theta}_1, \theta)$, then we have

$$f(\hat{\theta}_1, \hat{\theta}_2 \mid \theta) = p(\hat{\theta}_1 \mid \theta)\, q(\hat{\theta}_2 \mid \hat{\theta}_1, \theta). \tag{17.63}$$

If the distribution q is unrelated to θ, then $\hat{\theta}_2$ contains no information about θ. That is, all information about θ is contained in $\hat{\theta}_1$. Such a statistics $\hat{\theta}_1$ is called a sufficient statistics. If it is a sufficient statistics, then it holds that

$$f(\hat{\theta}_1, \hat{\theta}_2 \mid \theta) = p(\hat{\theta}_1 \mid \theta)\, q(\hat{\theta}_2 \mid \hat{\theta}_1). \tag{17.64}$$

3. **Properties of the maximum likelihood estimator[54][95]**

The maximum likelihood estimator described in Chapter 8 approximately satisfies all of the properties that make it a good statistics as described in the previous subsection. That is, the maximum likelihood estimator $\hat{\theta}_N$ has the following properties;

(a) If an efficient estimator exists, it is consistent with the maximum likelihood estimator.

(b) If a sufficient estimator exists, it can be expressed as a function of the maximum likelihood estimator.

(c) If an estimator is efficient, it is sufficient, but the converse is not necessarily true.

(d) The maximum likelihood estimator is a consistent estimator.

(e) The maximum likelihood estimator is an asymptotically efficient estimator.

(f) $\sqrt{N}\,(\hat{\theta}_N - \theta)$ asymptotically follows a normal distribution with mean 0 and variance $N\sigma_0^2$ [20][95], where

$$\sigma_0^2 = -\frac{1}{N}E\left[\frac{\partial^2 \log p(x \mid \theta)}{\partial \theta^2}\right]. \tag{17.65}$$

18

Kalman Filter

In this chapter, we derive the Kalman filter used in Chapter 10 and other chapters. The Kalman filter first demonstrated its power in the calculation of satellite orbits[67]. It is also used in navigation to estimate ship's position. In this chapter, the Kalman filter is derived from the standpoint of estimation.

18.1 Calculations for Navigation and Kalman Filter

Assume now that the ship is sailing at position P at time $s-1$, as shown in Figure 18.1. The position mentioned here means a state vector containing the ship's longitude $l_x(s-1)$ and latitude $l_y(s-1)$, plus its speed $v_x(s-1)$ and $v_y(s-1)$ in each direction and its acceleration $\alpha_x(s-1)$ and $\alpha_y(s-1)$,

$$x(s-1) = (l_x(s-1),\, l_y(s-1),\, u_x(s-1),\, u_y(s-1),\, \alpha_x(s-1),\, \alpha_y(s-1))^t. \qquad (18.1)$$

Next, suppose that time has advanced Δt in this state and the observation vector of latitude and longitude $y(s) = (L_x(s), L_y(s))$ with observation error is obtained using **global positioning system** (**GPS**) and so on. We shall estimate the state $x(s)$ of this "position" (called dead reckoning position in nautical science) using the latitude and longitude information at the observed time s.

For the state at time s after time Δt has elapsed from time $s-1$, let $u_x(s-1)$ be the velocity and $\alpha_x(s-1)$ be the acceleration at time $s-1$. For example, the position in longitude direction is inferred as

$$l_x(s) = l_x(s-1) + u_x(s-1)\Delta t + \frac{1}{2}\alpha_x(s-1)\Delta t^2.$$

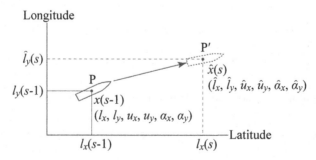

FIGURE 18.1
Estimation of ship's location

DOI: 10.1201/9781003428565-18

The l_y direction is also expressed by the same equation. In addition, the velocity in the l_x direction is estimated by acceleration as

$$u_x(s) = u_x(s-1) + \alpha_x(s-1)\Delta t.$$

The acceleration is constant when considered over a short time period, but random changes of $v_x(s)$ are added due to changes in weather and sea conditions, and is given by

$$\alpha_x(s) = \alpha_x(s-1) + v_x(s).$$

The same is true for the l_y direction. To represent such a change of state representing the dead reckoning navigation from time $s-1$, using the transition matrix F

$$F = \begin{bmatrix} 1 & 0 & \Delta t & 0 & \frac{1}{2}\Delta t^2 & 0 \\ 0 & 1 & 0 & \Delta t & 0 & \frac{1}{2}\Delta t^2 \\ 0 & 0 & 1 & 0 & \Delta t & 0 \\ 0 & 0 & 0 & 1 & 0 & \Delta t \\ 0 & 0 & 0 & 0 & 1 & 0 \\ 0 & 0 & 0 & 0 & 0 & 1 \end{bmatrix}, \tag{18.2}$$

it can be expressed as

$$x(s) = Fx(s-1) + v(s), \tag{18.3}$$

where $v(s) = (0,0,0,0,v_x(s),v_y(s))^t$ and $v_x(s)$ and $v_y(s)$ are Gaussian white noises with mean 0 and variances σ_x^2 and σ_y^2, respectively.

These are the internal movements of the navigation system. In this internal system, $x(s)$, especially velocity and acceleration, cannot be measured directly. Suppose now that only longitude $L_x(s)$ and latitude $L_y(s)$ information, containing the observation error related to the actual position state at s time, can be measured using an observer such as GPS. In this case, if we put

$$H = \begin{pmatrix} 1 & 0 & 0 & 0 & 0 & 0 \\ 0 & 1 & 0 & 0 & 0 & 0 \end{pmatrix}, \tag{18.4}$$

then the observation process is expressed as

$$y(s) = Hx(s) + w(s), \tag{18.5}$$

where $w(s)$ is a vector of observation errors in the latitude and longitude directions whose elements are Gaussian white noise with mean 0 and variance τ_x^2 and τ_y^2, respectively.

The Kalman filter is a method of **recursive estimation** of the mean and variance of $x(s)$ using the observed $y(s)$ in this situation.

18.2 Optimal Estimation

The objective of the task of **estimation** in signal processing considered below is to extract the most plausible signal at time t_1 from the observed values obtained up to time t [39]. Therefore, it is natural to consider the estimated value as a function of the observed process y. Therefore, consider to obtain the most plausible estimate $\hat{x}(t_1)$ of the signal $x(t_1)$ at time t_1 by using an appropriate function

$$\hat{x}(t_1) = f(y). \tag{18.6}$$

Considering this probabilistically, the problem is to evaluate the conditional probability density function that yields the value $x(t_1)$ at time t_1, given the observation of y by time $t = t_1$

$$p(x \mid y), \tag{18.7}$$

and to find the estimate with the highest conditional probability.

Therefore, the first thing that must be done in this task is to establish evaluation criterion to measure the goodness of the estimation. That is, in the task of estimation, it must be clarified what criterion is used to determine the **best estimation**. In such a case, the commonly used method is to use **squared error**, $e(t_1) = x(t_1) - \hat{x}(t_1)$, as the estimation error, which is the expected value of the (Figure 18.2),

$$l(e) = e^2. \tag{18.8}$$

This evaluation function (cost function) has the following properties:

1. It is zero or positive, i.e., $l(e) \geq 0$.

2. It is a symmetric function, i.e., $l(-e) = l(e)$.

3. It is a non-decreasing function for $|e| > 0$.

Since the original process of the evaluation function defined in this way is a stochastic process, the estimation error $x(t_1) - \hat{x}(t_1)$ is a random variable and $l(x(t_1) - \hat{x}(t_1))$ is also a random variable. In this case, if the expected value

$$E[l(x - \hat{x})] = \int_{-\infty}^{\infty} p(x) \int_{-\infty}^{\infty} (x(t_1) - f(y))^2 p(y \mid x) dy \, dx \tag{18.9}$$

is obtained, an estimator of the squared error evaluation function is obtained. This expected value is called **Bayes risk**. Here $p(x)$ is the prior probability density function of x and $p(y \mid x)$ is the probability of the observed value y appearing under the given conditions of the probability distribution of x (see Section 17.2). The estimate $f(y)$ that minimizes this Bayes risk is called the **Bayes estimate** with squared error as the evaluation function.

By the way, it is considered that the above perspective is the probability distribution of the observations assuming that the true signal is given. In other words, the results are estimated under the condition that the cause is known. However, in actual situation, the true distribution is unknown. Therefore, let us change the viewpoint and try to estimate the cause from the obtained result, namely the observations (Figure 18.3).

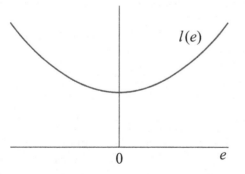

FIGURE 18.2
Evaluation function

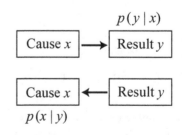

FIGURE 18.3
Cause and result perspective

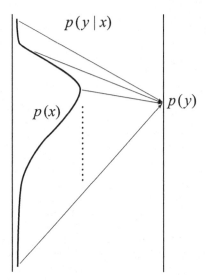

FIGURE 18.4
Bayes' formula

Bayes' theorem for the posterior probability $p(x \mid y)$ comes into play in such a case (equation 17.18),

$$p(x \mid y) = \frac{p(y \mid x)\,p(x)}{p(y)}. \tag{18.10}$$

Here, $p(x \mid y)$ is the probability of the appearance of x under the condition that the observed value y is obtained, and $p(x)$ is the prior probability of the event x occurring. $p(y)$ is the marginal probability of y given by (Figure 18.4),

$$p(y) = \int_{-\infty}^{\infty} p(y \mid x)\,p(x)dx.$$

Note that, $p(y)$ is fixed and constant because the observed value y is already given.

To minimize l using this theorem, we can use

$$
\begin{aligned}
E[l(x - \hat{x})] &= \int_{-\infty}^{\infty} p(x) \int_{-\infty}^{\infty} (x(t_1) - f(y))^2\, p(y \mid x)dy\, dx \\
&= \int_{-\infty}^{\infty} p(y) \int_{-\infty}^{\infty} (x(t_1) - f(y))^2\, p(x \mid y)dx\, dy
\end{aligned}
$$

and since the data y are already available, we know that the Bayes estimator in this case should minimize the inner integral of the second equation in the above equation, namely $f(y)$, which minimizes

$$\int_{-\infty}^{\infty} (x(t_1) - f(y))^2\, p(x \mid y)dx = E\left[(x(t_1) - f(y))^2 \mid y\right]. \tag{18.11}$$

Now, let $x(y)$ be the conditional expectation of the signal value $x(t_1)$ given the observed value y. In other words

$$x(y) = E[x \mid y] = \int_{-\infty}^{\infty} x\, p(x \mid y)dx. \tag{18.12}$$

In this case, we obtain the following inequality with respect to equation (18.11):

$$\int_{-\infty}^{\infty} (x(t_1) - f(y))^2 \, p(x \mid y) dx$$

$$= E[(x(t_1) - f(y))^2 \mid y]$$
$$= E[(x(t_1) - x(y) + x(y) - f(y))^2 \mid y]$$
$$= E[(x(t_1) - x(y))^2 \mid y] + 2(x(y) - f(y)) \, E[x(t_1) - x(y) \mid y] + (x(y) - f(y))^2$$
$$= E[(x(t_1) - x(y))^2 \mid y] + (x(y) - f(y))^2$$
$$\geq E[(x(t_1) - x(y))^2 \mid y].$$

The equality holds when

$$f(y) = x(y), \tag{18.13}$$

and in that case the minimum value is attained. Thus, the Bayes estimator when the evaluation function is chosen as the squared error is given by the "conditional expectation of $x(t_1)$ given the observed value y (equation (18.12))."

This is called the **minimum mean squared estimate** of $x(t_1)$ and is written as $\hat{x}_{\text{MMSE}}(y)$. In this case, it is obvious that

$$E_x[e] = E_x[x - \hat{x}_{\text{MMSE}}(y)] = E_x\left[x - E_y[x \mid y]\right] = 0, \tag{18.14}$$

and $\hat{x}_{\text{MMSE}}(y)$ is an unbiased estimator of x. The best estimate used in the Kalman filter is this minimum variance estimate.

Up to this point, we have chosen a quadratic function as $l(e)$ for the explanation, but it is generally true if the following condition 1 is satisfied as $l(e)$. Including that, we can say the following:

1. This conclusion holds in any case if the conditional probability density satisfies the given conditions, as long as the evaluation function is symmetric and non-decreasing with respect to positive variables.

2. The normal distribution satisfies the above conditions. Also, since the linear function of the normal random variable y is normally distributed, it is expected that the computation of the expected value $E[x \mid y]$ of the conditional probability distribution $p(x \mid y)$ is easier if it is given as an estimate by a linear function of y.

3. As we will see later, in general, when both signal and noise processes are described by second-order stochastic processes, the estimation problem can be formulated as minimization of the evaluation function, $E[(x - \hat{x})^2] \rightarrow \min$.

18.3   Linear Least Squares Estimation of the Observation Process of a Time Series

Now assume in general that the measurement y is given by a p-dimensional observation process, the signal process x is m-dimensional, and in observing the signal process,

p-dimensional noise process w is added[10]. Namely, they are expressed as

$$y = \begin{pmatrix} y_1 \\ y_2 \\ \vdots \\ y_p \end{pmatrix}, \quad x = \begin{pmatrix} x_1 \\ x_2 \\ \vdots \\ x_m \end{pmatrix}, \quad w = \begin{pmatrix} w_1 \\ w_2 \\ \vdots \\ w_p \end{pmatrix}. \tag{18.15}$$

The observation process is expressed as

$$y = Hx + w, \tag{18.16}$$

where H is the $m \times p$ observation matrix given as

$$H = \begin{pmatrix} H_{11} & H_{12} & \cdots & H_{1m} \\ H_{21} & H_{22} & \cdots & H_{2m} \\ \vdots & \vdots & \ddots & \vdots \\ H_{p1} & H_{p2} & \cdots & H_{pm} \end{pmatrix}. \tag{18.17}$$

Let $\bar{x}$ and $\bar{w}$ denote the expected values of x and w, respectively. Assume also that x and w are statistically independent (Figure 18.5).

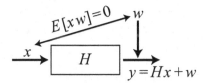

FIGURE 18.5
Observation process

Namely,

$$\bar{x} = E[x], \qquad E\left[(x - \bar{x})(x - \bar{x})^t\right] = \Sigma_{xx} \tag{18.18}$$
$$\bar{w} = E[w] = 0, \quad \left[(w - \bar{w})(w - \bar{w})^t\right] = R \tag{18.19}$$

The problem here is to obtain an estimator $\hat{x}$ of the unknown state x using this vector measurement y.

Thus, we consider constructing the estimate of x by a function $f(y)$ of y

$$\hat{x} = f(y), \tag{18.20}$$

and finding the minimum mean squared error of the estimator. Here, as its function, we choose the simplest linear expression

$$\hat{x} = \alpha y + \beta, \tag{18.21}$$

and choose the squared error as the evaluation function to estimate the α and β that minimize it.

First, the value of the signal x cannot be known directly. Also, the difference between its estimator, i.e., the error

$$e = x - \hat{x} \tag{18.22}$$

cannot also be known directly, but assuming that the mean and variance of random variables x and w are known and that x and w are independent, the mean value of e is

$$
\begin{aligned}
E[e] &= E[x - (\alpha y + \beta)] \\
&= (I - \alpha H)\bar{x} - \beta,
\end{aligned} \tag{18.23}
$$

and the variance-covariance matrix is given by

$$
\begin{aligned}
&E\left[(e - E[e])(e - E[e])^t\right] \\
&= E\left[\{(I - \alpha H)(x - \bar{x}) + \alpha(w - \bar{w})\}\{(I - \alpha H)(x - \bar{x}) + \alpha(w - \bar{w})\}^t\right] \\
&= (I - \alpha H)\, E\left[(x - \bar{x})(x - \bar{x})^t\right](I - \alpha H)^t + \alpha E\left[(w - \bar{w})(w - \bar{w})^t\right]\alpha^t \\
&= (I - \alpha H)\,\Sigma_{xx}(I - \alpha H)^t + \alpha R\alpha^t.
\end{aligned} \tag{18.24}
$$

The variance-covariance matrix depends only on α and is independent of β. Therefore, we adjust β so that the mean of the error e is zero in equation (18.23). For this purpose, we put

$$
\beta = \bar{x} - \alpha H\bar{x}.
$$

In this case, the expected value of the squared error is given by

$$
\begin{aligned}
E\left[e^2\right] &= \alpha H\Sigma_{xx}H^t\alpha^t + \alpha R\alpha^t - \alpha H\Sigma_{xx} - \Sigma_{xx}H^t\alpha^t + \Sigma_{xx} \\
&= \alpha(H\Sigma_{xx}H^t + R)\,\alpha^t - \alpha H\Sigma_{xx} - \Sigma_{xx}H^t\alpha^t + \Sigma_{xx} \\
&= \left\{\alpha - \Sigma_{xx}H^t(H\Sigma_{xx}H^t + R)^{-1}\right\}(H\Sigma_{xx}H^t + R)\left\{\alpha - \Sigma_{xx}H^t(H\Sigma_{xx}H^t + R)^{-1}\right\}^t \\
&\quad + \Sigma_{xx} - \Sigma_{xx}H^t(H\Sigma_{xx}H^t + R)^{-1}H\Sigma_{xx}.
\end{aligned} \tag{18.25}
$$

Thus the α that minimizes $E[e^2]$ is obviously obtained as

$$
\alpha = \Sigma_{xx}H^t(H\Sigma_{xx}H^t + R)^{-1}. \tag{18.26}
$$

That is, the linear estimator with zero mean errors and minimum mean squares errors is obtained by

$$
\begin{aligned}
\hat{x} &= \alpha y + \beta \\
&= \bar{x} + \Sigma_{xx}H^t(H\Sigma_{xx}H^t + R)^{-1}(y - H\bar{x}).
\end{aligned} \tag{18.27}
$$

The variance of the error when $\hat{x}$ is used as the estimate of x can be obtained from equation (18.25) as

$$
E\left[(\hat{x} - x)(\hat{x} - x)^t\right] = \Sigma = \Sigma_{xx} - \Sigma_{xx}H^t(H\Sigma_{xx}H^t + R)^{-1}H\Sigma_{xx}. \tag{18.28}
$$

In this case, since

$$
\Sigma H^t R^{-1} = \Sigma_{xx}H^t(H\Sigma_{xx}H^t + R)^{-1}, \tag{18.29}
$$

equation (18.27) can also be written as

$$
\hat{x} = \bar{x} + \Sigma H^t R^{-1}(y - H\bar{x}). \tag{18.30}
$$

If x and y are vector Gaussian processes, the conditional probability of x given y can be calculated directly as follows.

First, let x be an n-dimensional Gaussian vector with mean vector $\bar{x}$ and variance–covariance matrix Σ, then the n-dimensional Gaussian distribution is given by (equation (17.55))

$$
p(x) = \frac{1}{(2\pi)^n}|\Sigma|^{-1/2}\exp\left\{-\frac{1}{2}(x - \bar{x})^t\,\Sigma^{-1}(x - \bar{x})\right\}. \tag{18.31}
$$

By the way, suppose we are given two random variables represented by an m-dimensional Gaussian vector x and a p-dimensional Gaussian vector y. Further assume that these two vectors are related to each other.

In this case, considering the linear transformation into variables z_1 and z_2 that are independent of each other,

$$z_1 = x + My \tag{18.32}$$

$$z_2 = y, \tag{18.33}$$

the following must hold

$$
\begin{aligned}
0 &= E\left[(z_1 - E[z_1])(z_2 - E[z_2])^t\right] \\
&= E\left[\{(x - E[x]) + M(y - E[y])\}(y - E[y])^t\right] \\
&= \Sigma_{xy} + M\Sigma_{yy}.
\end{aligned}
\tag{18.34}
$$

Therefore, if we adopt

$$M = -\Sigma_{xy}\Sigma_{yy}^{-1}, \tag{18.35}$$

z_1 and z_2 are independent.

After this transformation, since the vector $z = (z_1 \ z_2)^t$ is a linear combination of these two vectors,

$$
z = \begin{pmatrix} z_1 \\ z_2 \end{pmatrix} = \begin{pmatrix} I & -\Sigma_{xy}\Sigma_{yy}^{-1} \\ 0 & I \end{pmatrix} \begin{pmatrix} x \\ y \end{pmatrix}, \tag{18.36}
$$

the distribution of this vector also follows a Gaussian distribution.

Then the mean vector is given by

$$
\begin{aligned}
E\begin{pmatrix} z_1 \\ z_2 \end{pmatrix} &= E\left[\begin{pmatrix} I & -\Sigma_{xy}\Sigma_{yy}^{-1} \\ 0 & I \end{pmatrix}\begin{pmatrix} x \\ y \end{pmatrix}\right] \\
&= \begin{pmatrix} I & -\Sigma_{xy}\Sigma_{yy}^{-1} \\ 0 & I \end{pmatrix}\begin{pmatrix} \bar{x} \\ \bar{y} \end{pmatrix} \\
&= \begin{pmatrix} \bar{x} - \Sigma_{xy}\Sigma_{yy}^{-1}y \\ \bar{y} \end{pmatrix} = \begin{pmatrix} \bar{z}_1 \\ \bar{z}_2 \end{pmatrix} \equiv \bar{z}.
\end{aligned}
\tag{18.37}
$$

And since

$$
\begin{aligned}
E\left[(z_1 - \bar{z}_1)(z_1 - \bar{z}_1)^t\right] &= E\left[\{(x - \bar{x}) - \Sigma_{xy}\Sigma_{yy}^{-1}(y - \bar{y})\}\{(x - x) - \Sigma_{xy}\Sigma_{yy}^{-1}(y - \bar{y})\}^t\right] \\
&= \Sigma_{xx} - \Sigma_{xy}\Sigma_{yy}^{-1}\Sigma_{yx},
\end{aligned}
\tag{18.38}
$$

the variance-covariance matrix is obtained by

$$
\begin{aligned}
E[\{z - \bar{z}\}\{z - \bar{z}\}]^t &= \begin{pmatrix} E[(z_1 - \bar{z}_1)(z_1 - \bar{z}_1)^t] & E[(z_1 - \bar{z}_1)(z_2 - \bar{z}_2)^t] \\ E[(z_2 - \bar{z}_2)(z_1 - \bar{z}_1)^t] & E[(z_2 - \bar{z}_2)(z_2 - \bar{z}_2)^t] \end{pmatrix} \\
&= \begin{pmatrix} \Sigma_{xx} - \Sigma_{xy}\Sigma_{yy}^{-1}\Sigma_{yx} & 0 \\ 0 & \Sigma_{yy} \end{pmatrix}.
\end{aligned}
\tag{18.39}
$$

Thus, the joint distribution of the two transformed independent Gaussian random variables $z_1 = x - \sigma_{xy}\sigma_{yy}^{-1}y$ and $z_2 = y$ is expressed as the product of two Gaussian distributions

$$N(z_1 \mid \bar{x} - \Sigma_{xy}\Sigma_{yy}^{-1}\bar{y}, \Sigma_{xx} - \Sigma_{xy}\Sigma_{yy}^{-1}\Sigma_{yx}) \cdot N(z_2 \mid y, \Sigma_{yy}).$$

From this result, it can be seen that the distributions of x and y are obtained by substituting $x - \Sigma_{xy}\Sigma_{yy}^{-1}y$ for z_1 and y for z_2 (from equation (18.36), the Jacobian of this transformation is 1).

First, the joint distribution of x and y is given by

$$
p(x, y) = \frac{1}{(2\pi)^{m/2}} |\Sigma_{xx \cdot y}|^{-1/2}
$$

$$
\times \exp \left\{ -\frac{1}{2} \left[(x - \bar{x}) - \Sigma_{xy} \Sigma_{yy}^{-1} (y - \bar{y}) \right] \Sigma_{xx \cdot y}^{-1} \left[(x - \bar{x}) - \Sigma_{xy} \Sigma_{yy}^{-1} (y - \bar{y}) \right]^t \right\}
$$

$$
\times \frac{1}{(2\pi)^{p/2}} |\Sigma_{yy}|^{-1/2} \exp \left[-\frac{1}{2} (y - \bar{y}) \Sigma_{yy}^{-1} (y - \bar{y}) \right], \tag{18.40}
$$

where we put

$$
\Sigma_{xx \cdot y} = \Sigma_{xx} - \Sigma_{xy} \Sigma_{yy}^{-1} \Sigma_{yx}. \tag{18.41}
$$

Since the distribution of y is given by

$$
p(y) = \frac{1}{(2\pi)^{p/2}} |\Sigma_{yy}|^{-1/2} \exp \left\{ -\frac{1}{2} (y - \bar{y}) \Sigma_{yy}^{-1} (y - \bar{y})^t \right\},
$$

from the Bayes' theorem

$$
p(x \mid y) = \frac{p(x, y)}{p(y)},
$$

the conditional probability distribution of x given y is obtained as

$$
p(x \mid y) = \frac{1}{(2\pi)^{m/2}} |\Sigma_{xx \cdot y}|^{-1/2}
$$

$$
\times \exp \left\{ -\frac{1}{2} \left[(x - \bar{x}) - \Sigma_{xy} \Sigma_{yy}^{-1} (y - \bar{y}) \right] \Sigma_{xx \cdot y}^{-1} \left[(x - \bar{x}) - \Sigma_{xy} \Sigma_{yy}^{-1} (y - \bar{y}) \right] \right\}. \tag{18.42}
$$

In other words, the following conclusions can be drawn;

[**Conclusion 1**]   *The conditional probability distribution of x given y is an m-dimensional normal distribution, namely;*

Mean vector	$\hat{x} = E[x \mid y] = \bar{x} + \Sigma_{xy} \Sigma_{yy}^{-1} (y - \bar{y})$	(18.43)
Variance-covariance matrix	$\Sigma_{xx \cdot y} = \Sigma_{xx} - \Sigma_{xy} \Sigma_{yy}^{-1} \Sigma_{yx}.$	(18.44)

Next, we compute $E[\hat{x} e^t]$. From $E[e] = 0$, we obtain

$$
\begin{aligned}
E \left[\hat{x} e^t \right] &= E \left[(\bar{x} + \Sigma H^t R^{-1} \{ y - H\bar{x} \}) e^t \right] \\
&= \Sigma H^t R^{-1} E \left[y(\hat{x} - x)^t \right].
\end{aligned}
$$

Since, from the conditions $E[xw^t] = 0$ and $E[w(x - \bar{x})^t] = 0$, equations (18.16) and (18.30) yield

$$
\begin{aligned}
E \left[y(\hat{x} - x)^t \right] &= E \left[(Hx + w)(\bar{x} + \Sigma H^t R^{-1} \{ y - H\bar{x} \}^t \right] \\
&= E \left[(Hx + w)(\Sigma H^t R^{-1} H - I)(x - \bar{x}) + \Sigma H^t R^{-1} w) \right] \\
&= HE \left[x(x - \bar{x}) \right] (\Sigma H^t R^{-1} H - I)^t + E \left[ww^t \right] (\Sigma H^t R^{-1})^t.
\end{aligned}
$$

Here, since

$$
\begin{aligned}
E \left[x(x - \bar{x})^t \right] &= E \left[(x - \bar{x})(x - \bar{x})^t + \bar{x}(x - \bar{x})^t \right] \\
&= \Sigma_{xx} \\
E \left[ww^t \right] &= R \\
\left(\Sigma_{xx}^{-1} + H^t R^{-1} H \right) \Sigma &= I,
\end{aligned}
$$

we obtain

$$
\begin{aligned}
E\left[y(\hat{x}-x)^t\right] &= H\,\Sigma_{xx}(\Sigma H^t R^{-1}H - I)^t + R(\Sigma H^t R^{-1})^t \\
&= H\,\Sigma_{xx}(-\Sigma\Sigma_{xx}^{-1})^t + H\,\Sigma^t \\
&= -H\,\Sigma^t + H\,\Sigma^t = 0.
\end{aligned}
$$

Thus, we obtain the following conclusions (Figure 18.6);

[**Conclusion 2**]   *The least squares estimator $\hat{x}$ and the estimation error $e = x - \hat{x}$ are independent. In other words, we have*

$$
E\left[\hat{x}\,e^t\right] = 0. \tag{18.45}
$$

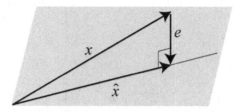

FIGURE 18.6
The orthogonal projection of x is $\hat{x}$

18.4   Kalman Filter

18.4.1   Prediction, Filtering, Smoothing and Sequential Estimation of Dynamic Systems

Up to the previous section, we have studied statistical methods for estimating true values from observed values. However, in the previous discussions, signal x does not necessarily change with time. In this section and thereafter, we will consider the case in which these signals change with time, i.e., a time-varying system called a time evolution system or a dynamic system. Let x be a time-varying multivariate random variable, and let $x_i(s)$ be the ith element at the sampling time $s\Delta t$. Such a variable that varies over time is called **state variable**. This variable transitions with time s. Here, the change is assumed to follows

$$
x(s) = f(x(s-1)) + v(s), \tag{18.46}
$$

where $v(s)$ is a noise orthogonal to the vector $f(x(s-1))$. In other words, the time transition of the state variable, which should really consider not only time $s-1$ but also more past state variables such as time $s-2$ and time $s-3$ in general, is aggregated into a function of only the previous state, i.e., value at time $s-1$. This can be expressed in terms of a statistical distribution that we assume the following for the conditional probability

$$
p(x(s) \mid x(s-1), x(s-2), \cdots) = p(x(s) \mid x(s-1)).
$$

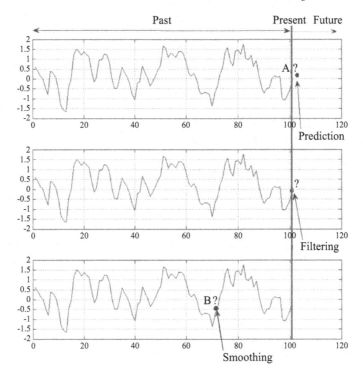

FIGURE 18.7
Prediction, filtering and smoothing of a time series

This way of thinking is called the **Markov representation model**, and it is the basic assumption on which the following discussion is based.

The problem here is to estimate the state of the system, which internally varies in time with such stochastic model, from the observed time series $y(0), y(1), \cdots, y(N)$ associated with these state variables. That is, to estimate the past, present and future values of the time series $x(s)$ using the observed $\{y(t), t = 0, 1, 2, \cdots, N\}$. In this case, when $s = N$, we call it **filter**, when $s = N + i$, we call it **prediction**, when $s = N - i$, we call it **smoothing** (Figure 18.7). Viewing this as a time evolution of the statistical distribution, it means following the time evolution of the statistical distribution $p(x(s)|Y_N)$ of the state x, given the observations $Y_N = \{y(s), s = 0, 1, \cdots, N\}$ obtained to date. At this point, the Kalman filter described below can be viewed as a **sequential estimation** that can estimate the unknown internal state of the system each time an observation is obtained for this estimation problem.

18.4.2 State-Space Representation of Dynamic Systems and Problem Formulation

The representation of a system by a state-space model introduced by Kalman is very convenient in that it allows a unified treatment of system-represented models. In this section, we show how to perform sequential state estimation of a dynamic system represented by a stationary **linear state-space model**.

The state-space model is a structure that combines the two processes described so far: the process of changing the state of the system (equation (18.46)) and the observation process (equation (18.16));

$$
\begin{aligned}
x(s) &= Fx(s-1) + Gv(s), &\text{(system model)} &\qquad(18.47)\\
y(s) &= Hx(s) + w(s) &\text{(observation model).} &\qquad(18.48)
\end{aligned}
$$

$x(s)$ is an unknown m-dimensional state variable representing the interior of the system, and $y(s)$ is a p-dimensional time series that is actually observed. The $w(s)$ and $v(s)$ are m-dimensional and p-dimensional **Gaussian white noise** satisfying

$$
\begin{cases}
E[v(s)] = 0, & E[v(s)\,v^t(s)] = Q\\
E[w(s)] = 0, & E[w(s)\,w^t(s)] = R \qquad\qquad(18.49)\\
E[w(s)\,v^t(s)] = 0.
\end{cases}
$$

Also, the initial state is given as

$$
E[x(0)] = \bar{x}, \quad E\left[\{x(0) - \bar{x}\}\{x(0) - \bar{x}\}^t\right] = \Sigma_0. \qquad(18.50)
$$

The Kalman filter is characterized by this separation of the system's **internal and external representations**.

The problem here is that given the observations $\{y(s-1), \cdots, y(1)\}$ and a new observation $y(s)$, find an estimate of $x(s)$ that minimizes the evaluation function

$$
E[l(x(s) - \bar{x})]. \qquad(18.51)
$$

Assume that the function l is symmetric and nondecreasing with respect to positive real numbers. Denote the estimate of $\hat{x}(s \mid s-1)$ obtained by minimizing equation (18.51). The estimation error is denoted as

$$
\tilde{x}(s \mid s-1) = x(s) - \hat{x}(s \mid s-1). \qquad(18.52)
$$

18.4.3 Derivation of the Kalman Filter

The solution to this problem is given by the Bayes risk estimator, i.e., the conditional expectation, as described in Section 18.2,

$$
\hat{x}(s \mid s-1) = E[x(s) \mid y(1), \cdots, y(s-1)]. \qquad(18.53)
$$

To simplify the description, we introduce the notation

$$
Y_{s-1}^t = \left(y^t(1), \cdots, y^t(s-1)\right). \qquad(18.54)
$$

According to this notation, equation (18.53) can be written as

$$
\hat{x}(s \mid s-1) = E[x(s) \mid Y_{s-1}]. \qquad(18.55)
$$

Since the stochastic processes $x(s)$ and $y(s)$ are Gaussian processes, the conditional probability is given by the results of Section 18.3. However, this computation is difficult. Therefore, we will consider **sequential estimation method**. First, it holds that

$$
\begin{aligned}
E[x(s) \mid Y_{s-1}] &= E[Fx(s-1) + Gv(s) \mid Y_{s-1}]\\
&= Fx(s-1 \mid s-1). \qquad(18.56)
\end{aligned}
$$

This is the prediction at the stage Y_{s-1} of the observation.

Also, the variance of the prediction error at this stage is given by

$$
\begin{aligned}
\Sigma(s \mid s-1) &= E\left[x(s) - x(s \mid s-1)(x(s) - x(s \mid s-1))^t\right] \\
&= E\big[\{F(x(s-1) - x(s-1 \mid s-1)) + Gv(s)\} \\
&\quad \times \{F(x(s-1) - x(s-1 \mid s-1)) + Gv(s)\}^t\big] \\
&= FE\left[(x(s-1) - x(s-1 \mid s-1))(x(s-1) - x(s-1 \mid s-1))^t\right] F^t \\
&\quad + GE\left[v(s)\, v^t(s)\right] G^t \\
&= F\,\Sigma(s-1 \mid s-1)\, F^t + GQG^t.
\end{aligned} \tag{18.57}
$$

Equations (18.56) and (18.57) are collectively referred to as the **time update algorithm**.

Next, compute the conditional expectation of $x(s \mid s)$ when a new observation $y(s)$ is obtained and the data set is $Y_s = \{Y_{s-1}, y(s)\}$,

$$
\hat{x}(s \mid s) = E[x(s) \mid Y_s] = E[x(s) \mid Y_{s-1}, y(s)]. \tag{18.58}
$$

First, Y_{s-1} and the transition error of the observed variable at time s

$$
\begin{aligned}
\tilde{y}(s) &\equiv y(s) - E[y(s) \mid Y_{s-1}] = y(s) - E[Hx(s) + w(s) \mid Y_{s-1}] \\
&= Hx(s) + w(s) - H\hat{x}(s) = H\tilde{x}(s) + w(s)
\end{aligned} \tag{18.59}
$$

are independent. Here $\tilde{x}(s) = x(s) - \hat{x}(s) = x(s) - E[x(s) \mid Y_{s-1}]$. Since such $\tilde{y}(s)$ is part of the observation signal and contains information that has not been used before, it is called **innovation** at time s.

These show that Y_{s-1} and $y(s)$ are equivalent to the set of innovations $\tilde{Y}_{s-1}$ up to time $s-1$ and $\tilde{y}(s)$ at time s for computing the conditional expectation of $x(s)$. Therefore, we obtain

$$
\begin{aligned}
\hat{x}(s \mid s) &= E[x(s) \mid \tilde{Y}_{s-1}, \tilde{y}(s)] \\
&= E[x(s) \mid \tilde{Y}_{s-1}] + E[x(s) \mid \tilde{y}(s)] - E[x(s)].
\end{aligned} \tag{18.60}
$$

The last equality follows from the following result. That is, considering $E[x(s) \mid \tilde{Y}_{s-1}, \tilde{y}(s)]$, we generally have vector random variables $x\ (= x(s))$, $u\ (= \tilde{Y}_{s-1})$, $v\ (= \tilde{y}(s))$, where u and v are independent and have variance-covariance matrices Σ_{uu} and Σ_{vv}, respectively, while u and v are correlated with x. It is equivalent to computing the expected value of $x(s)$ in the case where the variance-covariance matrices of u and v with x are Σ_{xu} and Σ_{xv}, respectively. That is, when the joint vector of u and v is assumed to be

$$
y = \begin{bmatrix} u \\ v \end{bmatrix},
$$

we have

$$
\Sigma_{yy} = \begin{bmatrix} \Sigma_{uu} & 0 \\ 0 & \Sigma_{vv} \end{bmatrix}
$$

$$
\mathrm{Cov}[x, y] = E\left[(x - \bar{x})(y - \bar{y})^t\right] = E\left[(x - \bar{x}) \begin{bmatrix} u - \bar{u} \\ v - \bar{v} \end{bmatrix}^t\right] = (\Sigma_{xu}, \Sigma_{xv}).
$$

Then we obtain

$$
\Sigma_{xy}\,\Sigma_{yy}^{-1} = (\Sigma_{xu}, \Sigma_{xv}) \begin{bmatrix} \Sigma_{uu} & 0 \\ 0 & \Sigma_{vv} \end{bmatrix}^{-1} = (\Sigma_{xu}, \Sigma_{xv}) \begin{bmatrix} \Sigma_{uu}^{-1} & 0 \\ 0 & \Sigma_{vv}^{-1} \end{bmatrix} = (\Sigma_{xu}\,\Sigma_{uu}^{-1}, \Sigma_{xv}\,\Sigma_{vv}^{-1}).
$$

Therefore, the from Conclusion 1 in Section 18.3,

$$
\begin{aligned}
E[x \mid u, v] &= \bar{x} + (\Sigma_{xu} \Sigma_u^{-1}, \Sigma_{xv} \Sigma_{vv}^{-1}) \begin{bmatrix} u - \bar{u} \\ v - \bar{v} \end{bmatrix} \\
&= \bar{x} + \Sigma_{xu} \Sigma_{uu}^{-1} (u - \bar{u}) + \Sigma_{xv} \Sigma_{vv}^{-1} (v - \bar{v}) \\
&= E[x \mid u] + E[x \mid v] - E[x] \quad\quad (18.61)
\end{aligned}
$$

and we obtain equation (18.60).

The first term in equation (18.61) is equal to $\hat{x}(s \mid s - 1)$. Again, from equation (18.43) in Conclusion 1 of Section 18.3, the second term in equation (18.60) is obtained as

$$
E[x(s) \mid \tilde{y}(s)] = E[x(s)] + \text{Cov}[x(s), \tilde{y}(s)] \text{Cov}[\tilde{y}(s), \tilde{y}(s)]^{-1},
$$

where

$$
\begin{aligned}
\text{Cov}[x(s), \tilde{y}(s)] &= E\left[(\hat{x}(s) + \tilde{x}(s) - E[x(s)])(H\tilde{x}(s) + v(s))^t\right] \\
&= E\left[\tilde{x}(s)\,\tilde{x}(s)^t\right] H^t \\
&= \Sigma(s \mid s - 1)\,H^t. \quad\quad (18.62)
\end{aligned}
$$

Also, we have

$$
\begin{aligned}
\text{Cov}[\tilde{y}(s), \tilde{y}(s)] &= E\left[\{H\tilde{x}(s) + w(s)\}\{H\tilde{x}(s) + w(s)\}^t\right] \\
&= H\left\{E[\tilde{x}(s)\,\tilde{x}^t(s)]\right\} H^t + R \\
&= H\,\Sigma(s \mid s - 1)\,H^t + R. \quad\quad (18.63)
\end{aligned}
$$

Thus, by putting

$$
K(s) = \Sigma(s \mid s - 1)\,H^t(H\,\Sigma(s \mid s - 1)\,H^t + R)^{-1}, \quad\quad (18.64)
$$

we end up with equation (18.60) by putting

$$
\hat{x}(s \mid s) = \hat{x}(s \mid s - 1) + K(s)(y(s) - H\hat{x}(s \mid s - 1)). \quad\quad (18.65)
$$

Next, we compute $\Sigma(s \mid s)$. From equation (18.65), since we get

$$
\begin{aligned}
\hat{x}(s \mid s - 1) &= \hat{x}(s \mid s) - K(s)(y(s) - H^t\hat{x}(s \mid s - 1)) \\
&= \hat{x}(s \mid s) - K(s)\,\tilde{y}(s),
\end{aligned}
$$

we obtain

$$
x(s) - \hat{x}(s \mid s - 1) = \{x(s) - \hat{x}(s \mid s)\} + K(s)\,\tilde{y}(s).
$$

By the way, in this equation, $x(s) - \hat{x}(s \mid s)$ in { } is orthogonal to its next term, $\tilde{y}(s)$, so in the operation of covariance of $x(s) - \hat{x}(s \mid s - 1)$, when taking expectation, the product on the right side is { } and the term of $\tilde{y}(s)$ is zero, and the expected value of the sum of the product of the terms of $\tilde{y}(s)$ and $\tilde{y}(s)$ is zero, and we end up with

$$
\begin{aligned}
\Sigma(s \mid s - 1) &= \Sigma(s \mid s) + K(s)\,E[\tilde{y}(s)\,\tilde{y}^t(s)]\,K^t(s) \\
&= \Sigma(s \mid s) + K(s)\,H^t\,\Sigma(s \mid s - 1).
\end{aligned}
$$

From this, we have

$$
\Sigma(s \mid s) = \Sigma(s \mid s - 1) - K(s)\,H^t\,\Sigma(s \mid s - 1). \quad\quad (18.66)
$$

Equations (18.65) and (18.66) are called **observation update algorithm**. We also refer to equation (18.64) as **Kalman gain**.

Summing up these results, we obtain the following conclusions[12].

[Conclusion 3]   Kalman filter

For the dynamic system given by equations (18.47) and (18.48), the estimate that minimizes the evaluation function (equation (18.51)) at time $s + 1$ given the observations $y(1), \cdots, y(s)$ and the initial distribution $x(1 \mid 0) = \bar{x}$, $\Sigma(1 \mid 0) = \Sigma_0$, are obtained by the following sequential procedure;

- *Observation Update*

$$K(s) \;=\; \Sigma(s \mid s-1)\, H^t \,(H\, \Sigma(s \mid s-1)\, H^t + R)^{-1} \tag{18.67}$$

$$\hat{x}(s \mid s) \;=\; \hat{x}(s \mid s-1) + K(s)(y(s) - H^t\, \hat{x}(s \mid s-1)) \tag{18.68}$$

$$\Sigma(s \mid s) \;=\; \Sigma(s \mid s-1) - K(s)\, H^t\, \Sigma(s \mid s-1), \tag{18.69}$$

- *Time Update*

$$\hat{x}(s \mid s-1) \;=\; F\, x(s-1 \mid s-1) \tag{18.70}$$

$$\Sigma(s \mid s-1) \;=\; F\, \Sigma(s-1 \mid s-1)\, F^t + G Q G^t. \tag{18.71}$$

19

Statistical Optimal Control Theory

In this chapter, we derive the control law in the statistical optimal control of ship motions using the control type autoregressive model described in Chapter 15. There are various methods for obtaining the control law, but here we present a method using dynamic programming. In this case, the cost function is assumed to be the usual quadratic form, but the control law can be obtained by proceeding in the same approach even when constraints are added to smooth rudder motion, etc.

19.1   Preparation for Understanding Optimal Control Theory

19.1.1   Optimal Routing Problem and Dynamic Programming

Once the control model and cost function are determined, the next step is to establish control method to maintain the responding system in the desired state or to move it along the desired trajectory. In general, when a mathematical model of the target system is built in some way and an appropriate cost function is set, the theory that finds a control method to optimize the cost function is called **optimal control theory**.

The following three methods are known as typical mathematical methods for obtaining such optimal control laws;

1. Method by dynamic programming.
2. Variational method.
3. Method by the maximum principle.

In this section, we adopt the dynamic programming method proposed by Richard Bellman to solve the optimal problem.

19.1.2   Example for Understanding Dynamic Programming: Shortest Route Problem

Dynamic programming (DP), advocated by Richard Bellman, is a principle for the optimal design of dynamic systems, and the design of optimal control systems proceeds according to the following **principles of optimality**.

- **Principle of Optimality**
  An optimal policy has the property that whatever the initial state and initial decision are, the remaining decisions must constitute an optimal policy with regard to the state resulting from the first decision[15]. (Figure 19.1)

That is, in Figure 19.1, the optimal strategy must be optimal not only from the starting point A, but also from any point C leading to the target point B.

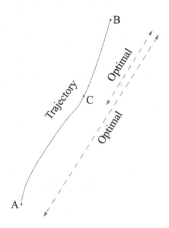

FIGURE 19.1
Principle of optimality

This principle may seem obvious at first glance, but in fact it is not. For example, it is not generally true to generalize it to say that "any part of the trajectory chosen by the optimal strategy is also the optimal trajectory." For example, how would a marathoner distribute his power in a long-distance race? If the runner follows the policy of always running at full speed in every section of the race, the runner will tire before reaching the finish line. The runner always comes out on top at the final point by rationally allocating his efforts toward the final goal. What Bellman is saying is that short-sightedly running fastest in any part of the race is not ultimately the best strategy.

To understand **dynamic programming** (abbreviated as **DP**), it is easy to explain it using the following shortest route problem described by Kaufman[40]. Although backward solving is the usual method of solving dynamic programming, the two coincide in decision problems. In stochastic problems, the problem is solved backward, and at run time, the stored optimal strategy is advanced forward.

The principle of dynamic programming is to always anticipate to the last step and make the current strategy. However, there is one exceptional stage. It is the last stage. Only the last stage should be optimized by considering only that stage. After this is done, another stage is added, the second to last stage and so on. Obviously, in a new two-stage process, one cannot plan the last stage without knowing how the second to last stage will end. On that point, we must make various assumptions about the state in which the second-to-last stage will end, and for each possible assumption, we must decide on a strategy for the last stage. In dynamic programming, this procedure is optimized while moving backward.

Therefore, as an example, consider the problem of finding the shortest route as shown in Figure 19.2. The problem here is to find the shortest route from A to N in the figure. The distance to each point at each step is shown above the connecting line.

To solve this problem, we define the decision variables x_0, x_1, x_2, x_3, x_4 and x_5 as variables representing the subsequent optimal routes determined at each stage. Candidates for the contents of the decision variables in each stage are;

x_0 : A
x_1 : B, C, D
x_2 : E, F, G
x_3 : H, I, J, K
x_4 : L, M
x_5 : N

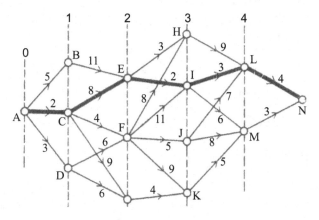

FIGURE 19.2
Problem of finding the shortest route

The problem here is to find the shortest route from point A to N by optimally selecting measures x_0, x_1, x_2, x_3, x_4 and x_5 at each stage. The cost function F that assesses the cost of all selected routes (in this case, total distances) is given by

$$F(x_0, x_1, x_2, x_3, x_4, x_5) = v_1(x_0, x_1) + v_2(x_1, x_2) + v_3(x_2, x_3) + v_4(x_3, x_4) + v_5(x_4, x_5),$$

where $v_1(x_0, x_1)$ is the cost of section 1, $v_2(x_1, x_2)$ is the cost of section 2 and so on.

The cost of each route in the fourth stage, one step down from the final point N, has two possible initial points, L and M, in the fourth stage. Thus, there is no choice, and in each case they are given by

$$f_5(L) = v_5(L, N) = 4, \quad f_5(M) = v_5(M, N) = 3.$$

Next, we consider the route from the third stage to the fourth stage, going back one step. At this point, since the principle of optimality requires that the route must be additively optimal up to the point where it reaches the final step, the following must hold;

$$f_{4,5}(H) = \min_{x_4 = L, M} [f_5(x_4) + v_4(H, x_4)]$$

$$f_{4,5}(I) = \min_{x_4 = L, M} [f_5(x_4) + v_4(I, x_4)]$$

$$f_{4,5}(J) = \min_{x_4 = L, M} [f_5(x_4) + v_4(J, x_4)]$$

$$f_{4,5}(K) = \min_{x_4 = L, M} [f_5(x_4) + v_4(K, x_4)].$$

Note here that, if there is no direct route connecting them, such as (H,M) or (K,L), the distance between them is denoted as infinity. If we put in concrete numbers, we get

$$
\begin{aligned}
f_{4,5}(H) &= \min[4 + 9, 3 + \infty] = 13, &\text{with} \quad x_4 &= L \\
f_{4,5}(I) &= \min[4 + 3, 3 + 6] = 7, &\text{with} \quad x_4 &= L \\
f_{4,5}(J) &= \min[4 + 7, 3 + 8] = 11, &\text{with} \quad x_4 &= L \text{ or } M \\
f_{4,5}(K) &= \min[4 + \infty, 3 + 5] = 8, &\text{with} \quad x_4 &= M.
\end{aligned}
$$

Therefore,

If we stop at H, then NLH is the shortest and has a cost of 13.
If we stop at I, then NLI is the shortest and has a cost of 7.
If we stop at J, then NLJ or NMJ is the shortest with a cost of 11.
If we stop at K, then NMK is the shortest and the cost is 8.

Then proceed to the third, fourth and fifth steps. From the same principle of optimality, it must additively hold that

$$f_{3,4,5}(\text{E}) = \min_{x_4=\text{H,I,J,K}}[f_{4,5}(x_3) + v_3(\text{E}, x_3)]$$

$$f_{3,4,5}(\text{F}) = \min_{x_4=\text{H,I,J,K}}[f_{4,5}(x_3) + v_3(\text{F}, x_3)]$$

$$f_{3,4,5}(\text{G}) = \min_{x_4=\text{H,I,J,K}}[f_{4,5}(x_3) + v_3(\text{G}, x_3)].$$

Substituting numerical values, we obtain

$$f_{3,4,5}(\text{E}) = \min[13+3, 7+2, 11+\infty, 8+\infty] = 9, \quad \text{with} \quad x_3 = \text{I}$$

$$f_{3,4,5}(\text{F}) = \min[13+8, 7+11, 11+5, 8+9] = 16, \quad \text{with} \quad x_3 = \text{J}$$

$$f_{3,4,5}(\text{G}) = \min[13+\infty, 7+\infty, 11+\infty, 8+4] = 12, \quad \text{with} \quad x_3 = \text{K}.$$

Therefore, it can be seen that

If we stop at E, then NLIE is the shortest and the cost is 9.
If we stop at F, then NLJF or NMJ is the shortest and the cost is 16.
If we stop at G, then NMGK is the shortest and the cost is 12.

Similarly, considering the second to the last stage, it must hold that

$$f_{2,3,4,5}(\text{B}) = \min_{x_4=\text{E,F,G}}[f_{3,4,5}(x_2) + v_3(\text{B}, x_2)]$$

$$f_{2,3,4,5}(\text{C}) = \min_{x_4=\text{E,F,G}}[f_{3,4,5}(x_2) + v_3(\text{C}, x_2)]$$

$$f_{2,3,4,5}(\text{D}) = \min_{x_4=\text{E,F,G}}[f_{3,4,5}(x_2) + v_3(\text{D}, x_2)].$$

Substituting the numbers from the first stage, we get

$$f_{2,3,4,5}(\text{B}) = \min[9+11, 16+\infty, 12+\infty] = 20, \quad \text{with} \quad x_2 = \text{E}$$

$$f_{2,3,4,5}(\text{C}) = \min[9+8, 16+4, 12+9] = 17, \quad \text{with} \quad x_2 = \text{E}$$

$$f_{2,3,4,5}(\text{D}) = \min[9+\infty, 16+6, 12+6] = 18, \quad \text{with} \quad x_2 = \text{G}.$$

Therefore, we obtain

If it stops at B, then NLIEB is the shortest and the cost is 20.
If it stops at C, then NLIEC is the shortest and the cost is 17.
If it stops at D, then NMKGD is the shortest and the cost is 18.

Finally, let f be the function that gives the minimum value of $F(x_0, x_1, x_2, x_3, x_4, x_5)$, then we obtain

$$f = \min_{x_1=\text{B,C,D}}[f_{2,3,4,5}(x_1) + v_1(\text{A}, x_1)].$$

In other words, it holds that

$$f = \min[20+5, 17+2, 18+3] = 19, \quad \text{with} \quad x_1 = \text{C}.$$

Thus, the shortest route is NLIECA. This solution was obtained by optimizing from N to A in the reverse direction. This is called the **backward search**.

19.1.3 Formulation of Optimal Control Problem and Dynamic Programming

The dynamic programming method described in the previous section is applied to the case of optimal control theory for the dynamic system under consideration here. The general formulation is as follows:

System in Problem

First, the dynamic motion of the system in problem that we are about to consider is divided into appropriate **stages**. Then, number the initial stage as 0 and the terminal stage as N stage. Let us assume that the **state variable** representing the state of each stage is defined as a scalar variable $z(s)$. Then, the rule for transitioning from any step l to step $l+1$ of $z(s)$ is assumed to be deterministic or stochastically determined by some **control** (**strategy, policy**) $u(l)$.

Cost Function

Next, we evaluate the cost, or energy, of reaching the N-step for this problem, using the **cost function** defined as

$$J_N = \sum_{s=1}^{N} f_s(z(s), u(s-1)), \tag{19.1}$$

where it is assumed that this function is additive.

The optimal problem we are considering here is to determine the **optimal control law** or **optimal strategy** that $u(l)$ at each stage should take so that the above cost function is optimal as a whole. That is, if the cost function (equation (19.1)) is minimized, then the goal here is to optimize the cost function,

$$\min_{u(0), u(1), \cdots, u(N-1)} J_N. \tag{19.2}$$

Application of Bellman's Principle of Optimality

What Bellman's principle of optimality states, as explained at the beginning of the previous section, is not simply that the overall optimal (in this case, minimum) cost value can be obtained by optimizing the cost function from the middle stage l to the next stage $l+1$ of this process, but the claim is to always minimize the cost function leading to the target point, namely, the N-step.

Now consider a scalar state $z(s)$ representing a change of state (Figure 19.3). Then, the process is divided into two stages: the first stage from the initial stage to the l-stage and the second stage from the $l+1$-stage onward. At this point, the cost function is additive, and thus it follows that

$$J_N = \sum_{s=1}^{l} f_s(z(s), u(s-1)) + \sum_{s=l+1}^{N} f_s(z(s), u(s-1)). \tag{19.3}$$

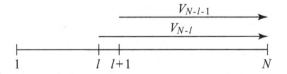

FIGURE 19.3
Optimization interval

From the principle of optimality, the *remaining steps must be optimal* both *from the step l to the last step* and *from the step l+1 to the last step*. Therefore, consider the cost function of the remaining stages from the *l*-stage to the final stage

$$J_{N-l} = \sum_{s=l+1}^{N} f_s(z(s), u(s-1)),\tag{19.4}$$

and define the minimum value of the cost function of this *remaining number of stages $N-l$* stages as

$$V_{N-l}(z(l)) = \min_{u(l),\cdots,u(N-1)} J_{N-l}.\tag{19.5}$$

Then, it must hold

$$\begin{aligned}V_{N-l}(z(l)) &= \min_{u(l)}[f_{l+1}(z(l+1), u(l)) + \min_{u(l+1),\cdots,u(N-1)} J_{N-l-1}]\\ &= \min_{u(l)}[f_{l+1}(z(l+1), u(l)) + V_{N-l-1}(z(l+1))].\end{aligned}\tag{19.6}$$

This is the functional statement of Bellman's principle of optimality, and equation (19.6) is called **Bellman's functional equation**.

19.1.4   Prior Knowledge of Extrema of Quadratic Functions

By the way, the idea used in solving this problem is the same as obtaining the minimum value *u* of the quadratic function with respect to *u*,

$$f(u) = au^2 + 2buz + cz^2.\tag{19.7}$$

As can be seen by transforming this equation as

$$f(u) = a\left(u^2 + 2\frac{b}{a}uz\right) + cz^2 = a\left(u + \frac{b}{a}z\right)^2 - \frac{b^2}{a}z^2 + cz^2,$$

the minimum value of the quadratic function $f(z)$ is given by the minimum value $-\frac{b^2}{a}z^2 + cz^2$ when $u = -\frac{b}{a}z$.

Let us extend this to the case where x and u are multi-dimensional. In this case, consider the following quadratic form

$$F(u) = u^t au + z^t bu + u^t b^t z + z^t cz,\tag{19.8}$$

where a is a symmetric $n \times n$ positive definite matrix, b is a $p \times n$, c is a $p \times p$ matrix, u is an n-dimensional vector and z is a p-dimensional vector. Transforming the above equation as

$$\begin{aligned}F(u) &= u^t au + z^t b^t u + u^t bz + z^t cz + z^t b^t a^{-1} bz - z^t b^t a^{-1} bz\\ &= (u + a^{-1}bz)^t a(u + a^{-1}bz) - z^t b^t a^{-1} bz + z^t cz,\end{aligned}$$

the minimum value is attained at

$$u = -a^{-1}bz,\tag{19.9}$$

and its value is obtained by

$$F_{\min} = z^t(c - b^t a^{-1} b)z.\tag{19.10}$$

19.1.5 Preliminary Considerations on the Expected Value of a Random Variable in Quadratic Form

We show how to compute the expected value of a quadratic form $E[x^t S x]$, in general, when x is a Gaussian random variable with mean m and variance-covariance matrix R. Let S be a non-negative definite matrix. Since $E[x] = m$, we have

$$
\begin{aligned}
E[x^t S x] &= E[(x-m)^t S(x-m)] + E[m^t S x] + E[x^t S m] - E[m^t S m] \\
&= E[(x-m)^t S(x-m)] + m^t S m.
\end{aligned} \tag{19.11}
$$

Furthermore, for the quadratic form, it holds that

$$
(x-m)^t S(x-m) = \mathrm{tr}[S(x-m)(x-m)^t], \tag{19.12}
$$

where $\mathrm{tr}[\cdot]$ is the trace of the matrix. Thus, taking the statistical expectation yields

$$
\begin{aligned}
E[(x-m)^t S(x-m)] &= \mathrm{tr}\left[S \cdot E[(x-m)(x-m)^t]\right] \\
&= \mathrm{tr}[SR] = \mathrm{tr}[RS].
\end{aligned} \tag{19.13}
$$

From this we obtain

$$
E[x^t S x] = m^t S m + \mathrm{tr}[RS]. \tag{19.14}
$$

19.2 Optimal Control Methods for Linear Systems

19.2.1 Optimization of Stochastic Fully Informed Linear Discrete Systems

With the above preparations, let us solve the optimal control problem for linear systems described in Section 15.2. This problem, commonly referred to as the **LQG control problem** (**linear quadratic Gaussian control problem**), is an optimization problem for a complete information discrete stochastic system with unknown Gaussian noise input as shown below[58][14].

Linear Stochastic Control Model

$$
\begin{aligned}
Z(s+1) &= \Phi Z(s) + \Gamma U(s) + W(s+1) \tag{19.15} \\
E[Z(0)] &= Z_0, \tag{19.16}
\end{aligned}
$$

where $Z(s)$ is an r-dimensional state vector representing the system state, $U(s)$ is an l-dimensional operating variable, Φ is an $r \times r$ transition matrix and Γ is an $r \times l$ driving matrix. Z_0 is the mean of the initial values of $Z(s)$ and is assumed to be known. Also the structure of the noise system in $W(s)$ is given by

$$
E[W(s)] = 0, \quad E[W(s) W^t(s)] = R_1. \tag{19.17}
$$

where $W(s)$ is assumed uncorrelated with $Z(s)$ and $U(s)$.

Complete State Information System

If the observation equation is not disturbed by noise, then we have

$$
y(s) = Z(s). \tag{19.18}
$$

Cost function

The cost function is given by

$$J_N = E\left[\sum_{s=1}^{N}\left\{Z^t(s)\,QZ(s) + U^t(s-1)\,RU(s-1)\right\}\right], \tag{19.19}$$

where Q and R are symmetric non-negative definite and positive definite weight function matrices, respectively.

Objective

The objective here is to achieve

$$\min_{U(0),U(1),\cdots,U(N-1)} J_N. \tag{19.20}$$

That is, to minimize the cost function J by optimizing the control input $U(0), U(1), \cdots, U(N-1)$.

Bellman's Functional Equation

Suppose that we have now reached any l-stage (remaining number of stages is $N-l$) of the first through N-stages, and we divide the cost function up to this stage and the cost function after that, as follows

$$E\left[\sum_{s=1}^{N}\left\{Z^t(s)\,QZ(s) + U^t(s-1)\,RU(s-1)\right\}\right]$$

$$= E\left[\sum_{s=1}^{l}\left\{Z^t(s)\,QZ(s) + U^t(s-1)\,RU(s-1)\right\}\right]$$

$$+ E\left[\sum_{s=l+1}^{N}\left\{Z^t(s)\,QZ(s) + U^t(s-1)\,RU(s-1)\right\}\right]. \tag{19.21}$$

Here $E[\cdot]$ indicates that we take the expectation of the value in [ ] due to stochastic variation in $W(s)$. The first term in the second equation does not depend on $U(l), \cdots, U(N)$ because it is a term related to the l-stage or earlier. Therefore, the goal of the minimization problem in equation (19.21) at this point is to minimize the second term. Considering this in the third stage of Figure 19.2, this means that the possibilities in this stage, H, I, J or K, are to formulate subsequent measures under given conditions. Thus, the second term in equation (19.21) can be obtained by computing the minimum value with respect to $U(s)$ after it, under the condition that the value of $Z(s)$ in the l-stage already exists. Thus, with respect to the second term, it holds that

$$\min_{U(l),\cdots,U(N-1)} E\left[\sum_{s=l+1}^{N}\left\{Z^t(s)\,QZ(s) + U^t(s-1)\,RU(s-1)\right\}\right]$$

$$= \min_{U(l),\cdots,U(N-1)} E\left[\sum_{s=l+1}^{N}\left\{Z^t(s)\,QZ(s) + U^t(s-1)\,RU(s-1)\right\} \mid Z(l)\right]$$

$$= E\left[V_{N-l}(Z(l))\right], \tag{19.22}$$

where

$$V_{N-l}(Z(l)) = \min_{U(l),\cdots,U(N-1)} E\left[\sum_{s=l+1}^{N}\left\{Z^t(s)\,QZ(s) + U^t(s-1)\,RU(s-1)\right\} \mid Z(l)\right]$$

$$\tag{19.23}$$

is the expected value of the total cost of the remaining stages after the lth stage. Therefore, we have

$$V_0(Z(N)) = 0, \qquad (19.24)$$

which is the expected value of the total cost of the remaining stages after the l-stage.

The $E[\,\cdot\mid Z(l)]$ in equation (19.22) represents the conditional expectation of the stochastic variable in $\{\ \}$ due to the variation of $W(l)$ in each subsequent interval to the final stage, given the value of Z in the lth stage. The min on its left implies that the optimal control law is adopted for subsequent $U(l), \cdots, U(N-1)$ and taken so that the expected value on the right is minimized.

Next, suppose that the value of Z at the $l+1$th stage (remaining number of stages is $N-l-1$) is given as $Z(l+1)$. $V_{N-l}(Z(l))$ assumes that $Z(l+1)$ has been determined after time $l+1$ from the principle of optimality, and after adopting the optimal measure $U(l+1)$ that minimizes the entire cost up to the Nth stage after that, the optimal value of $U(l)$ is adopted and $V_{N-l}(Z(l))$ is minimized. Therefore, it must hold

$$
\begin{aligned}
V_{N-l}(Z(l)) = \ &\min_{U(l)} E\big[Z^t(l+1)\,QZ(l+1) + U^t(l)\,RU(l) + \\
&\min_{U(l+1)} E\{Z^t(l+2)\,QZ(l+2) + U^t(l+1)\,RU(l+1) + \cdots + \\
&\min_{U(N-1)} E\big(Z^t(N)\,QZ(N) + U^t(N-1)\,RU(N-1)\big) \mid Z(l+1)\} \mid Z(l)\big].
\end{aligned}
$$

Considering the previous definition of $V_{N-l}(Z(l))$, the second bracket $\{\cdot\}$ is given by

$$
\begin{aligned}
&\min_{U(l+1)} E\{[Z^t(l+2)\,QZ(l+2) + U^t(l+1)\,RU(l+1) + \cdots \\
&\quad + \min_{U(N-1)} E\big(Z^t(N)\,QZ(N) + U^t(N-1)\,RU(N-1)\big) \\
&= \min_{U(l+1),\cdots,U(N-1)} E\left[\sum_{s=l+2}^{N} \{Z^t(s)\,QZ(s) + U^t(s-1)\,RU(s-1)\} \mid Z(l+1)\right] \\
&= V_{N-l-1}(Z(l+1)).
\end{aligned}
$$

From these, it follows that

$$
\begin{aligned}
V_{N-l}(Z(l)) = \ &\min_{U(l)} \big[E[\{Z^t(l+1)\,QZ(l+1) + U^t(l)\,RU(l) + V_{N-l-1}(Z(l+1))\} \mid Z(l)]\big] \\
= \ &\min_{U(l)} \big[Z^t(l+1)\,QZ(l+1) + U^t(l)\,RU(l) + E\{V_{N-l-1}(Z(l+1)) \mid Z(l)\}\big]
\end{aligned}
$$

$$(19.25)$$

The above equation is the **Bellman's functional equation** for this problem.

Solution

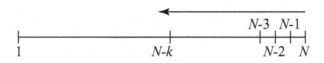

FIGURE 19.4
Backward optimization

Next, let us consider solving Bellman's functional equation backward[7]. First, in Figure 19.4, at the $N - 1$st stage (with 1 step remaining), from the principle of optimality, we obtain

$$V_1(Z(N-1)) \;=\; \min_{U(N-1)} \big[E[Z^t(N)\,QZ(N) + U^t(N-1)\,RU(N-1)$$
$$+\, V_0(Z(N)) \mid Z(N-1)] \big]. \tag{19.26}$$

The first term can be written as

$$E[Z^t(N)\,QZ(N) \mid Z(N-1)] \;=\; E[(\Phi Z(N-1) + \Gamma U(N-1) + W(N))^t Q$$
$$\times\, (\Phi Z(N-1) + \Gamma U(N-1) + W(N)) \mid Z(N-1)].$$

However, since $U(s)$ and $W(s)$, and $Z(s)$ and $W(s)$ are uncorrelated, expanding the above equation, considering equation (19.13), we obtain

$$E[Z^t(N)\,QZ(N) \mid Z(N-1)]$$
$$= E[U^t(N-1)\,\Gamma^t Q\Gamma U(N-1) + U^t(N-1)\,\Gamma^t Q\Phi Z(N-1)$$
$$+\, Z^t(N-1)\,\Phi^t Q\Gamma U(N-1) + Z^t(N-1)\,\Phi^t Q\Phi Z(N-1)] + \mathrm{tr}[R_1 Q]. \tag{19.27}$$

Therefore, V_1 can be written as

$$V_1(Z(N-1)) \;=\; \min_{U(N-1)} \big[E[U^t(N-1)(\Gamma^t Q\Gamma + R)\,U(N-1)$$
$$+\, U^t(N-1)\,\Gamma^t Q\Phi Z(N-1) + Z^t(N-1)\,\Phi^t Q\Gamma U(N-1)$$
$$+\, Z^t(N-1)\,\Phi^t Q\Phi Z(N-1)] + \mathrm{tr}[R_1 Q] \big]. \tag{19.28}$$

In equation (19.9), if $a = \Gamma^t Q\Gamma + R$ and $b = \Gamma^t Q\Phi$, the minimum value in [ ] of the quadratic function (equation (19.28)) is attained by putting $U(N-1)$ as

$$U(N-1) = -L(N-1)\,Z(N-1), \tag{19.29}$$

where

$$L(N-1) = (\Gamma^t Q\Gamma + R)^{-1}\,\Gamma^t Q\Phi. \tag{19.30}$$

As a result, when this procedure is adopted, $V_1(Z(N-1))$ can be regarded as $c = \Phi^t Q\Phi$ in equation (19.10), we obtain

$$V_1(Z(N-1))$$
$$= E[Z^t(N-1)\{\Phi^t Q\Phi - \Phi^t Q\Gamma(\Gamma^t Q\Gamma + R)^{-1}\Gamma^t Q\Phi\}Z(N-1)] + \mathrm{tr}[R_1 Q]$$
$$= E[Z^t(N-1)\{\Phi^t (Q - Q\Gamma(\Gamma^t Q\Gamma + R)^{-1}\Gamma^t Q)\Phi\}Z(N-1)] + \mathrm{tr}[R_1 Q], \tag{19.31}$$

which is a quadratic function of $Z(N-1)$ at this stage. Here, if we put

$$\begin{cases} M(N-1) = Q - Q\Gamma(\Gamma^t Q\Gamma + R)^{-1}\Gamma^t Q \\ g(N-1) \;= \mathrm{tr}[R_1 Q], \end{cases} \tag{19.32}$$

then equation (19.31) can be written as

$$V_1(Z(N-1)) = E[Z^t(N-1)\,\Phi^t M(N-1)\,\Phi Z(N-1)] + g(N-1).$$

Considering the $N - 2$nd step (2 steps remaining), it follows from Bellman's functional equation that

$$
\begin{aligned}
V_2(Z(N-2)) &= \min_{U(N-2)} \left[E[Z^t(N-1) Q Z(N-1) \right. \\
&\quad \left. + U^t(N-2) R U(N-2) + V_1(Z(N-1)) \mid Z(N-2)] \right] \\
&= \min_{U(N-2)} E[Z^t(N-1)\{Q + \Phi^t M(N-1)\,\Phi\} Z(N-1) \\
&\quad + U^t(N-2) R U(N-2) + g(N-1) \mid Z(N-2)].
\end{aligned}
\tag{19.33}
$$

Here, by putting

$$
S(N-1) = Q + \Phi^t M(N-1)\,\Phi,
\tag{19.34}
$$

equation (19.33) can be expressed as

$$
\begin{aligned}
V_2(Z(N-2)) &= \min_{U(N-2)} E[Z^t(N-1)\{S(N-1)\} Z(N-1) \\
&\quad + U^t(N-2) R U(N-2) + g(N-1) \mid Z(N-2)].
\end{aligned}
$$

This means that in equation (19.26), by putting

$$
S(N) = Q, \quad g(N) = 0,
\tag{19.35}
$$

and by substituting N to $N-1$, we obtain the same cost function except that $\mathrm{tr}[R_1 Q]$ is added to the final term. Thus, the optimization methods of equations (19.26) through (19.30) can be used in this stage as well, simply by changing the corresponding Q to $S(N)$. In other words, if we put

$$
\begin{cases}
M(N-1) = S(N) - S(N)\,\Gamma(\Gamma^t S(N)\,\Gamma + R)^{-1}\Gamma^t S(N) \\
g(N-1) \;= g(N) + \mathrm{tr}[R_1 S(N)],
\end{cases}
\tag{19.36}
$$

then equation (19.34) becomes

$$
\begin{aligned}
S(N-1) &= Q + \Phi^t M(N-1)\,\Phi \\
&= Q + \Phi^t\{S(N) - S(N)\,\Gamma(\Gamma^t S(N)\,\Gamma + R)^{-1}\Gamma^t S(N)\}\Phi.
\end{aligned}
\tag{19.37}
$$

This equation is a backward recursive formula for computing $S(N-1)$ from the known $S(N)$, which can be computed since all the other coefficients are known. Equation (19.37) is called the **Riccati equation**.

The optimal control strategy is then given by, from equation (19.9)

$$
U(N-2) = -L(N-2)\,Z(N-2),
\tag{19.38}
$$

where

$$
L(N-2) = (\Gamma^t S(N-1)\,\Gamma + R)^{-1}\Gamma^t S(N-1)\,\Phi.
\tag{19.39}
$$

Note that equation (19.37) can be written using equation (19.30) as

$$
S(N-1) = Q + \Phi^t S(N)\,\Phi - L^t(N-1)(\Gamma^t S(N)\,\Gamma + R) L(N-1).
\tag{19.40}
$$

The optimization procedure for the following steps is a repetition of this procedure. Suppose now that the result of stage $N - 2$ is applied to stage $N - 3$ and $N - 3$ to $N - 4$, etc., resulting in $N - k$ stages in general. Then the relational equation corresponding to equation (19.36) is given by

$$\begin{cases} M(N - k) = S(N - k + 1) \\ \qquad\qquad - S(N - k + 1)\,\Gamma(\Gamma^t S(N - k + 1)\,\Gamma + R)^{-1}\Gamma^t S(N - k + 1) \\ g(N - k) \; = g(N - k + 1) + \mathrm{tr}[R_1 S(N - k + 1)]. \end{cases} \qquad (19.41)$$

Furthermore, the relation corresponding to equation (19.39) becomes

$$L(N - k) = (\Gamma^t S(N - k + 1)\,\Gamma + R)^{-1}\Gamma^t S(N - k + 1)\,\Phi, \qquad (19.42)$$

and the corresponding relation in equation (19.37) is

$$\begin{aligned} S(N - k) &= Q + \Phi^t M(N - k)\,\Phi \qquad\qquad\qquad\qquad\qquad (19.43) \\ &= Q + \Phi^t\{S(N - k + 1) \\ &\qquad - S(N - k + 1)\,\Gamma(\Gamma^t S(N - k + 1)\,\Gamma + R)^{-1}\Gamma^t S(N - k + 1)\}\Phi. \end{aligned}$$

Thus, using these, the optimal control law in $N - k$ stages is given by

$$U(N - k) = -L(N - k)\,Z(N - k). \qquad (19.44)$$

Also, since $V_N(Z(0))$ equals $\min J_N$ by definition, the result of optimal control over all stages is given by

$$\begin{aligned} V_N(Z(0)) &= \min_{U(0)} E[Z^t(1)\,QZ(1) + U^t(0)\,RU(0) + V_{N-1}(Z(1)) + g(0) \mid Z(0)] \\ &= E[Z(0)\,M(0)\,Z(0)] + g(0) \\ &= m_z^t M(0)\,m_x + \mathrm{tr}[R_1 M(0)] + \sum_{s=0}^{N-1} \mathrm{tr}[R_1 S(s + 1)]. \qquad (19.45) \end{aligned}$$

Now, in equations (19.41), (19.43) and (19.42), to change the optimization direction from backward to forward, we replace $M(N - k) \to M(k)$, $S(N - k) \to S(k)$, $L(N - k) \to L(k)$. Then, starting from $S(0) = Q$, we have

$$\begin{cases} M(k) = S(k - 1) - S(k - 1)\,\Gamma(\Gamma^t S(k - 1)\,\Gamma + R)^{-1}\Gamma^t S(k - 1) \\ S(k) \; = Q + \Phi^t M(k)\,\Phi. \end{cases} \qquad (19.46)$$

Proceeding this computation sequentially, equation (19.44) becomes

$$L(k) = (\Gamma^t S(k - 1)\,\Gamma + R)^{-1}\Gamma^t S(k - 1)\,\Phi.$$

In this case, the optimal control law is determined as

$$U(N - k) = -L(k)\,Z(N - k), \qquad (19.47)$$

where, considering $k = N$, we have

$$L(N) = (\Gamma^t S(N - 1)\,\Gamma + R)^{-1}\Gamma^t S(N - 1)\,\Phi, \qquad (19.48)$$

which is the optimal control law to adopt considering the state of the system N points ahead from $S = 1$.

If the system is stationary, $L(k)$ will be constant if N is sufficiently large. Thus, in effect we obtain

$$U(s) = -LZ(s) \qquad (19.49)$$

and thus optimal control with fixed gains becomes possible.

Bibliography

[1] Akaike, H. (1968), On the Use of a Linear Model for the Identification of Feedback Systems, *Ann. Inst. Statist. Math.*, Vol.20, No.1, 425–439.

[2] Akaike, H. (1969a), Fitting Autoregressive Models for Prediction, *Ann. Inst. Statist. Math.*, Vol.21, 243–247.

[3] Akaike, H. (1969b), A Method of Statistical Identification of Discrete Time Parameter Linear System, *Ann. Inst. Statist. Math.*, Vol.21, 225–242.

[4] Akaike, H. (1973), Information Theory and an Extension of the Maximum Likelihood Principle, *Proc. 2nd International Symposium on Information Theory*, B. N. Petrov and F. Csaki, (eds.), Akademiai Kiado, Budapest, 267–281.

[5] Akaike, H. (1974), A New Look at the Statistical Model Identification, *IEEE, Trans. Automat. Contr.*, AC-19, 716–723.

[6] Akaike, H. and Kitagawa, G. (eds.) (1998), *The Practice of Time Series Analysis*, Statistics for Engineering and Physical Science, Springer, New York.

[7] Akaike, H. and Nakagawa, T. (1988), *Statistical Analysis and Control of Dynamic Systems*, Mathematics and Its Applications, Kluwer Academic Publishers, Tokyo.

[8] Akaike, H. and Yamanouchi, Y. (1962), On the Statistical Estimation of Frequency Response Function, *Ann. Inst. Statist. Math.*, Vol. 14, 23–56.

[9] Amerongen, J. V. and Udink Cate, A. J. (1975), Model Reference Adaptive Autopilots for Ships, *Automatica*, Vol.11, No.1, 441–449.

[10] Anderson, B. D. O. and Moore, J. E. (1979), *Optimal Filtering*, Prentice Hall.

[11] Aoki, M. (1972), *Mathematical Programming II — Nonlinear Programming*, Kyoritsu Publishing, (in Japanese).

[12] Åström, K. J. (1970), *Introduction to Stochastic Control Theory*, Academic Press Inc.

[13] Åström, K. J. and Källström, C. G. (1976), Identification of Ship Steering Dynamics, *Automatica*, Vol.12, No.1, 9–22.

[14] Åström, K. J. and Wittenmark B. (2011), *Computer Controlled Systems: Theory and Design*, Dover Publications.

[15] Bellman, R. (1955), *Introduction to the Mathematical Theory of Control Processes*, RAND. Corp.

[16] Bellman, R. (1957), *Dynamic Programming*, Princeton, N. J., Princeton University Press.

[17] Blackman, R. B. and Tukey, J. W. (1958), *The Measurement of Power Spectra*, Dover, NY.

[18] Box, G. E. P. and Jenkins, G. M. (1970), *Time Series Analysis: Forecasting and Control*, Holden-Day, California.

[19] Bryson Jr., A. E. and Ho, Y. C. (1975), *Applied Optimal Control, Optimization, Estimation and Control*, Hemisphere Publishing Corporation.

[20] Cramer, H. (1999), *Mathematical Methods of Statistics*, Princeton University Press, Princeton.

[21] Denis, M. St. and Pierson Jr., W. J. (1953), On the Motions of Ships in Confused Seas, *Trans. Soc. Nav. Arch. Mar. Eng.*, No.61, 280–357.

[22] Feller, W. (1950), *An Introduction to Probability Theory and Its Applications*, 1, John Wiley & Sons.

[23] Fossen, T. I. (2002), *Marine Control Systems: Guidance, Navigation and Control of Ship's, Rigs and Underwater Vehicles*, Marine Cybernetics AS, Trondheim, Norway.

[24] Fukuda, H., Ohtsu, K., Tasaki, T. and Okazaki, T. (2001), A Study of Ship's Guidance Control by Time-Varying LQ Controller, *Journal of the Society of Naval Architects of Japan*, No.190, 715–722, (in Japanese).

[25] Gersch, W. (1981), Nearest Neighbor Rule Classification of Stationary and Nonstationary Time Series, in *Applied Time Series II*, D. F. Findley Ed., Academic Press, New York, 221–270.

[26] Hino, M. (eds.) (2004), *Spectrum Analysis Handbook*, Asakura Publishing, (in Japanese).

[27] Hirano, M. (2006), *Note on Anti-Rolling*, privately printed, (in Japanese).

[28] Hsu, H. P. (1970), *Fourier Analysis*, Simon & Schuster, New York.

[29] Ibaraki, T. and Fukushima, M. (1992), *Optimized Programming for FORTRAN 77*, (Computer Science Series), Iwanami Publishing, (in Japanese).

[30] Imamura, T. (1994), *Physics and Fourier Transformation*, Physics and Mathematics Series, **3**, Iwanami Publishing, (in Japanese).

[31] Iseki, T., Ishizuka, M. and Ohtsu, K. (1994), Effect of Hull Bending Stress on Main Engine Loads, *Journal of the Society of Naval Architects of Japan*, No. 90, 215–221, (in Japanese).

[32] Ishizuka, M., Ohtsu, K., Hotta, T. and Horigome, M. (1991), Statistical Identification and Optimal Control of Marine Engine System (Part 1), *Journal of the Society of Naval Architects of Japan*, No.170, 211-220, (in Japanese).

[33] Ishizuka, M., Ohtsu, K., Hotta, T. and Horigome, M. (1992), Statistical Identification and Optimal Control of Marine Engine System (Part 2), *Journal of the Society of Naval Architects of Japan*, No.171, 211-220, (in Japanese).

[34] Isobe, T. (eds.), (1968), *Correlation Functions and Spectra*, The University of Tokyo Press, Tokyo, (in Japanese).

[35] Jenkins, G. M. and Watts, D. G. (1968), *Spectral Analysis and its Applications*, Holden-Day, San Francisco.

[36] Johnson, R. A. and Wichern, D. W. (2002), *Applied Multivariate Statistical Analysis*, Prentice Hall, 5th ed.

[37] Källström, C. G. and Åström, K. J. (1979), Adaptive Autopilots for Tankers, *Automatica*, Vol.15, No.3, 241–254.

[38] Kalman, R. E. (1960), A New Approach to Linear Filtering and Prediction Problems, *Trans. ASME, J. Basic Eng.*, 83, 34–45.

[39] Katayama, T. (1983), *Applied Kalman Filter*, Asakura Publishing, Tokyo, (in Japanese).

[40] Kaufman, A. (1967), *Graphs, Dynamic Programming, and Finite Games*, Academic Press.

[41] Kayano, J. and Ohtsu, K. (2003), On Berthing Maneuvering Using Optimal Variable Gain Control, *Journal of Kansai Society of Naval Architects of Japan*, No.240, 221–225, (in Japanese).

[42] Kayano, J., Ohtsu, K., Tasaki, T. and Shouji, K. (2003), Study on Optimal Variable Gain Steering through a Discrete Linear Model, *The Journal of Japan Institute of Navigation*, No.108, 225–230, (in Japanese).

[43] Kimura, H. (1993), *Fourier Laplace Analysis*, Iwanami Publishing, Applied Mathematics Series, Method No.4.

[44] Kitagawa, G. (1983), Changing Spectrum Estimation, *J. Sound and Vib.*, Vol.89, No.3, 443–445.

[45] Kitagawa, G. (2021), *Introduction to Time Series Modeling with Applications in R*, Monographs on Statistics and Applied Probability 166, CRC Press, Chapman & Hall, New York.

[46] Kitagawa, G. and Akaike, H. (1978), A Procedure for the Modeling of Non-Stationary Time Series, *Ann. Inst. Statist. Math*, Vol.30, 351–363.

[47] Kitagawa, G. (1981), A Nonstationary Time Series Model and its Fitting by a Recursive Filter, *J. Time Ser. Anal.*, Vol.2, No.2, 103–116.

[48] Kitagawa, G. and Gersch, W. (1984), A Smoothness Priors-State Modeling of Time Series with Trend and Seasonality, *J. Am. Stat. Assoc.*, Vol.79, No.386, 378–389.

[49] Kitagawa, G. and Gersch, W. (1985), A Smoothness Priors Time-Varying AR Coefficient Modeling of Nonstationary Covariance Time Series, *IEEE Trans., Autom. Control* AC-30, 48–56.

[50] Kitagawa, G. and Gersch, W. (1996), *Smoothness Priors Analysis of Time Series*, Lecture Notes in Statistics, 116, Springer.

[51] Kohari, A. (1972), *Introduction to Probability and Statistics*, Iwanami Publishing. (in Japanese)

[52] Konishi, S. (2014), *Introduction to Multivariate Analysis: Linear and Nonlinear Modeling*, Texts in Statistical Science, Chapman & Hall/CRC, New York.

[53] Konishi, S. and Kitagawa, G. (2010), *Information Criteria and Statistical Modeling*, Springer Series in Statistics, Springer, New York.

[54] Kudo, H. (1973), *Calculation of Probability*, Iwanami Publishing, (in Japanese).

[55] Kullback, S. (1959), *Information Theory and Statistics*, Wiley, New York.

[56] Kullback, S. and Leibler, R. A. (1951), *On Information and Sufficiency*, Ann. Math. Statist., Vol.22, No.1, 79–86.

[57] Kvam, K., Ohtsu, K. and Fossen, T. I., (2000), Optimal Ship Maneuvering Using Bryson and Ho's Time Varying LQ Controller, *5th IFAC Conference on Maneuvering and Control of Marine Crafts (MCMC2000)*, Aalborg, Denmark.

[58] Kwakernaak, H. and Sivan, R. (1972), *Linear Optimal Control Systems*, Wiley-Interscience.

[59] Lewis, E. V. (1966), *Principles of Naval Architecture*, SNAME.

[60] Mehra, R. (1970), Maximum Likelihood Identification of Aircraft Parameters, *Joint Automatic Control Conference*, Atlanta, No.8, 442-444.

[61] Minorsky, N. (1922), Directional Stability of Automatically Steered Bodies, *J. Ameri. Soc. of Naval Engineers*, Vol.34, No.2, 280–309.

[62] Miyoshi, S., Hara, Y. and Ohtsu, K. (2007), A Study on Optimum Tracking Control with Kalman Filter for Vessel, *The Journal of Japan Institute of Navigation*, No.118, 47–53, (in Japanese).

[63] Motora, S. (ed.) (1982), *Kinematics of Hull and Offshore Structures*, Seizando Publishing, (in Japanese).

[64] Nagai, M., Kageyama, I. and Tagawa, Y. (1995), *General Introduction to Vibration Engineering*, Sangyotosho Publishing, (in Japanese).

[65] Nakatani, T., Ohtsu, K. and Horigome, M. (1988), A Statistical Analysis of Ship's Main Engine's Propeller Revolution Control and its Optimal Control, *The Journal of Japan Institute of Navigation*, No.79, 157–167, (in Japanese).

[66] Neumann, G. (1953), On Ocean Wave Spectra and a New Method of Forecasting Wind-Generated Sea, *Technical Memo*, No.43, Beach Erosion Board.

[67] Nishimura, T. (1982), Application of Kalman Filters to Space Systems, *Probability System Theory*, Sunahara, Y. (ed.), Asakura Publishing, (in Japanese).

[68] Nomoto, K. (ed.) (1956), On Steering Qualities of Ships (1) (2), *Proc. of The Society of Naval Architects of Japan*, No.99 and No.100, (in Japanese).

[69] Oda, H., Ohtsu, K., Sasaki, M. and Seki, Y. (1991), Roll Stabilization by Rudder Control through Multivariate Autoregressive Model, *Journal of Kansai Society of Naval Architects of Japan*, No.216, 165–173, (in Japanese).

[70] Oda, H., Yamanouchi, Y., Oda, T. and Takanashi S. (1983), Application of New Statistical Methods to Offshore Structures Dynamics, *Mitsui Zosen Technical Review*, No. 117, 1–10, (in Japanese).

[71] Ohgushi, M. (1999), *Theory of Marine Engineering* (revised edition), Kaibundo Publishing, (in Japanese).

[72] Ohtsu, K. (1982), The Study on a Statistical Optimal Control of Ship's Motion (Part 1), *Journal of the Society of Naval Architects of Japan*, No.152, 216–228, (in Japanese).

[73] Ohtsu, K. (1983), The Study on a Statistical Optimal Control of Ship's Motion (part 2), *Journal of the Society of Naval Architects of Japan*, No.153, 135–141, (in Japanese).

[74] Ohtsu, K. (2006), Research and Development on Coasting Vessel's Environmental Harmony Type Shipping Schedule Supporting System, Ministry of Economy, Trade and Industry, New Energy and Industrial Technology Development Organization (NEDO), (in Japanese).

[75] Ohtsu, K. and Hasegawa, K. (1981), Autopilot Evaluation and the Prospects, *The Society of Naval Architects of Japan*, Motion Performance Research Committee, The 3rd Symposium, (in Japanese).

[76] Ohtsu, K., Horigome, M. and Kitagawa, G. (1976), Statistical Identification of Ship's Course Keeping Motion and Optimal Control, *Journal of the Society of Naval Architects of Japan*, No.139, 31–43, (in Japanese).

[77] Ohtsu, K., Horigome, M. and Kitagawa, G. (1978), Statistical Identification of Ship's Course Keeping Motion and Optimal Control—On the result of actual tests under on-line controlling, *Journal of the Society of Naval Architects of Japan*, No.143, 216–224, (in Japanese).

[78] Ohtsu, K., Horigome, M. and Kitagawa, G. (1979a), A Robust Autopilot System against the Various Sea Conditions—Noise Adaptive Autopilot, *Proceed. ISSOA*, Tokyo, 119–124.

[79] Ohtsu, K., Horigome, M. and Kitagawa, G. (1979b), A New Ship's Auto Pilot Design through a Stochastic Model, *Automatica*, Vol.15, No.3, 255–268.

[80] Ohtsu, K. and Iseki, T. (1994), Recent Developments in Time Series Analysis, *The Society of Naval Architects of Japan, Motion Performance Research Committee, 11th Symposium, 'Applications of seaworthiness theory to design'*, 375–379, (in Japanese).

[81] Ohtsu, K. and Iseki, T. (1995), Study on the Onboard Predictive Ship Motion Analyser-I., *Japan Institute of Navigation*, No.92, 177–184, (in Japanese).

[82] Ohtsu, K. and Iwai, A. (1986), Weekly Adjustment of Time History of In-Bay Vessels in Port Yokohama, *The Journal of Japan Institute of Navigation*, No. 75, 45–53, (in Japanese).

[83] Ohtsu, K. and Kitagawa, G. (1989) Full Scale Data Depended Statistical Estimate of the Parameters in the Equation of Ship's Oscillation—Application of a Continuous Auto Regressive Model, *The Journal of Japan Society of Naval Architecture*, No.165, 181–191, (in Japanese).

[84] Ohtsu, K., Oda, H. and Iida, T. (1997), Challenges to the Advanced Optimization of Ship Motion Control Systems, *The Society of Naval Architects of Japan, Motion Performance Research Committee, 13th Symposium, 'Ship motion and its control and sea conditions'*, Tokyo, Japan, (in Japanese).

[85] Oishi, G. (2019), *Fourier Analysis, An Introductory Course of Mathematics for Science and Engineering*, Iwanami Publishing, (in Japanese).

[86] Ozaki, T. and Kitagawa, G. (eds.) (1998), *Method of Time Series Analysis*, Selected Works of Statistical Science, **5**, Asakura Publishing, (in Japanese).

[87] Ozaki, T. and Tong, H. (1975), On the Fitting of Non-stationary Autoregressive Models in Time Series Analysis, *Proceed. of 8th Hawaii International Conference on System Science*, Western Periodical Co., Hawaii.

[88] Pandit, S. M. and Wu, S. M. (1983), *Time Series and System Analysis with Applications*, John Wiley and Sons.

[89] Park, J. S., Ohtsu, K. and Kitagawa, G. (2000), Batch Adaptive Ship's Autopilots, *Inter. Journal of Adaptive Control and Signal Processing*, Vol.14, No.4, 427–439.

[90] Pierson Jr., W. J. (1954), An Interpretation of the Observable Properties of Sea Waves in Term of the Energy Spectrum of the Gaussian Record, *Trans. American Geo. Union*, Vol.35, No.5, 747–757.

[91] Sakamoto, Y., Ishiguro, M. and Kitagawa, G. (1986), *Akaike Information Criterion Statistics*, Mathematics and Its Applications, Kluwer Academic Publishers, Tokyo.

[92] Shoji, R., Nishiyama, H. and Ohtsu, K. (2010), A Study on Prediction Method of Zonal Index for Weather Routing, *The Journal of Japan Institute of Navigation*, No.123, 29–37, (in Japanese).

[93] Takahashi, Y. (1968), *System and Control*, first and second volume, Iwanami Publishing, (in Japanese).

[94] Togawa, H. (1971), *Numerical Computation of the Matrix*, Ohm Publishing, (in Japanese).

[95] Tokumaru, H. (1987), *Handbook of Statistical Engineering*, Baihukan Publishing, (in Japanese).

[96] Tokumaru, H., Soeda, T., Nakamizo, T. and Akizuki, K. (1982), *Counting and Measurements: Theory and Applications of Random Data Processing*, Baihukan Publishing, (in Japanese).

[97] Tsuda, T. (1969), *Monte Carlo Methods and Simulation: Stochastic Applications for Computers*, Baihukan Publishing, Tokyo, (in Japanese).

[98] Watanabe, R., Natori, R. and Oguni, R. (eds.) (1990), *Numerical Software with Fortran 77*, Maruzen Publishing.

[99] Wiener, N. (1948), *Cybernetics: or Control and Communication in the Animal and the Machine*, The MIT Press, MA.

[100] Wiener, N. (1949), *Extrapolation, Interpolation and Smoothing of Stationary Time Series*, The MIT Press, MA.

[101] Wiener, N. (1956), *I am a Mathematician: The Later Life of a Prodigy*, Doubleday, NY.

[102] Woud, H. and Stapersma, D. (2002), Design of Propulsion and Electric Power Generation Systems, *IMarEST*, Institute of Marine Engineering. Science and Technology, London, 263–271.

[103] Yamanouchi, Y. (1961), Methods for Analysing the Response of Ships to Motion in Waves, No.1 and No.2, *Proc. of The Society of Naval Architects of Japan*, No. 109, 169–183 and No. 110, 19–29, (in Japanese).

[104] Yamanouchi, Y. (1969a), Response in Ocean Waves. *Sea-Keeping Symposium*, The Society of Naval Architects of Japan, No.2, 53–97, (in Japanese).

[105] Yamanouchi, Y. (1969b), On the Application of the Multiple Input Analysis to the Study of Ship's Behaviour and an Approach to the Non-linearity of Responses, *Journal of the Japan Society of Naval Architecture*, No.125, 73–87, (in Japanese).

[106] Yamanouchi, Y. (ed.) (1986), *Theory of Irregular Phenomena for Ship and Marine Engineers*, Kaibundo Publishing, (in Japanese).

[107] Yamanouchi, Y., Ohtsu, K., Kitagawa, G. and Oda, H. (1978), Trends in Data Analysis (1), (2), *Journal of the Japan Society of Naval Architecture*, No.589, 337–344 and No.591, 436–447, (in Japanese).

[108] Yule, G. U. (1927), On a Method of Investigating Periodicities in Disturbed Series, with Special Reference to Wolfer's Sunspot Numbers, *Phil. Trans.*, A226, 267–298.

Index

Printed in the United States
by Baker & Taylor Publisher Services